"A must-read for every parent, educator, and physician. In *Kids and COVID*, Dr. Elizabeth Mumper, a dedicated pediatrician with decades of experience, examines the consequences of hurried and often unscientific pandemic policies on children's health and well-being. With a sharp, evidence-based approach, she challenges the narratives that shaped public health decisions and exposes the long-term risks that were unfortunately ignored despite outcries from many of us in the physician community.

From flawed testing strategies to the unintended harm of school closures and the devastating impacts of mandating barely tested, novel gene therapy vaccines, Dr. Mumper presents her meticulous research and first-hand clinical experience. Her book is an urgent call for accountability— and a roadmap to ensure we never make the same mistakes again. *Kids and COVID* is an eye-opening, essential read that will leave you questioning everything you thought you knew."

— **Pierre Kory, MD, MPA, Leading Edge Clinic**

"Dr. Elizabeth Mumper delivers a searing exposé on the grave failures of the COVID response and its impact on children. With precision, compassion, and unwavering integrity, she dismantles the flawed trials, suppressed data, and unethical policies that put millions of kids at risk. Drawing on decades of frontline experience and global research, Dr. Mumper reveals how science was censored, medicine was politicized, and children became collateral damage in a system driven by control rather than care. *Kids and COVID* is more than a book—it's a wake-up call. A must-read for parents, doctors, and policymakers alike. Ignore it at your peril."

— **Crisanna Shackelford, PhD, advocate for service members and veterans harmed by countermeasures, biologics and vaccines; and non-linear warfare expert**

"As we've seen in graphic detail, doctors and patients can no longer put blind faith into guidance from once-trusted authorities, including CDC, medical associations, and public health experts. And when medical doctors exercised independent thought, reason, and common sense, they were censored and punished in unprecedented ways. It's not a simple thing to determine which information and studies are untouched by ethical and

financial conflicts of interest. Dr. Mumper has taken on the job in order to arrive at truths that Big Pharma and their media partners worked so hard to keep so many from learning."
— **Sharyl Attkisson, investigative journalist and *New York Times* bestselling author of *Follow the $cience***

"Dr. Mumper is the pediatrician you want in your living room (bookshelf). She has years of experience providing patient-centered care. She is compassionate—the eternal advocate for her patients. She understands the true science and will give you the advice she would give her own family. Her book *Kids and COVID* is essential reading for all responsible parents. Dr. Mumper is not afraid of tackling the difficult issues (i.e., COVID and vaccines), however, she does this with a strong knowledge base providing the best practical course of action for your children. She details in depth the complications associated with the COVID vaccines and provides an explanation why the ACIP added COVID 'vaccines' to the childhood immunization schedule. She explores the role propaganda and censorship played in the rollout of the vaccines and the lack of true informed consent. *Kids and COVID* is essential reading in order to learn from our mistakes and to ensure they never happen again."
— **Paul Marik, MD, chief scientific officer for the Independent Medical Alliance**

"It is a delight to see this groundbreaking contribution in print about what happened during the COVID era to kids—and why it must never happen again. Dr. Mumper has had a distinguished career as pediatrician and Cassandra about the serious problems with vaccines. This comprehensive book shines bright with Dr. Mumper's intelligence, ethics, humility, and profound concern for today's and tomorrow's children."
— **Mary Holland, CEO of Children's Health Defense**

"Elizabeth Mumper, MD, is a compassionate and brilliant clinician with decades of pediatrics experience. She is also an accomplished lecturer and writer, making complex topics understandable and digestible. In this book, she again has tackled a complex and controversial topic, COVID, and provided well-researched and evidence-based information in a thoughtful and comprehensible manner. This is a must-read for parents and practitioners caring for our most vulnerable and precious assets, our children."

— **Nancy O'Hara, MD, MPH, author of *Demystifying PANS/ PANDAS: A Functional Medicine Desktop Reference on Basal Ganglia Encephalitis* and faculty at the Medical Academy of Pediatrics and Special Needs**

"COVID forever changed the way the once-trusting public views health authorities. Fear mongering, lies, and threats to education and employment convinced many to participate in the greatest medical experiment ever conducted. Dr. Mumper exposes the propaganda and deceit with expert analysis of scientific data and indisputable facts. Our children, who were among the most harmed by the policies implemented, would have been the least affected if calmer and more honest heads prevailed. *Kids and COVID* is an enlightening, thought-provoking review of the harrowing events of the past few years that should prompt open discussion and debate of these important topics.

— **Cindy Schneider, MD, medical director at the Center for Autism Research and Education**

"Dr. Mumper so diligently and precisely explains the issues surrounding our indiscriminate use of COVID vaccination in kids. The issues she brings to the forefront are a mirror of the issues we face in the way we deliver and decide on health care for our children. We desperately need this information to make informed choices for our future generations. Thank you, Liz."

— **Anju Usman-Singh, MD, medical director at True Health Medical Center and faculty at Medical Academy of Pediatrics and Special Needs**

"If you are a critical thinker—someone who questions the realities of what we lived through during the COVID-19 pandemic—then Dr. Mumper's new book is sure to challenge and engage you. A seasoned pediatrician and fearless advocate for common sense, Dr. Mumper peels back the layers of pandemic policies that defied logic: school closures, mask mandates for toddlers, universal mRNA vaccinations for children, and more.

We live in an era where questioning the status quo makes people uncomfortable—but that is precisely what great leaders do. They question. They ask. They think. Dr. Mumper embodies this spirit of inquiry, making her a voice worth listening to in the ongoing conversation about health, science, and policy."

— **Chris Magryta, MD, managing partner at Salisbury Pediatric Associates**

"For those of you who want to understand the truth of the COVID-19 pandemic and our flawed response to it, I urge you to read this book. *Kids and COVID* will take you on a science-based journey of our often non-science-based response to the pandemic. In the end as Elizabeth Mumper states, 'You Are The Judge.'"

— **Joel Wallskog, MD, co-chair, React19**

"This thoroughly researched guide to Covid and the pandemic challenges the mainstream narrative about what really happened and reveals what our government and medical establishment got wrong. More importantly, it explains why it's no longer taboo to question today's pediatric healthcare policies and why more and more parents are choosing to raise their kids naturally."

— **Dr. Bob Sears, pediatrician and author of *The Vaccine Book***

"Elizabeth Mumper understands the principle that battles may be won through power, but wars are won through love. It is her love for individual struggling children that has made her an essential general in the war to restore all children's health, one child at a time. She brings her decades of experience, in nurturing the most difficult pediatric cases back to health and in improving their quality of life, to the new challenges that COVID has presented to the health of all Americans. She gently makes the complex digestible, and the seemingly impossible attainable, as medicine moves from the symptom-drug paradigm to the deeper applied understanding of individual metabolomics and their role in the new epidemics of the 21st

century. This book will be of great benefit to both the practicing physician and the mother healer."
— **Ginger Taylor, MS, autism activist and author** *In Many Words* **Substack**

"Dr. Elizabeth Mumper's book is an insightful exposé of COVID. Her courage to stand against Big Pharma empowers evidence-based reasoning that the shots are dangerous in fact, and the alternative remedies objectively superior. The paradox presented at the end of the book is cathartic, like a well-earned sigh of relief that the future is bright for our children."
— **Greg Glaser, JD, general counsel of Physicians for Informed Consent**

"You want the truth? The science, the facts? Dr. Mumper is unflinching in her quest to share what you need to know as you make healthcare choices for your young family. Her book is a beacon of truth in the fog of misinformation and disinformation surrounding COVID vaccines for children."
— **Janet Lintala, DC, author of** *The Un-Prescription for Autism*

"*Kids and COVID* is hands-down the most comprehensive, researched-backed writing addressing the long-term physical and mental health implications of the COVID-19 pandemic, particularly concerning the pediatric population. From treatment options to masking to vaccines, Dr. Mumper does a brilliant job of breaking down complex information to be easily digestible. This book is for parents, grandparents, practitioners, and anyone who knows a child- hence, virtually everyone! If you care about the future of children and families, you need to read this book!"
— **M. Buerger, BA, DC, founder of** *Intersec4Life*

"I have known Dr. Elizabeth Mumper for over twenty years, primarily as a dedicated, truly caring doctor who worked with children with autism. I have had the great honor and pleasure of listening to her speak at many of our shared conferences. She is the most competent, reliable, and trustworthy pediatrician I know. *Kids and COVID* is many things. *It is an eye-opener. It is a fountain of information. It is a source of wisdom. And it is an inspiration.* This book should be read not just by scientists, medical practitioners, and their patients; it should be read by anyone who works with children, who has children, or is even planning to have children, and that means it should be read by everybody. I whole-heartedly endorse this book.

— **Aristo Vojdani, PhD, MSc, CLS, CEO of Immunosciences Labs and chief scientific officer at Cyrex Labs**

"Dr. Mumper provides clear, factual, and evidence-based information that every parent yearns for in making decisions regarding their children's health care. The knowledge she shares is based on her clinical experience as a pediatrician, listening to parents, and witnessing a dramatic decline in the health of our children over the last four decades.

Kids and COVID exposes our federal health agencies' and medical professional associations' financial conflicts of interests in their policies and cautions parents to be skeptical of any new medical interventions that do not have long-term safety data, but large profit motives. She includes recommendations for what parents and grandparents can do to protect their children from being monetized by pharmaceutical companies and captured health agencies who have put profits over the people they are tasked to protect.

Don't be misled by the title of this book. It is so much more than a book about kids and covid. Dr. Mumper takes a deep dive into the real science and statistics that every person who is faced with a vaccine decision needs to know to be able to exercise true informed consent. This book is a must read for everyone!"

— **Lyn Redwood, RN, MSN, FNP, president emerita, Children's Health Defense**

"If we want to keep our health, freedom, and sanity intact the next time we encounter a pandemic, we can't simply erase the memories of the COVID debacle. Instead, we could learn from the author's treasure trove of revelations and lessons that deepen our understanding of the tools and perspectives necessary for successfully navigating through future global health emergencies.

As a holistic, pediatric registered nurse for over forty years, I appreciate Dr. Mumper's well-honed medical detective skills, unique ability to bring her years of experience as a pediatrician and teacher of clinicians and her unyielding commitment to researching and telling the truth to skillfully illuminate the costly mistakes we made and must never repeat again."

— **Maureen McDonnell, BSN, founder of Millions Against Mandates**

"*Kids and COVID* is a must-read for all health authorities, pediatricians, and parents. Dr. Liz Mumper draws upon her long, successful medical career to zero in on the medical mistakes made during the COVID epidemic in hopes that shining a bright light on the events surrounding COVID and children will ensure that the world never suffers from the lack of evidence-based science again. Dr. Mumper discusses COVID from a historical perspective and outlines how countermeasures hurt children. She examines the science and data, events and policies that created "Operative Warp Speed" and how the cries of warning from seasoned doctors and scientists were ignored. She reminds parents that parental instincts have been honed over millennia to ensure the survival of the species and urges them to trust themselves when it comes to the health and well-being of their children."

— **Laura Bono, executive vice president emerita, Children's Health Defense**

"A triumph. This book is an exquisitely written account of the COVID war years, 2020–2024, woven with a double thread that deploys both an empathic narrative voice, and a crystalline lens on the 'science.' Dr. Mumper's emphasis is on the attack on children, and through her careful weaving of fact, scientific literature, and personal anecdotes, we are left with one of *the* definitive accounts of this crime of the ages. Four years of expensive propaganda lays in ruins, gathered at the end of each chapter, in her voluminous, careful annotations. We also get a living history of those doctors, scientists, parents, and citizens, who stood against it. A must read."

— **Celia Farber, author of *Serious Adverse Events: An Uncensored History of AIDS* and *Sacrifice: How the Deadliest Vaccine in History Targeted the Most Vulnerable***

"*Kids and Covid* is a powerful post-op on how a one-size-fits-all public health approach, which prioritizes profits at the expense of patient outcomes, has lasting consequences for future generations. Dr. Mumper has a unique voice that offers sharp analysis of the data, wise insights on effective therapies, and empathetic reckoning for colleagues and patients who have been sidelined for decades."
   — **Leslie Embersits, executive director of MINDD Foundation, Sydney, Australia**

"As a pediatrician with decades of experience treating kids with neurodevelopmental disorders, Dr. Elizabeth Mumper's perspective is extremely valuable in this thorough—and necessary—debriefing on how COVID policies affected kids. To protect children from future harm, we must listen and change course—especially in light of the Wuhan Institute of Virology's recent announcement of a new viral threat!"
   — **Zoey O'Toole, co-editor of *Turtles All the Way Down: Vaccine Science and Myth* and editor of *Vax Facts***

"This book is a powerful wake-up call, shedding light on the healthcare system, offering a vision for a better, safer future. It validates the struggles so many of us face while making healthcare choices for our children."
   — **Shannon Kenitz, president of International Hyperbarics Association**

"I have gotten to know Dr. Elizabeth Mumper over the past few years as part of organizations such as MAPS and FLCCC (now the IMA). She is certainly an advocate for her patients and their families. The pandemic response has brought to the surface deficiencies and dangers in our healthcare system. Now Dr. Mumper, a trusted and honest pediatrician, has something to tell the public in her new book."
   — **JP Saleeby, MD**

"There are those who believe that humanity (excluding themselves of course) has overstayed its welcome, and have thus waged war on the populace. The preferred target is the young with the aim that they shall remain, in the words of Bob Dylan, forever young. In my field, this involves exposing infants to significant amounts of aluminium through unnecessary vaccines and convenient cheap nutrition (infant formulas). Dr. Elizabeth Mumper exposes the latest weapon against children in her seminal tome *Kids and COVID*. If I might once again borrow from Bob Dylan, the times they are a changin' and imminent arrivals to power will only benefit from learning from the lessons so carefully spelled out by Dr. Mumper."

— **Dr Chris Exley, FRSB, author of** *Imagine You Are an Aluminum Atom*

"*Kids and COVID* is a book every parent should own and read. You will learn that children survive COVID 99.9987 percent of the time! And you will learn about ivermectin and repurposed medications. You will have a clear understanding that COVID shots are made with synthetic mRNA, contain dangerous lipid nanoparticles and are associated with increased autoimmune diseases, clotting and strokes, myocarditis, sudden death, severe neurological issues, pregnancy loss, menstrual issues, cancer and disabilities.

Dr. Mumper concludes "as a pediatrician with over forty years of experience both in the private and academic sectors, I strongly oppose recommending COVID vaccines for children or teenagers." I join Dr. Mumper and thousands of informed doctors in calling for a moratorium on COVID vaccines.

This book makes one thing very clear. The past few years we have been subjected to propaganda, manipulation, and censorship like never before. Is there any doubt when Pfizer and the FDA requested that the clinical data stay secret for seventy-five years!"

— **Paul Thomas, MD, host of** *Pediatric Perspectives* **on CHDTV**

KIDS AND
COVID

KIDS AND COVID

Costly Mistakes That Must Never Happen Again

Elizabeth Mumper, MD

Skyhorse Publishing

Children's
Health Defense

Skyhorse Publishing books may be purchased in bulk at special discounts for sales promotion, corporate gifts, fund-raising, or educational purposes. Special editions can also be created to specifications. For details, contact the Special Sales Department, Skyhorse Publishing, 307 West 36th Street, 11th Floor, New York, NY 10018 or info@skyhorsepublishing.com.

Skyhorse® and Skyhorse Publishing® are registered trademarks of Skyhorse Publishing, Inc.®, a Delaware corporation.

Visit our website at www.skyhorsepublishing.com.
Please follow our publisher Tony Lyons on Instagram @tonylyonsisuncertain.

10 9 8 7 6 5 4 3 2 1

Library of Congress Cataloging-in-Publication Data is available on file.

ISBN: 978-1-64821-103-4
eBook ISBN: 978-1-64821-104-1

Cover design by David Ter-Avanesyan

Printed in the United States of America

To the vaccine risk-aware healers,
who have courageously accepted the challenge of becoming
specialists in vaccine injury.
The world needs your expertise and compassion.

To those who lost cherished careers, essential jobs, or family businesses
because of COVID injection mandates.
I stand with you in empathy and solidarity.

To the vaccine injured and the loved ones who support them,
your pain is real, your voices matter, and I believe you.

Contents

PART III – RISKS AND BENEFITS OF COVID INJECTIONS

PART IV – LESSONS FROM THE PANDEMIC

Acknowledgments

This book is a result of what I have learned over forty years working in the worlds of medical education, nutrition, vaccine injury, medical treatments for autism, and research about environmental causes of the ever-increasing rates of chronic illness in childhood. Much of what I know I have learned from the families of my patients, who have shared their insights with clinicians, researchers, and advocates to generate hypotheses and move the science forward. Something new and terrible impacted the current generation of children, and I will always be grateful for my experiences with families of children with autism, vaccine injuries, and chronic illness.

I am grateful to my colleagues at the Autism Research Institute, especially Jim Adams, PhD; Sid Baker, MD; Mark Blaxill; Richard Deth, PhD; Steve Edelson, PhD; Jill James, PhD; Dan Rossignol, MD; Richard Frye, MD; Bob Hendron, DO; Mary Megson, MD; Maureen McDonnell, RN; Mary Megson, MD; Jon Pangborn, PhD; Bernie Rimland, PhD; and many more colleagues who challenged and enlightened me during our many think tanks. My colleagues at the Medical Academy of Special Needs have my undying gratitude for persevering in the quest to help children with chronic illnesses and neurologic disorders. Special thanks to David Dornfeld, DO; Michael Elice, MD; John Gaitanis, MD; Sheila Kilbane, MD; Vicki Kobliner, RD; Bob Naviaux, MD, PhD; James Neuenschwander, MD; Nancy O'Hara, MD; Dan Rossignol, MD; Cindy Schneider, MD; Sue Swedo, MD; Anju Usman Singh, MD; and our intrepid administrators, Honey Rinicella and Kathy Morris.

My colleagues at the Institute for Functional Medicine have worked tirelessly to change the paradigm of medicine toward promotion of health and the importance of lifestyle changes. Special thanks to Jeff Bland, PhD; Kara Fitzgerld, ND; Patrick Hanaway, MD; Bethany Hays, MD; Mark Hyman, MD; David Jones, MD; Liz Lipski, PhD; Robert Luby, MD;

Dan Lukaczer, ND; Marion Owen, MD; Joe Pizzorno, ND; Michael and Leslie Stone, MDs. The staff at Children's Health Defense have worked tirelessly to bring true science to the public despite massive criticism. I am grateful for collaborations with Laura Bono; Rolf Hazelhurst, JD; Mary Holland, JD; Brian Hooker, PhD; Robert Kennedy Jr., JD; Meryl Nass, MD; Polly Tommy; Larry Palevski, MD; Madhava Setty, MD; Crisanna Shackleford, PhD; Paul Thomas, MD; and Naomi Wolf, PhD. To my colleagues at the Front Line COVID Critical Care Organization (now Independent Medical Alliance), whose efforts to "let doctors be doctors" and follow clinical wisdom instead of governmental dictates brought scorn and derision from on high, thanks for bringing me into the fold. Many thanks to Kelly Bumann; Ryan Cole, MD; Suzanne Gazda, MD; Pierre Kory, MD; Paul Marik, JP; Saleeby, MD; Joe Varon, MD; Jordan Vaughn, MD; and Fred Wagshul, MD among others.

To those at the Informed Consent Action Network, especially Cat, Del, Jefferey, and Aaron, I am grateful for the incredible work. To Brianne, Joel, Candace, Ernest, and others at React 19, I applaud your efforts to prevent the terrible things that happened to you and those you love from happening to others.

My travels abroad to teach clinicians have enriched my life immensely. Many thanks to my international colleagues, including Nicola Antonucci, MD; Judith Bowman, MD; Ruth Edwards, MD; Kelly Francis, MD; Frank Golnik, MD; Albert Mensah, MD; Kaye Napalinga, MD; Christina Reyes, MD; Paul Shattock, Pharm: Annabel Stucky, MD: Richard Stuckey, MD: Lourdes Bernadette Tanchanco, MD (Tippy); Bill Walsh, PhD; and Andy Wakefield, MD.

I support and applaud the work of American Citizens for Health Choice, the Autism Action Network, The Canary Party, Documenting Hope, Green Med Info, the Institute for Pure and Applied Knowledge, Millions Against Mandates, Moms Across America, the National Autism Association, the National Vaccine Information Center, Physicians for Informed Consent, The Autism Community in Action, and Stand for Health Freedom. Small groups of concerned citizens banded together to save our children.

To my staff at the Rimland Center for Integrative Medicine, thanks for keeping me on track during more than two decades of treating children with special needs and providing general pediatric care designed to decrease the risks of chronic illness for our patients. Special thanks to Amber Chenault, RN; Lisa Coleman, RN; Vickie Gardner; Melissa Hudson, NP; Susan Robinson, MN; and Ginna Petrikonis, NP. To friends and family who have supported me through tough times, my gratitude to Mike, Leigh, William, Jim, Pamela, Larry, Nancy, Judy, Lynn, Connie, Valerie, and Kenny. Special thanks to the ladies at my aqua aerobics class and book club.

Thanks to tireless advocate Ginger Taylor for helping me format almost a thousand footnotes. Thanks to my editors Hector Carosso and Zoey O'Toole for helping me shape my messages.

To the vaccine risk-aware pioneers who paved the way over many decades, the heroes who emerged during the COVID crisis and are documented in this book, as well as many other healers I did not write about, know that you were on the right side of history.

PROLOGUE

Cassandra: A Cautionary Tale

In the days of Ancient Greece, Cassandra was a beautiful Trojan priestess. Her brother was Hector, the hero of the Greek-Trojan war. Apollo sought to win her love, which means sexual favors, by giving her the divine power of being able to see the future. When her romantic attention did not materialize, Apollo added a curse that no one would believe her prophecies. She was doomed to see terrible events due to happen in the future and be powerless to intervene.

Three millennia later, Lissandra was a hard-working healer of children in the kingdom of the Virgin Queen. She dispensed the potions recommended by the American Academy of Pharmacopedia to the babies in her care. Once, a beautiful Black baby reacted badly to the mixed potion of the MM&R. He started having running of the bowels and lost his ability to speak. How can this be, Lissandra thought? These potions are deemed by those on high to be safe and effective. However, since Lissandra was taught to believe what she sees instead of seeing what she believes, she questioned the authorities and the scholars. She told the authoritative scholars that the potions given to every child seemed to act like a poison for some children. They dismissed her concerns and banned her from the ivory tower.

She was told by a mystic that her destiny was to speak for the children who had no voices. She began gathering the stories of mothers and fathers of the voiceless children. She tried to figure out what was happening. She thought the potions given to babies in their infancy might be connected somehow with the strange loss of speaking sickness. She went to the Center for Disease Counting, but they did not believe her. She went to the National Institutes of Wealth, but they did not believe her. She went to the American Academy of Pharmacopedia, but they did not believe her. The Institute of Medicine said there was not enough data to see if what she said was true and waited another two decades to see what might happen. Lissandra watched in horror as the speaking sickness increased from one in 5,000 children to one in 160 to one in 36.

Then a great plague came upon the land. The plague was dangerous for the old and sick. Lissandra was relieved that children were usually not hurt. The feudal lords of the land in cahoots with the aristocrats of the new potions decreed that a totally new type of potion would save the day. They declared that existing effective potions could not be used, not even the magic potion found in the rich soil from the land of Japan. Sir Pierre of Kory, Sir Paul of Marik, and others healed many people with the magic potion from the Earth itself, but the feudal lords overruled them.

Lissandra was sure that the new potion was not needed for young children. She watched in horror as the potion merchants came for the young men, causing their hearts to scar. Then they came after young women, interfering with the tidal moon's effects on their womanly cycles and fertility. Not to be impeded, the potion merchants then came for the children and babies. Sir Edward of Dowd made a parchment of all the young people of the kingdom who were dropping dead of unknown causes. But Lissandra and Sir Edward of Dowd knew what the cause could be. Lissandra heard of a group of rebels, led by Sir Pierre of Kory and Sir Paul of Marik and assisted by the fair maidens—Kelly of Bumann, Kristina of Morris, and Betsy of Ashton. Lissandra knew that these rebels were her people. Together they would take their discoveries to the people of the Grassy Roots and the Healers of the Front Line. Mary of Holland, Polly of Thom, and Tessa of Laurie would spread the truth far and wide. Together they would fight the feudal lords to defend children's health!

PART I
THE CRISIS

CHAPTER 1

Why Should You Believe Me?

For, in the final analysis, our most basic common link is that we all inhabit this small planet. . . . We all cherish our children's future.
—John F. Kennedy

Women always figure out the truth. Always.
—Han Solo, *The Force Awakens*

The Role of Perception and Life Experiences

How did you react to the COVID pandemic? Was your first instinct to be altruistic and comply with recommendations of the authorities for the greater good? Were you skeptical about the promises made by government officials? Did widespread business lockdowns and school closures make sense to you? Did you have faith in the Warp Speed injections?

"We do not see things as they are, we see them as we are." Whether that insight has its roots in Talmudic wisdom, Immanuel Kant's musings, or the mind of an anonymous person long since lost to history, it has always rung true to me. The maxim is mentioned in H. M. Tomlinson's 1931 collection of short stories called "The Gift," in the 1991 book *Think and Grow Rich*, by Napolean Hill, and in Stephen Covey's *The 7 Habits of Highly Effective People* (which I read during my rather prolonged period of reading self-improvement bestsellers). When one of my functional medicine mentors, Dr. Sid Baker, said it to me, it reinforced my feeling that he was a leader of my tribe.

If I am to develop any credibility with my readers, it is fair for you to know what has shaped the perspectives I will share in this book. If you are old enough, think back to the 1960s, which was the time when much of my world view was shaped. I was born in 1954 in the foothills of the Blue Ridge Mountains. I was in New Haven briefly as a toddler when my dad was a Fulbright scholar at Yale. Then we moved to Charlottesville, Virginia, where we lived in converted army barracks at the site of the current John Paul Jones arena, home to the 2019 national basketball champions. My dad was getting a PhD in history at UVA and writing his dissertation on the "Jeffersonian Administration from the Federalist Point of View." My preschool year memories included seeing my mom type his dissertation on an old-fashioned typewriter, which meant that every time she made a mistake, she had to start the page over. How that memory makes me appreciate my access to word processing!

During my elementary school years we lived in Bridgewater, a small town of around three thousand people, where my father was chairman of the college history department. JFK's assassination was one of the defining events of my childhood. We were dismissed from school early and I walked home. From a nine-year-old's perspective, I watched Walter Cronkite's coverage and the devastation on my parents' faces. I was on the banks of the North River in my hometown in the Shenandoah Valley with my friend Libby on the day Robert Kennedy was shot in 1968. My future husband was at Georgetown Prep (where RFK Jr. went to school) when Father Dugan came to tell Bobby that his father had been shot after winning the California primary for president. I vividly remember the television coverage of people lined up along the train tracks in the Northeast paying their respects as RFK's body was transported to DC for burial at Arlington National Cemetery. JFK, RFK, MLK, and Malcolm X—all assassinated in the same decade. It seemed like a lot for all of us, and profoundly affected a young girl from a small town in Virginia.

The Civil Rights Movement was a consequential force in shaping my outlook on race relations and the importance of the quest for equality and justice. One of my most consistent memories of the dog days of summer during my adolescence was sweating as my brother and I played in our yard instead of going to the community swimming pool that was just up the hill. We could hear splashing and squeals of laughter from other kids. My father and mother refused to let us join the pool because it was segregated. A beloved coach at the college, Carlyle Whitelow, was refused membership because he was black. In protest, my family and others boycotted the pool until it was integrated. Coach Whitelow was nicknamed "Cotton"—a moniker he seemed to embrace with his usual jovial demeanor. What can I say to those of you reading today who find that as horrific as I do now? It was a different time. . . . Coach Whitelow was BC's first athlete of African American descent and Virginia's first African American college athlete who did not attend a historic black college or university. He coached and taught at Bridgewater for many years, influencing thousands of college students. I am so grateful to my parents for teaching us the value of sacrifice for the cause of civil rights. In retrospect, days in the pool were an embarrassingly small price to pay. In one of those weird quirks of history, I discovered many years after moving to my current town of Lynchburg that local citizens poured concrete into many city pools rather than integrate them. A local theatre group, Endstation Theatre, produced a compelling play about it. By the way, Lynchburg was named for John Lynch, who founded the local ferry, not for the acts of lynching that undoubtedly occurred there.

I have always been a bookworm. I remember reading biographies and historical fiction under the covers after bedtime. My parents nurtured our love for books, history, and current events during discussions around the dinner table. My love of books served me well during my medical career. When I left my position as a medical educator and embraced functional medicine, I read over seventy-five books on integrative medicine, physiology, and cell biology in just a few years. I have many colleagues who have written books, which I like to buy from independent stores (like Givens Book and Toy Store in Lynchburg) or discount warehouses (like Green Valley Book Fair in the Shenandoah Valley).

My mother was a woman raised in the South who put the needs of others before her own. Despite working at the college and putting dinner on the table every night (except for the two nights a year we went out to the Belle Meade restaurant) she drove with us for two hours from Bridgewater to Lynchburg most weekends for years to help take care of her elderly mother. Granny Flossie had Parkinson's disease and lived with my Uncle TJ and my Aunt Kitty on the family dairy farm. TJ lived in that home from birth until the day he died. When it was hard to get good childcare when my children were young, my mother moved to Lynchburg so she could take care of my kids while I worked as a pediatrician. As a new grandmother, I am paying that practice forward by taking care of my daughter's first baby in Washington state while she teaches.

By attending Bridgewater College as a faculty brat, I got free tuition. I took Western Civilization class from my dad because he was the only one teaching it. I remember some of his teaching aphorisms that have helped me in the medical lecture circuit like: "Tell them what you are going to tell them, tell them, then tell them what you told them." Another was "do not assume anyone listens after twenty minutes." My father's new onset diabetes in his early forties undoubtedly influenced my brother and me to become interested in medicine. In college, I got good grades, chaired clubs like "Public and Cultural Affairs" and the Honor Society and studied to be a nurse. It was the early 1970s and I had never seen a woman physician or read feminist literature. I will be forever grateful to Dr. Harry "Doc" Jopson, who coached track, taught biology for forty-five years and lived to be one hundred years old. One day after class he took me aside and said: "You are just as smart as the guys in this class. You should go to medical school." After a summer in nursing classes at UVA, during which time family legend reports I wanted to be the one to order enemas instead of giving them, I returned to Bridgewater to major in General Science and prep for med school.

A quick trip through my CV will reveal that I went to medical school at the Medical College of Virginia in Richmond, did my internship at the University of Massachusetts in Worcester, then did two years of pediatric residency and a year as chief resident at the University of Virginia. In 1984 (yes, the year in Orwell's book), I was recruited to be one of the two first female pediatricians to come to Lynchburg. I joined a group practice of five wonderful men who saved up cases of vaginitis, teenage angst, and behavior problems for me. I found it difficult to keep up the pace of a busy group practice. Dr. Tom Albertson, the senior partner, saw an average of sixty-five patients a day. I struggled to see forty to forty-five kids a day, often feeling like I was backing out of the room while the parents still had questions. During those five years of group practice, my dad died and I had five miscarriages.

When my first-born son was a baby, I joined the local Family Practice Residency Program as Director of Pediatric Education. I loved to teach and still do. I remember taking my son and three-week-old daughter to a faculty development conference at Duke, where my mom helped care for my children while I learned to become a better medical educator. My favorite memory of that trip is when we succumbed to my son's cranky requests for a snack as we crossed the border into North Carolina and got him his first ever doughnut at Dunkin Donuts. For years he asked to go back to North Carolina because that was where he could get doughnuts. Since I was trying to condition my kids to eat healthier than I do, it took a few years for us to "fess up" that donuts were widely available.

Over the years I have received many awards, like being chosen "Woman of the Year in Science" in Central Virginia, being named "Alumni of the Year" at my college, getting a lifetime achievement award from the Front Line COVID Critical Care Alliance, and being voted "Best Teacher" at the Residency program (competing against eighty other doctors). Objectively, one could argue that I had a successful career and have done some good work for my community. However, as you will come to understand in this book, I am haunted by the patients I could not help enough. I am my own worst critic.

I taught pediatrics at the residency until I got swept up in the autism tsunami and started a solo private practice. A beautiful fifteen-month-old African American baby regressed into autism with chronic diarrhea after I gave him MMR vaccine. This happened before I was aware of the 1998 paper in *The Lancet* in which Andrew Wakefield and colleagues wrote about a novel form of colitis in a series of patients with autism at the Royal Free Hospital in London. I had noticed an increase in the percentage of children with neurodevelopmental problems and thought somebody should investigate it. As Attorney General Robert F. Kennedy said, "If not us, who? If not now, when?"

It took me sixteen months in my new solo practice before I gave myself a paycheck—a whopping $200 for that month's work. As a pediatrician who likes to spend extra time with patients and hire people to provide extra services like nutrition counseling, I managed to work for twenty-three years cracking a six-figure salary from my practice only a few times. My point is that I put my money where my mouth was trying to fill what I perceived as a gap in care for children in my community. We saw patients from twenty states and five countries. We found that treating the medical problems of children with autism could improve their behavior and development. Some children we treated early in life lost their diagnosis of autism.

I started attending Autism Research Institute conferences, then got invited to their think tanks, then served as medical director for five years. During that time, under the tutelage of Dr. Sid Baker, the previous medical director, and watchful eye of Maureen McDonnell, the conference coordinator, my colleague Dr. Nancy O'Hara and I started a clinician training series.

One thing led to another and before long, people who heard me lecture for ARI started inviting me to lecture and mentor abroad. I love to travel, so this was a dream come true for me! One of my favorite things to do is upgrade myself to a business class seat on a lecture trip to Australia and have my every whim fulfilled by wonderful Qantas flight attendants with Aussie accents. My profound thanks to my colleagues in Austria, Australia, the Czech Republic, Denmark, England, Finland, Germany, Hong Kong, Italy, Japan, Mexico, New Zealand, Norway, the Philippines, Poland, Thailand, Sweden, and Switzerland. Your hospitality enriched my life immeasurably! Over the years, I attended over five hundred hours of think tanks, seven hundred hours of lectures on international trips, and thousands of hours of lectures at the Medical Academy of Special Needs and the Autism Research Institute. In addition, I completed fellowship training at the Institute for Functional Medicine, which included a weeklong course, six modules lasting three days each plus a challenging exam. Later I developed the first pediatric curriculum for IFM. Along the way I continued to see patients and managed to publish eleven clinical research projects with my colleagues, more than most pediatricians in private practice take time to do.

Joys of Life

My children are the pride and joy of my life. Raising them with my husband and lots of help from my mom is the best thing I have ever done. We valued education highly in our household. Both of my kids graduated with highest honors from college and got postgraduate degrees. More importantly, they are highly ethical people who do meaningful work. They live on the left coast now; I miss them terribly. I am rearranging my life geographically to spend more time with my grandchildren.

Being a Mediator in Times of Conflict

Even though I have the personality of a peacemaker and live a life that values non-violent resolutions, I love good spy movies, conspiracy flicks, and adventure films. I have vivid memories of seeing the first Indiana Jones movie in Charlottesville during my pediatric residency. Heather Heyer would die within blocks of that theater many years later in August of 2017 during a "Unite the Right" rally which propelled Charlottesville into the national news.

For those of you who are Myers-Briggs fans, I am an INFP. The one-word summary for my personality type is "mediator." Allegedly less than 1 percent of physicians and medical students fall into this category. Perhaps this is why I often feel in the minority professionally. As Mark Twain allegedly said, "Whenever you find yourself on the side of the majority, it is time to pause and reflect." I am grateful for the gifts of my personality type, which reportedly include open-mindedness, creativity, and generosity. Empathy is one of the greatest strengths of my personality type, but for a pediatrician who sees a lot of children with neurodevelopmental problems, empathy can be a liability. I tend to over-empathize with the troubles of parents and find it hard to set boundaries and turn my mind off from obsessing over how to help my patients.

INFPs tend to feel compelled to speak their truth. My personality type wants to contribute to a world where everyone's voice is heard, and every person's needs are met. You see the problem here, right? Clearly that's not gonna happen. INFPs tend to be unrealistic, and I certainly had unrealistic expectations about how the medical world would respond once they were made aware of the exponential rise in chronic illnesses and neurodevelopmental problems like autism.

COVID Knowledge

But wait, you ask, I may know a thing or two about general pediatrics and autism, but why should we believe what you say about COVID and COVID shots? Beginning in March of 2020, I tried to learn all I could about COVID (to the point of obsession, my husband would say). The response to the COVID crisis did not seem right to me from the start. Many days in that first year I spent several hours per day reading medical literature, which at the time was contradictory and difficult to interpret. I quickly found leaders I thought were on target and became skeptical of pundits on the evening news. My practice never shut down during COVID. Frankly, it never occurred to me that we should. I was a doctor, patients could get sick, and it was my job to try to help them when they got COVID. As a pediatrician, I rejected the "stay home until you turn blue" approach and was interested in prevention strategies. As a former State Debate Champion, I learned how to analyze and argue both sides of multiple issues. Thanks to my debate coach, Ernie Deyerle, for training me to analyze both sides of each issue and see strengths and weaknesses inherent in each.

Since spring of 2021, I have been part of an international consortium that meets weekly by Zoom to share insights and research from around the world. It has been so informative to hear directly from a guy in Israel that ambulance sirens increased after the shots rolled out compared to during the first outbreaks. Or to hear a person from Sweden share that country's experience during COVID without lockdowns and compare it to the US experience where our mortality rates were among the worst in the world. Or to hear economists discuss the predictable impact of lockdowns on job loss, mental health problems and deaths from despair, which Dr. Anthony Fauci did not include in his decision-making.

I was an early adopter of the Front Line COVID Critical Care Alliance protocols. By my assessment, FLCCC does their homework, knows the science, and has the clinical acumen to apply it to help patients. My work in the world of autism has taught me to be skeptical of overblown vaccine promises, like novel injections developed at Warp Speed with poor regulatory insight. I helped edit the footnotes of *The Real Anthony Fauci*, which added to my knowledge base about vaccinology and vaccine politics. I took a public stand early on with concerns about the emerging vaccines, taping a video for Children's Health Defense on Halloween in 2020. I wrote an ebook in June 2021 explaining my concerns about spike proteins, lipid nanoparticles, and synthetic mRNA. Many of my concerns have sadly come to pass.

I read through the Pfizer clinical trial documents; I wonder how many primary care physicians did? After analyzing the trial data, I decided I could not recommend them to my patients and did not want them for myself. Many members of my immediate family were mandated to get the COVID jabs to keep their jobs or chose to get them since they had different viewpoints from me. I was in the minority again.

You Are the Judge

Whether or not you find me credible is for you to judge. Maybe you will think my arguments are well reasoned or maybe you will conclude I am wrong to recommend against COVID vaccines for kids. Let the analysis begin. Buckle your seat belt. You are not going to like everything you read.

FOOD FOR THOUGHT

Can you think of times in your life where you felt compelled to speak out or take an unpopular stand even though you suspected it would lead to criticism or ostracism? If you heard new evidence about COVID that contradicted what you thought, would you consider changing your mind?

CHAPTER 2

What Are the Real Risks of COVID Infection in Babies, Children, and Adolescents?

There are three types of lies—lies, damn lies, and statistics.
—Benjamin Disraeli

A single death is a tragedy; a million deaths is a statistic.
—Joseph Stalin

Children Survive COVID 99.9987 Percent of the Time

Within a few months of COVID-19 hitting our shores, we knew that risks were stratified according to age and underlying risk factors. Healthy people under age fifty were likely to experience COVID-19 in the same way as other coronaviruses—like a bad head cold or a flu that caused symptoms for less than a week. Since the beginning of the pandemic, children and adolescents (in the absence of underlying obesity, diabetes, or serious chronic illness) have faced risks that are vanishingly small. To interpret the data, it is important to know what infection fatality rate means. The IFR is calculated by dividing the number of COVID deaths by the number of COVID infections. For those of you who like math equations:

IFR = (COVID Deaths / COVID Infections) Let me show you some data released in November of 2020 and then we can deconstruct it.[1]

COVID Infection-Fatality Rates by Sex and Age Group (Numbers are shown as percentages)

Age group	Male	Female	Mean
0-4	0.003	0.003	0.003
5-9	0.001	0.001	0.001
10-14	0.001	0.001	0.001
15-19	0.003	0.002	0.003
20-24	0.008	0.005	0.006
25-29	0.017	0.009	0.013
30-34	0.033	0.015	0.024
35-39	0.056	0.025	0.040
40-44	0.106	0.044	0.075
45-49	0.168	0.073	0.121
50-54	0.291	0.123	0.207
55-59	0.448	0.197	0.323
60-64	0.595	0.318	0.456
65-69	1.452	0.698	1.075
70-74	2.307	1.042	1.674
75-79	4.260	2.145	3.203
80+	10.825	5.759	8.292

There are several caveats to interpreting the chart above. First, it is difficult to know the true number of cases because some people with COVID (and most children) are totally asymptomatic. That is quite a nice problem to have from the patient's viewpoint. Second, the possibility of undercounting the total cases means that the death risks could be even lower. Third, it is difficult to figure out which patients died "from COVID" versus "with COVID."[2] Hospitals were incentivized with extra money if they recorded COVID diagnoses. Many hospitals tested every patient for COVID. Some kids who had broken arms or were in the hospital after a car accident may have been counted in the COVID statistics if they had a positive test, even if they had no symptoms. Fourth, some kids had false positive COVID tests. In another chapter we dissect the problem of false positive and false negative COVID tests. These inaccuracies can certainly affect these results.

Data from Around the World

To get the best possible estimates, researchers looked at data from forty-five countries and nearly two dozen studies that looked at the percentage of the population who had antibodies against SARS CoV-2 and therefore had been infected prior to any vaccines. Important data emerges.

First, the least likely to die were in the five- to fourteen-year age group, the very group that was affected so severely by school closures. Their infection fatality rate was 0.001 percent; that corresponds to one in one hundred thousand children, usually those with underlying conditions. Pretty good odds, but obviously the family of any child who died would be devastated. Time would prove the real risk was even less.

Second, let's look at the data on babies. I was surprised that the babies with positive COVID tests in my practice did great! We usually worry about young infants with respiratory infections since they have such narrow airways. Respiratory syncytial virus and bacterial pneumonias can be quite serious in young infants. However, the infants in my practice who tested positive for COVID were often sick for just one or two days, and sometimes barely even had a runny nose. One baby had pink eye, which is easily treated with drops or gets better without any interventions. My clinical experience is supported by this data from forty-five countries.

Third, those same odds hold true in the adolescent age group of fifteen to nineteen years old as for babies.[3] Remember that college applicants were mandated by most colleges to get COVID vaccines to attend classes. As we examine potential risks of COVID shots later in this book, I hope you will think about the risk versus benefit calculations.

A seven-country analysis (France, Germany, Italy, South Korea, Spain, United Kingdom, and the United States) found that the COVID death rate in children was 1.7 per million (one in 588,000). Across the seven countries, as of February 2021, COVID-19 deaths in children comprised 0.48 percent of total mortality from all causes in a normal year. Another way of looking at that statistic is that, during the worst ravages of COVID, 99.5 percent of the children who died succumbed to something OTHER than COVID.[4] Those cold, hard numbers may help parents put in perspective the true risks facing children and adolescents as they make COVID vaccine decisions.

Deaths in Healthy Children Are Statistically Zero

A statistically zero chance of death in healthy children does not mean that no healthy children around the world died with COVID. It does mean that there is no difference in death rates between children dying with and without COVID. Marty Makary, MD, MPH worked with Fair Health, Inc. to analyze forty-eight thousand children with COVID-19. The mortality rate was zero for children under eighteen years old who did not have other conditions, like diabetes, obesity, or chronic cardiopulmonary conditions. This data was released in November 2020,[5] which was before COVID injections were distributed or recommended for children. By looking at the nation's largest private health-care claims database, Makary and his colleagues were able to find 467,773 patients diagnosed with COVID-19 from April 1, 2020, until August 31, 2020, and look for age effects. The overall mortality rate for all ages was 0.59 percent. Across all age groups, patients with developmental disabilities had the highest odds of dying with COVID-19, 3.06 times more likely. The odds of dying from or with COVID increased as the number of comorbid conditions increased. Dr. Makary, another of my COVID heroes, referred to this data in his July 19, 2021, editorial in the *Wall Street Journal*, titled "The Flimsy Evidence Behind the CDC's Push to Vaccinate Children."[6]

A study in *Nature*, which is usually considered one of the top five journals in the world, showed virtually no risk of death in children under eighteen who had no comorbidities.[7] A large study from Germany found the lowest risk for needing treatment for COVID was in children five through eleven without comorbidities, when one child in about thirty thousand needed intensive care and the case fatality rates could not be calculated due to the absence of deaths in that age group.[8]

According to CDC published data, not one death occurred in children aged six months through four years old associated with COVID-19 from December 2021 through March 2022, which was during the late Delta through early Omicron waves.[9]

Think About the Implications for the Future

Please, someone explain to me the rationale for giving an experimental vaccine to protect children from a disease that is often asymptomatic, is usually mild when there are symptoms, and for which death is less likely than many types of childhood accidents? It is very difficult to prove a statistical improvement from "vaccines" better than zero mortality! The studies did not prove injections lowered severity of disease. The studies did not prove vaccinated children protected their grandparents. But many children got pizza parties, gift cards, and free ice cream for taking the experimental shot! In a later chapter I will explain why I think those methods of coercion are highly unethical.

Deaths Due to COVID versus Deaths with Incidental COVID

A study published in July 2021 retrospectively examined nine months of records for 117 pediatric patients (< age eighteen) hospitalized "with COVID" (that is, young patients with a positive PCR test result). The researchers assessed severity of illness (asymptomatic, mild to moderate, severe, or critical) and whether the hospitalization was likely or unlikely to have been caused by SARS-CoV-2. They found that nearly four in ten (39.3 percent) children had no COVID-19 symptoms, and they categorized nearly half (45 percent) of the admissions as "unlikely to be caused by SARS-CoV-2."[10] The investigators concluded that diagnostic records based purely on positive PCR test results may overestimate pediatric SARS-CoV-2 hospitalizations. Nearly 40 percent hospitalized with COVID had no COVID symptoms. Nearly 50 percent of the admissions were not likely due to COVID itself. Let those numbers sink in.

CDC published COVID-19 mortality data as deaths "with COVID" exaggerating deaths that included those that were not actually caused by COVID-19 infections or complications.[11] Coroners have recorded deaths from gunshot wounds in which the patient had a positive COVID test as COVID deaths.[12] Clinicians in my region have reported that children with broken limbs in the Emergency Department have been counted in the COVID tally if they had a (potentially false) positive COVID test.

Kinder, Gentler Forms of the Virus

According to the CDC, as reported in the *New York Times*, data over a three-month period beginning on February 28, 2022, indicated fewer than one US child per one hundred thousand children were hospitalized daily for COVID-19.[13] Strong evidence was published in the *Journal of the American Medical Association* in April 2022 that newer variants of COVID-19 (Omicron) pose dramatically reduced risks to young children.[14] Evolution favors mutations that lead to less virulent forms, so that viruses can continue to find hosts to live in. If there are too many dead people, the virus has no place to thrive. Remember, viruses do not survive unless they take over a host's cellular machinery.

A huge medical database in the United States matched children under five years old who were infected with Delta versus Omicron. Children with Omicron (kinder, gentler than Delta) were only 35 percent as likely to need an ICU bed than the children with Delta. They were only 15 percent as likely to need a ventilator.[15]

Are Children Full of Germs and Disease That Will Kill Their Grandparents?

With other viral illnesses like influenza and measles, children are often viewed as major spreaders of infection. That does not seem to be the case with SARS, as "children do not seem to be major drivers of the COVID-19 pandemic" as noted in this article which was published in August 2020.[16] As early as June 2020, the evidence that children were unlikely to be "major transmitters" had already accumulated. During contact tracing in China, for example, the WHO identified zero episodes where COVID transmission went from child to adult.[17] In a subsequent multicounty analysis of studies that described "household transmission clusters," only three cases were identified in which the child was the index case, and in all three cases, the child had symptoms.[18] Another detailed analysis of 107 pediatric index cases and their household contacts showed definitive evidence of only one case of onward transmission from a teenager.[19]

In January 2021, two University of Vermont physicians thoughtfully looked back at COVID-19 household contact studies involving children, considering symptoms, infection rates and transmission dynamics.[20] Only thirty-three of the households they analyzed included children, so the small sample size merits caution in drawing conclusions. They found that having a parent sick with COVID was a clear risk factor for the child, but not the other way around. The overwhelming majority of children who got sick exhibited mild symptoms.[21]

Several recent meta-analyses suggest that children overall were less susceptible to severe infection than adults, especially if they were under ten to twelve years of age. The odds ratio for infection in children aged five to twelve versus adolescents aged thirteen to eighteen is 0.36, meaning that teens are three times more likely to get severely ill than elementary school children.[22] These data suggest that approaches should be tailored based on age. I could not find data to support the "children will kill granny" hypothesis that was used to encourage parents to vaccinate their children. Did the government weaponize our altruism to manipulate vaccine uptake?

What About Childcare, Camps and School?

Children's return to childcare, school, and summer camps has generated data that pointed to a very low prevalence of outbreaks in such settings early on. This has been confirmed in studies in Australia,[23] France,[24] Germany,[25] Ireland,[26] Singapore,[27] and Rhode Island.[28] In communities with widespread transmission of the COVID virus, adults are nearly always responsible for the spread in childcare settings.

Take-Home Messages

There are two take-home messages from this body of research. First, early data in the aggregate show that children have reduced susceptibility and infectivity compared to adults. Second, younger children and older teens have different susceptibility and infectivity profiles and should be treated differently when deciding about mitigation strategies and "vaccines."

Protective Mechanisms: Why Are Kids at Lower Risk?

There are a number of possible mechanisms that may be protecting our young children from infection with SARS-CoV-2. First, various types of coronaviruses are already responsible for 15 percent to 30 percent of seasonal upper respiratory infections. Several lines of research suggest that the frequent exposure to similar coronaviruses may induce some effective immunity in children (and perhaps pediatricians and elementary school teachers). Researchers found SARS-CoV-2 reactive antibodies, predominantly IgG, in patients uninfected with COVID-19, and more so in children and adolescents than in adults; the antibodies targeted the S2 subunit of the spike protein.[29]

T-cell responses are exquisitely orchestrated immune responses that are part of the adaptive immune system and, therefore, important in the development of long-term immunity. Several lines of research suggest T cells play a big role in protective immunity against SARS-CoV-2. Germany-based researchers found Spike-reactive CD4 T cells in 83 percent of COVID-19 patients but also in about a third of healthy blood donors who were SARS-CoV-2-unexposed.[30] In another important paper published in *Nature*, researchers in Singapore studied T-cell immune responses to the nucleocapsid protein (NP) and other regions of SARS-CoV-2 in patients with COVID-19.[31] In the twenty-three patients who had previously recovered from severe acute respiratory virus (SARS), the Singapore research group showed that even seventeen years after the

2003 SARS outbreak, SARS-recovered patients "still possess[ed] long-lasting memory T cells reactive to SARS-NP, which displayed robust cross-reactivity to SARS-CoV-2 NP."[32] Consider this—if you were one of those twenty-three patients who had SARS back in 2003, you had robust immunity in 2020, and I would wager you are still immune as I write this. The immune system is a spectacular thing if we do not get in its way! Furthermore, you have memory T cells that can cross-react to the current SARS-CoV-2 virus! The take-home message is that childhood exposure to certain coronaviruses is likely to induce strong and long-lasting immunity to the nucleocapsid protein that, in turn, is likely to protect your child from serious problems from COVID-19. I wish every official tasked with making COVID vaccine recommendations for young children—and parents trying to make vaccine decisions for or with their kids—knew about the results of this research.

Innate Immune System More Robust in Children Than Adults

Children mount effective, robust, and durable immunity when compared to adults.[33] Children tend to have strong innate immune systems. The innate arm of immunity is made up of "first responders," cells that rush to the scene when it is not yet clear what the problem is. Later, more specific immunity, called adaptive immunity kicks in. Adaptive immunity includes T cells and antibodies. The mRNA vaccines are intended to mobilize antibodies, which are only one part of one arm of the immune system.

Risks for the Elderly: Higher Than for Kids but Lower Than You Might Think

The elderly, in contrast—especially if they had underlying health problems like obesity, diabetes or chronic problems with their lungs, heart, or kidneys—were more likely to get very sick or die. The infection fatality rate sharply rises in the sixty-five- to sixty-nine-year age group. Even so, survival rates in the elderly are still 97 percent or better through age seventy-nine, with just over three deaths per hundred people ages seventy-five to seventy-nine infected. The survival rates fall to about 92 percent only for those age eighty and above.[34]

Patients older than sixty-nine years of age accounted for 4.82 percent of COVID diagnoses but 42.43 percent of total deaths.[35] Comorbid conditions like developmental disabilities, cancer, chronic kidney disease, and heart failure add to the odds of dying. An odds ratio of 3.06 means one is about three times more likely to die if they have that co-existing condition.

- Developmental disorders: odds ratio 3.06
- Cancer, especially lung cancer: odds ratio 6.74
- Chronic kidney disease: odds ratio 1.85
- Heart failure: odds ratio 1.58[36]

What About MIS-C (Multisystem Inflammatory Syndrome in Children)?

In May 2020, reports from Europe, the UK, and the US (specifically the San Francisco Bay area, New York, and Los Angeles) began reporting a pediatric inflammatory syndrome that made health professionals wonder if COVID was to blame. Up to five children in New York reportedly died. The symptoms of this ailment resembled Kawasaki disease and toxic shock syndrome. Kawasaki disease is an acute systemic vasculitis with characteristic findings: fever plus four of these symptoms—swollen lymph nodes, rash, bilateral non-purulent conjunctivitis, oral changes like cracked lips and strawberry tongue, and extremity changes like desquamation (peeling palms and soles). Atypical Kawasaki's presents with fever and two or three of the above.[37] Toxic shock syndrome can look similar, but is caused by bacteria, usually staphylococcal or streptococcal. According to the CDC, over five thousand children per year are hospitalized with Kawasaki disease.[38] Kawasaki's is important because it is a leading cause of acquired coronary artery disease, which carries the risk of aneurysms. In New York, only 40 percent of the kids hospitalized with the mysterious vasculitis tested positive for COVID.[39]

This led me to wonder if we were in an "all COVID all the time" mindset and were not recognizing that all the usual illnesses could still be around. We will learn about the high false positive rate for PCR COVID tests in a coming chapter, so the true infection rate in those children could be even lower. Doctors named the new syndrome initially "pediatric multisystem inflammatory syndrome temporally associated with COVID-19," which they changed to "pediatric inflammatory multisystem syndrome (PIMS) and later to MIS-C, multi-inflammatory syndrome in children.

Kawasaki disease was unheard of prior to the 1960s and 1970s and started about the same time as my career in pediatrics. Clinicians from around the world simultaneously noticed the characteristic symptoms and eventually named the illness after Dr. Tomisaku Kawasaki in Japan, not for the motorcycle. The children diagnosed with PIMS or MIS-C have some symptoms of Kawasaki's, but those symptoms can occur in other illnesses. In the spring of 2020, PIMS patients were typically older (75 percent above age five) than Kawasaki patients (usually under five).[40]

PIMS patients were more likely to present in shock, and a large number had severe abdominal pain. The heart effects were inflammation of muscle, not arteries. Knowing that the prevalence of influenza declined dramatically during COVID (or did it just get diagnosed as COVID instead?), I thought it would be interesting to look at CDC data on the prevalence of Kawasaki during 2021 and 2022. I wondered if there was diagnostic substitution such that cases of Kawasaki's would be much lower than baseline. I was disappointed to find that the Kawasaki page had not been updated since May 2020.[41]

Experience with MIS-C Accumulates

By 2021, clinicians were accumulating patient care experience with this seemingly new entity. A post-infectious immune dysregulation was suspected. Since patients could deteriorate rapidly, most were managed in pediatric intensive care units. Patients usually presented with fever and involvement of two or more organ systems. Some received specific therapies targeted at modulating the immune system and reducing inflammation.[42]

By 2021, a systemic review collating information from 1,726 published papers appeared which found thirty-five papers related to MIS-C cases. Between March and June 2020, there were 783 individual cases of MIS-C reported. Fifty-five percent were male. The median age was 8.6 years, but there were patients as young as three months and as old as twenty years.[43]

- Seventy-one percent of patients had gastrointestinal symptoms (34 percent abdominal pain and 27 percent diarrhea).
- Nearly 10 percent had respiratory distress
- About 5 percent had a cough.
- Fifty-nine percent had positive tests for SARS Co-V-2, but only forty-one percent had an abnormal chest x-ray.[44]

High white counts in 83 percent and high C reactive protein in 94 percent suggest infection and inflammation. Some kids got quite sick. Sixty-eight percent needed to be in the intensive care unit. Twenty-eight percent needed a ventilator, and four percent ended up on ECMO (a heart-lung machine).[45]

Treatments included intravenous immunoglobulin in 63 percent (which is also used in Kawasaki's) and IV steroids in 44 percent. Some kids were treated with monoclonal antibodies or interleukin receptor antagonists. Sadly, twelve out of the 783 children died, for a fatality rate of 1.5 percent. To put this in context, 1.5 percent of the subset of children who got SARS CoV-2 with MIS-C died.[46] This was during the time of the initial virus, which, remember, has mutated to become less serious over time.

According to the CDC, a child's risk in the US of getting MIS-C from 2020 to 2023 was about 1 in 7,443. By 2022, clinicians and scientists were able to summarize treatment strategies and make recommendations for management of MIS-C. The main strategy is to modulate the immune system with interventions like intravenous immunoglobulin (IVIG) and steroids as first-line therapy. IVIG comes from pooled plasma from multiple donors and contains antibodies. Patients must be watched carefully during the infusions to make sure they do not have an allergic reaction. Monoclonal antibodies are usually the next treatment strategy if patients do not respond to IVIG or steroids. Patients are at risk for blood clots, and some receive anti-coagulant therapy or anti-platelet treatments to stop clotting.[47]

MIS-C Over Time

The CDC has identified seventy-nine deaths in children from mid-May 2020 to mid-August 2023, who met the definition of MIS-C.[48] My heart goes out to those children and their families. To be more specific, the mortality risk for a child under eighteen from MIS-C is 0.0000011127 or one chance in 898,734. Unlike the side effects from various COVID injections, MIS-C risk is vanishingly small. I hope this math is helpful for those of you who will be making risk versus benefit decisions for your children and adolescents as COVID "vaccines" continue to be promoted and advertised.

Children are more likely to get Kawasaki disease than MIS-C during their childhood with an incidence from nine to twenty per 100,000 children under five years old, which works out to 1 in 6,666 in the vulnerable age group. Nearly twenty children per 100,000 under five years old (one in 5,000) are hospitalized. As an acute vasculitis, about 25 percent of untreated cases lead to coronary artery aneurysms.[49]

Oh, dear. For parents out there, I just mentioned one more thing for you to worry about. It is not easy being a parent in today's challenging times.

Conclusions

As intensive care doctor Pierre Kory testified to the Maryland legislature in March 2023, "given that CDC extrapolated that 95 percent of Americans already have partial to complete immunity, while we are at historic low levels for severe COVID-19 disease, it should be clear that there is no need to vaccinate anyone now." Makes sense to me—what do you think?

FOOD FOR THOUGHT

Now that you have seen the numbers from published research to compare to the level of concern you had from watching TV news during COVID, has your opinion on the need for COVID vaccines in kids changed? Were you surprised to learn that 92 percent of people older than eighty years survived COVID? Do you think people who had proven immunity to COVID-19 should have been pressured to get vaccinated?

CHAPTER 3

How Good Are Those PCR COVID Tests?

Medicine is a science of uncertainty and an art of probability.
—William Osler

Sell not virtue to purchase wealth, nor liberty to purchase power.
—Ben Franklin

Emergency Use Authorizations of Tests

The FDA authorized emergency use of Real-Time Reverse Transcriptase panels (hereafter RT-PCR or simply PCR testing) to detect SARS CoV-2 in March 2020.[50,51] The polymerase chain reaction (PCR) method was invented in 1983 by American biochemist Kary Mullis. PCR technology takes an infinitesimally small amount of biological material and multiplies it; the process is analogous to a xerox machine or copier. This is spectacular technology for studying biologic materials! You start with tiny amounts, then make a lot and go ahead with your experiments. Voilà!

Thermal cycling is used to expose reagents to repeated cycles of cooling and heating during PCR. This allows temperature-dependent reactions like DNA replication. Two main reagents are used: primers, short, single-strand DNA fragments, and DNA polymerases, enzymes which can act as scissors. By controlling the temperature, nucleic acids are denatured, then DNA strands become templates for DNA polymerase to assemble new DNA strands. These are replicated in a chain reaction in which the original DNA template is amplified exponentially. Hence, the voilà!

PCR Pitfalls

According to Mullis, the PCR method was never intended to be used for diagnosing illness. Let's figure out why. To put it simply, unless you know the number of amplification cycles used, it is difficult to differentiate between true illness and colonization with the virus in question. Since PCR amplifies genetic matter from a virus, the fewer cycles that are needed to show a viral footprint, the higher the viral load. Patients with large amounts of virus are likely to be more infectious than patients with small viral loads. Makes perfect sense. For any given patient, clinicians would find it useful to know how many replications (called the "cycle threshold") it took to make the patient's test turn positive. If the test turns positive in just a few cycles, the patient is likely to be more infectious, and more aggressive management and containment strategies might be indicated. But if a test turns positive after forty to forty-five cycles, it is likely to have picked up viral debris of no clinical consequence. The patient would not need to be isolated.

Several months into the pandemic, a pediatrician (OK, it was me) attended a virtual webinar with the local health department director and other community doctors. I asked what cycle thresholds were being used in the local community testing centers. The health department director could not tell me. A pathologist on the call said he thought we might never know, as the information might be considered proprietary. In spring of 2021, a national lab emailed a notice that they had the "best, most accurate" PCR testing. "Aha," I thought, "I can find out what cycle thresholds they are using." After waiting through a phone tree to get to the office of the company's medical director, I was told that they could not tell me. I made the case that, as a clinician, it was important to know if a patient's test was positive at fifteen to seventeen cycles (likely a high viral load) or not until forty to forty-five cycles (likely not truly infected and therefore not a threat to the community). My pleas fell on deaf ears. Yet the marketing persisted that this company has the "best, most accurate" PCR test. That is an absurd claim since those interpreting the test cannot know the cycle thresholds. I call foul.

False Positives and Negatives

Many months after PCR tests were widely used, a review by the *New York Times* on August 31, 2020 looked at three sets of testing data that included cycle thresholds and concluded that up to 90 percent of people who tested positive "carried barely any virus."[52] If we were forensic scientists looking for evidence that the virus was at the scene of a crime, we might want an ultrasensitive test that would not miss any fragments of the virus, living or dead. But as clinicians, we want a test that gives some indication of how much virus the patient is carrying so we can make judgments about how aggressively to treat them and how much they need to avoid being around vulnerable people. Throughout the pandemic, I was never able to find out how many cycle thresholds were being used in my community. From the standpoint of providing appropriate care and protecting vulnerable people, I hope you understand how absurd that is.

Profitable Enterprises

Despite the very limited clinical utility of PCR testing, it was quite profitable. When I went to visit my son in California, he requested a COVID PCR test to reassure them I was not a threat to his wife of child-bearing age or his infant daughter. I paid $180 for a PCR test that was collected while I sat in my car. Patient appointments were scheduled every five minutes, and three sets of patients were processed at the same time. I could not find the cycle threshold used for that test either.

We now know that many people who tested "positive" for COVID by PCR test had a false-positive test. So, did we really have a pandemic of COVID cases, or did we have an epidemic that included false-positive tests—with drastic consequences? To make judgments about that question it is important to know some basic statistical terms.

Statistics 101

If a person is infectious and the test is positive, that is a true positive. If a person is not infectious and the test is positive, that is a false positive. It is also possible to have the illness and have a negative test (false negative) or not be sick and have a negative test (true negative).

Next, let's go through the concepts of sensitivity and specificity. Sensitivity is the ability of a test to detect the virus when it is present. How sensitive a test is can depend on many factors, such as the competence of the people running the tests, how the specimens are collected (i.e., risk of contamination), whether the reagents are good or bad, and the number of cycle thresholds in the case of PCR testing. In the case of COVID, Emergency Use Authorizations were approved based on the ability of the test to find virus in samples known to have the virus. Specificity is a measure of the test's tendency to produce a negative result when the virus is indeed absent. An ideal test would have specificity and sensitivity rates of 100 percent and zero false positives or false negatives. Alas, we do not live in such an ideal world.

Scientists Tried to Warn Us

More testing is not always better. If the test is not specific it will detect a lot of false positives. False positives can happen with asymptomatic people, influenza, and other strains of coronaviruses. In fact, back in 2020, scientists published their concerns about false positives. Rahman and colleagues cautioned about inaccurate results with commercially available nucleic acid testing.[53]

A Connecticut pathologist published a paper in 2020 showing that, in COVID, the false positive rate of PCR testing was 30 percent and the false negative rate was 20 percent. Dr. Lee did RT-PCR testing, then confirmed or disproved the results by using a gold-standard Sanger DNA sequencing test.[54] He argued that the gold-standard testing should be widely used.

That information did not change policy. The false positive rates led to costs estimated in trillions of dollars. Positive tests kept adults who felt well from working, which followed quarantine guidelines but may have put some people at risk of being fired. Many people lost small businesses.

Fifty percent of businesses reported a large negative effect in spring of 2020, and even in late 2022, 25 percent were still experiencing a negative effect.[55] Around 17 percent of restaurants closed permanently.[56] In the Black community, many small businesses that were forced to close had been handed down through three or four generations and never reopened. Businesses owned by African Americans were disproportionally adversely affected by the pandemic.[57,58]

Many people who tested positive at high cycle numbers—including children—probably had only dead fragments or other protein sequences identified and were not a threat to their communities. Nevertheless, the outpatient protocols for handling PCR-positive tests often derailed children from being able to attend school. In another chapter we examine the disastrous consequences of keeping children and adolescents out of school for prolonged periods.

FOOD FOR THOUGHT

36 PCR tests per hour @$180 per test x 10 hours per day = $64,800 per day in charges at the lab I used in California.

If Dr. Lee's research is correct (and I think it is), that is $19,440 per day for false positives and $12,960 per day for false negatives.

How do those of you who lost your livelihoods or suffered significant mental health consequences due to COVID countermeasures feel about that?

CHAPTER 4

What's So Great About the Great Barrington Declaration?

People are generally better persuaded by the reasons which they have themselves discovered than by those which have come into the minds of others.

—Blaise Pascal

When we try to pick out anything by itself, we find it hitched to everything else in the universe.

—John Muir

Proposing a Rational Approach

I hope history eventually gives adequate credit to the authors of the Great Barrington Declaration. In that document, which I signed shortly after it was posted, sound, fundamental, and ethical principles were outlined. Many of my colleagues think that, if the principles of the Great Barrington Declaration had been applied worldwide, the social, economic, and medical devastation that occurred globally would have been averted. You can read it at GBDeclaration.org.[59]

However, I was outraged to read emails between Drs. Anthony Fauci (then director of the National Institute of Allergy and Infectious Diseases) and Francis Collins (then director of the NIH and a native of the Shenandoah Valley where I grew up) in the wake of the Great Barrington

Declaration. We have now learned from a Freedom of Information Act request that on October 8, 2020, Dr. Collins wrote to Dr. Fauci: "The proposal from the three fringe epidemiologists who met with the Secretary seems to be getting a lot of attention. . . . There needs to be a quick and devastating published take down of its premises . . ." "Don't worry, I got this," Fauci emailed back.[60]

Do These People Seem Like Fringe Kooks to You?

Let's look at the authors, and you can decide if they deserved the gaslighting and derision they received at the hands of our public health officials. Dr. Jay Bhattacharya is a professor of medicine, economics, and health research policy at Stanford University. Stanford is widely regarded as a leading university for law, medicine, and graduate studies. He has four—count 'em—four degrees from Stanford: Bachelor of Arts with honors, a Master of Arts, an MD degree, and a PhD in economics. Not too shabby in my book. He is the director of Stanford's Center for Demography and Economics of Health and Aging. He literally wrote a book about the economics of health, appropriately titled *Health Economics*. His research focuses on the role of government programs, biomedical innovations, and health economics. He is exactly the kind of expert I would want to weigh in on public health decisions that would affect the entire planet! Does he sound like a "fringe" doctor to you?

Sadly, Dr. Bhattacharya reported he experienced racist attacks and death threats during the pandemic, all because he applied his considerable ability to an extraordinarily complex problem.[61] He was also placed on a Twitter Trends blacklist in August 2021, which was revealed when the Twitter files were released in December 2022.[62]

Martin Kulldorff, PhD, is a Swedish biostatistician with experience on faculty at Harvard Medical School, at the US FDA Drug Safety and Risk Management Advisory Committee, and on the Advisory Committee on Immunization Practices at the CDC. He received a Fulbright fellowship and got a PhD from Cornell. He developed a free software program for disease surveillance and helped develop the statistics for the Vaccine Safety Data Link, used by the CDC. Does *he* sound "fringe" to you?

The third "fringe epidemiologist," Dr. Sunetra Gupta, is an infectious disease epidemiologist at Oxford. Her research on transmission of various infectious diseases includes influenza, malaria, and COVID-19. She received awards including the Scientific Medal of the Zoological Society of London and the Rosalind Franklin Award of the Royal Society. She has written five novels: *Memories of Rain, The Glassblower's Breath, Moonlight into Marzipan, A Sin of Colour,* and *So Good in Black.* She is developing a universal flu vaccine. Does she sound "fringe" to you?

The mainstream media promptly crucified them as Drs. Fauci and Collins wished, and as a result, their bios on the web contain allegations of spreading COVID misinformation. "Misinformation" was later redefined as anything that conflicted with the official government narrative, whether or not it was actually true.

Rational Strategies

The scientists above advocated "focused protection" to protect the most vulnerable people from COVID infection while avoiding restrictions that have been shown to damage the economy and the mental health of the population. They noted that lockdowns, as during 2020, would disproportionally harm the underprivileged. They pointed out that the risk from COVID-19 was more than a thousand times greater for the elderly with other chronic conditions than for children. They argued for making decisions based on risks stratified by age.[63] Time has borne out their predictions and given credence to their recommendations.

They argued that schools could re-open and young, healthy people could resume sports and cultural activities. They argued that as healthy people got COVID, most would experience it like other respiratory infections and their recoveries would contribute to herd immunity.[64] Their arguments were epidemiologically quite sound, in my opinion. History has proven the consequences they foretold.

They outlined several measures to keep the elderly and infirm safe from infection. They suggested meeting family members outside, having groceries delivered, and frequent hand washing.[65] They noted that 2020 protocols did not protect the elderly. The most glaring example of this was New York State Governor Cuomo's decision to send COVID-recovering patients back to their nursing homes. For elderly care homes, the Great Barrington authors suggested frequent testing of staff and visitors and less staff rotation to limit the number of people interacting with the vulnerable. Residents with active COVID needed to be sequestered to prevent transmission.

Based on epidemiologic data, they pointed out that if age-wide lockdown measures are used to try and suppress the disease, it could take up to three years to reach herd immunity. They argued that with a focused protection strategy, herd immunity could be achieved in three to six months. Shifting infection risk to young and healthy people would then result in fewer deaths overall.[66] And they were right. Though Sweden also made mistakes in its handling of the vulnerable elderly, the pandemic played out there with the virus circulating in the young and healthy in spring and summer of 2020, and the result was a much lower overall death rate than in the United States.

Conclusion

The consequences of draconian lockdowns and reported adverse events from COVID injections have given credence to the recommendations outlined in the Great Barrington Declaration. The Declaration was based on sound epidemiologic principles. Countermeasures were based on unreliable data and an unreasonable reliance on vaccines as the only way out of the crisis. I admire the authors of the Great Barrington Declaration and hope history treats them well.

FOOD FOR THOUGHT

Have you heard of the Great Barrington Declaration as an alternative approach to the COVID crisis? Are you disappointed that credible scientists were denigrated on social media and in the mainstream press? Was your family hurt by lockdown strategies?

CHAPTER 5

Did We Pick a New One-Size-Fits-All "Vaccine" Over Effective, Individualized Treatments?

Conformity is the jailer of freedom and the enemy of growth.
—John F. Kennedy

These aren't the droids you're looking for.
—Obi-Wan Kenobi, *A New Hope*

Different Strokes for Different Folks

It is ironic that at a time when we can customize chemotherapy for the particular genome of a tumor, and functional and integrative medicine are gaining traction in their strategies to evaluate patients individually, the world was asked to accept one-size-fits-all COVID injections that were initially studied only in extremely healthy adults. In fact, people with comorbid conditions, patients with autoimmune disease, and pregnant and breast-feeding women were specifically excluded from the initial clinical trials.

One-size-fits-all public health measures are both illogical and harmful. Most pediatric patients have robust innate immune systems, meaning that they have strong defenses against a variety of viruses. We have seen that their risk from COVID infections is dramatically less than the risk in the elderly. In contrast, many older people experience immunosenescence as we age, becoming more vulnerable to a variety of viral, bacterial, and fungal infections.[67]

Common Sense Approach to Early Treatment

When faced with what seems to be a new public health threat, tried and true strategies include enhancing immune protection for individuals and looking for repurposed drugs. Doctors in the trenches who were seeing sick patients came up with a lot of good ideas. Drs. Keith Berkowitz, Flavio Cadegiani, Suzanne Gazda, Meryl Nass, Tina Peers, Robin Rose, JP Saleeby, Mobeen Syed, Fred Wagshul, and others helped develop treatment protocols published by the Front Line COVID Critical Care Alliance under the watchful eyes of Paul Marik and Pierre Kory.[68,69] Research by Flavio Cadegiani from Brazil provided valuable treatment data.[70,71,72]

Dr. Didier Raoult published articles in July 2020 about the effectiveness of azithromycin (you may know it as a Z-pack) and hydroxychloroquine (a medicine so safe it is one of the few used widely in pregnancy). This was six months before the "vaccines" were available and only a few months into the declared pandemic.[73]

Vladimir "Zev" Zelenko, a beloved community doctor in upstate New York, developed life-saving protocols for his patients and treated thousands of people. He died in July 2022 at the age of forty-eight from cancer. After a lifetime of service and dedication to patients, his obituary headlines emphasized that he promoted "unproven treatments" and "spread misinformation." I have deep scorn for the so-called journalists who shamed him! By the way, his treatments proved to be extremely helpful. He saved many lives and deserved praise and gratitude, not to be vilified when he could not defend himself.

New vaccines cannot get emergency use authorization if there are effective existing treatments. Anyone who was philosophically, politically, or financially committed to COVID vaccines would have strong incentives to criticize alternative treatments. We'll cover the war on ivermectin in the next chapter. Hydroxychloroquine was another casualty of the propaganda wars, as were the careers of many dedicated health-care providers who offered patients more than waiting for the vaccine.

Cheap, Over-the-Counter, and Effective Treatments

Let's look at very simple strategies that did not get any press but could have been lifesaving. Vitamin C has long been known to be an excellent antiviral, and it has great antioxidant properties. In mid-March, when our medical staff had the first hospital meeting to discuss how we would respond to the coming onslaught of patients we had been told to expect, I was first to the microphone to suggest we look at giving vitamin C intravenously to sick patients. I knew it had a long history of effectiveness in a wide variety of ailments and admired the work Linus Pauling had pioneered.[74,75] According to nurses in my local hospital, vitamin C levels were not checked in patients and IV vitamin C infusions were not given.

Early in the COVID crisis, it was demonstrated that low vitamin D levels were a risk factor for worse outcomes.[76,77,78] Yet, the nurses at our local hospital tell me that vitamin D levels were not routinely checked and supplementation to increase vitamin D in patients hospitalized with or for COVID did not happen. It made me furious that folks in our community did not have access to these safe, effective, and cheap treatments. My efforts to bring attention to simple, proven nutritional supplements that would save lives led to eye rolling and criticism.

Were Vaccines Really the Cavalry?

It seemed that, early on, the decision was made to pursue novel vaccines as the "way out" of the pandemic. One reason I came to that conclusion is because so much effort and taxpayer money went to denigrating already existing treatments. The ivermectin trial that allegedly showed ivermectin did not work used too little, too late, but that was the study that got lots of press.

A close analysis of the methods led me to believe those trials were *designed* to fail. A group of scientists and clinicians called "Doctors for COVID Ethics" agreed with me and have written voluminously and in exquisite detail about the pathophysiology of COVID the illness and the dangers of the COVID "vaccines."[79] I am grateful for their work and guidance and reference their work here.

One of the many excellent points these doctors make is that "hasty approvals for COVID vaccines were not justified" because by mid-2020 (months before the injections were rolled out) several epidemiological studies "showed that the infection fatality rate of COVID-19 was on the order of 0.15 percent to 0.2 percent across all age groups, with a very strong bias towards elderly people who had comorbidities."[80,81,82] In other words, thoughtful scientists whom I respected found that COVID-19 was not more deadly than the flu in a typical year.

Epidemiology Done Well

I have an inherent fondness for scholars from Stanford since my son went there for law school, but that was not why I believed Dr. John Ioannidis's calculations over the predictions made by Neal Ferguson, which turned out to be wildly off-target but were used to justify lockdowns. I remembered Dr. Ioannidis from his essay in 2005, "Why Most Published Research Findings Are False," which by the time the pandemic started was the most accessed article in the Public Library of Science (PLOS). Dr. Ioannidis was raised in Athens and has had an exemplary career, much like my other favorite Greek scientist, Theo Theoharides, MD, PhD, who taught me most of what I know about mast cell activation. Dr. Ioannidis has served on the editorial board of over twenty scientific journals, including *JAMA* and *The Lancet*. He has expertise in sniffing out truth from fraud. He was an early critic of Theranos, the micro blood-sample company headed by Elizabeth Holmes (who is now in prison for fraudulent claims). Ioannidis has an h-index of 239 on Google Scholar, which means his publications are very frequently cited in good journals. Having achieved top honors at his medical school, scored an internal medicine residency at Harvard, done an infectious disease fellowship at Tufts, and gotten a PhD in biopathology, he is the kind of guy who seems to know what he is talking about. He is a professor of medicine, epidemiology, and population health at Stanford and has expertise in biomedical data science and statistics. He is co-director of the Meta-Research Innovation Center at Stanford (nick-named METRICS—isn't that clever?). On *March 17, 2020*, he wrote that COVID-19 might be a "Once in a century evidence *fiasco*."[83,84]

He cautioned that policy makers needed to know if they were doing more harm than good with "Draconian countermeasures" that had been adopted by many governments. He wrote: "Reported case fatality rates, like the official 3.4% rate from the World Health Organization, cause horror—and are **meaningless**. Patients who have been tested for SARS-CoV-2 are disproportionately those with severe symptoms and bad outcomes. As most health systems have limited testing capacity, selection bias may even worsen in the near future."[85] He pointed out that, on the Diamond Princess, a closed cruise ship population of mostly elderly, higher-risk passengers, the mortality rate was 1 percent. He reasoned, correctly, that mortality in an entire population that included younger people would be much less. He wrote all this on March 17, 2020, just days after the pandemic was declared. How I wish more people in positions of power had listened to him! Doctors for COVID Ethics believed him. They correctly point out that "This [mortality] rate does not exceed the range commonly observed with annually recurring waves of influenza."[86]

Another reason Doctors for COVID Ethics thought hasty approval and inadequate review of COVID vaccines were not justified is that COVID-19 can be treated. They do not recommend the hospital protocols developed by committees, which are influenced by financial incentives to use remdesivir and ventilators. They do recommend thoughtful treatment strategies developed by experienced clinicians who started publishing their guidelines in 2020. Peter McCullough was one such doc.[87]

FLCCC Treatment Protocols

Experienced ICU doctors, who founded the Front Line COVID Critical Care Alliance, mentioned in the prologue and prior chapter, developed and published treatment options. They emphasized early in the pandemic that there were two stages of COVID disease—the early stage of viral replication and the late stage during which inflammation predominated. This basic understanding of the pathology guided their treatment suggestions—antivirals early on, and anti-inflammatories in the late stage.[88] Doctors for COVID Ethics pointed out that ivermectin is on the World Health Organization's list of essential medicines. It is widely used for tropical parasitic diseases. Yet, when COVID-19 arrived, and novel vaccines were being developed at Warp Speed, "the WHO saw fit to warn against the use of this very same well-known and safe drug outside of clinical trials."[89]

Come on, y'all—to borrow a phrase from that horrid FDA tweet that inspired my local pharmacies to deny my ivermectin prescriptions but was later described by FDA lawyers in court as "just a quip." Let me make it clear. In a courtroom when the argument was made that the FDA interfered with doctors who wanted to prescribe a life saving medication with very few side effects, FDA lawyers stated (with a straight face) that their tweet about ivermectin being for horses and not for people was "just a quip" and not intended as medical advice. We had a safe, existing drug that is ubiquitous, cheap, and effective as both an antiviral and an anti-inflammatory, but its use was limited to those in clinical trials! Think about this logically. For ivermectin to have its best effect, it should be given early, ideally in the first twenty-four to forty-eight hours of symptoms. How is the average person going to find and enroll in a clinical trial when they are getting symptoms of an illness they have been told to fear greatly? The idea is absurd. Once again, Doctors for COVID Ethics agree. They write: "Such a policy cannot be rationally justified, and it has quite appropriately been overridden by national or regional health authorities and ignored by individual physicians worldwide."[90]

They Tried to Warn Us

If you have seen videos by Doctors for COVID Ethics, you know that they are earnest and super-smart. Sucharit Bhakdi is a seventy-seven-year-old retired Thai-German microbiologist with 314 PubMed references to his credit. He recognized early on the dangers of novel vaccines deployed worldwide. In his gentle voice, he made several videos that are worth watching, paying attention to the dates on which he conveyed his insights. Dr. Michael Palmer was an associate professor of biochemistry at the

University of Waterloo (Canada) teaching *Pharmacology* and *Toxicology*. Dr. Palmer's advanced research in biochemistry focuses on the interaction of peptides and proteins with biological membranes, which makes him uniquely qualified to opine about actions of novel mRNA "vaccines."

From their website, here is their origin story: "D4CE got started with three open letters to the European Medicines Agency, written shortly after the EMA had issued the emergency approval for the gene-based COVID vaccines. In these letters, which were signed by hundreds of physicians and scientists, we urgently warned of short-term and long-term dangers from COVID-19 vaccines, including clotting, bleeding, and autoimmune-like inflammation. We also demanded the immediate withdrawal of all experimental gene-based COVID-19 vaccines. Unfortunately, the vaccines were not withdrawn, and all of those adverse effects soon became manifest in abundance."[91]

They call it as they see it, and they see it through a perspective firmly grounded in science and the humanities. This is the kind of statement that makes me so proud of their work: "But not only was there no valid rationale for even contemplating such EUAs [Emergency Use Authorizations]—the issuance was based on incomplete and patently fraudulent documentation provided by the manufacturers. Some evidence of such fraud, which should have been caught by the regulators but apparently wasn't, is presented here."[92] Bazinga! These doctors are particularly upset about the use of untested genetic products for pregnant and breastfeeding women and bemoan the consequences.[93]

D4CE also bemoans the fact that, as reports of "grave adverse events mounted rapidly in VAERS and other major databases, the EUAs have since been extended to ever younger age groups and now apply even to babies."[94] Another bemoaned fact: COVID vaccine quality and manufacturing standards have not been met as proven by various contaminations detected in numerous batches of vaccines when tested by third-party scientists.[95]

That is a lot of bemoaning for a lot of good reasons. But the wonderful thing about Doctors for COVID Ethics is that they sprang into action early and tried to warn us. They wrote letters to authorities, made videos explaining the science, and created an ebook for wide distribution for free.[96] They are exemplary, dedicated scientists and deserve our appreciation. They certainly have my undying gratitude.

Fauci Acknowledges That COVID Vaccines Did Not Work

Dr. Fauci is listed as the "quarterback" for an article published in January 2023, in which he and the authors acknowledge the failure of COVID vaccines. In their words: "In this review, we examine challenges that have impeded development of effective mucosal respiratory vaccines, emphasizing that all of these viruses replicate extremely rapidly in the surface epithelium and are quickly transmitted to other hosts, within a narrow window of time before adaptive immune responses are fully marshaled. . . . Past unsuccessful attempts to elicit solid protection against mucosal respiratory viruses and to control the deadly outbreaks and pandemics they cause have been a scientific and public health failure that must be urgently addressed. We are excited and invigorated that many investigators and collaborative groups are rethinking, from the ground up, all of our past assumptions and approaches to preventing important respiratory viral diseases and working to find bold new paths forward."[97] Gee whiz—did you see that on the news?

Conclusion

Recent advances in medicine provide evidence for individualized care. People respond differently to a novel virus (and novel vaccines) based on their baseline health, genetics, environmental exposures, and social networks. This is not a new concept. We had access to existing, cheap, and effective treatments that could have prevented countless hospitalizations and deaths. We did not know how well the novel injections would work or how bad the side effects would be. Those in charge should have been aware that a one-size-fits-all strategy would harm some people and fail to help others. But the potential profits for a universally administered "vaccine" were huge. Infuriating!

FOOD FOR THOUGHT

Did you wonder why there was so little talk about good nutrition, vitamins C and D, and stress management from governmental officials? Did your instincts make you doubt that a universal novel "vaccine" would be the best solution to the crisis? Did you know about Doctors for COVID Ethics or the Front Line COVID Critical Care Alliance before this book?

CHAPTER 6

Ivermectin: Why Did the Government Call It Horse Paste?

The art of medicine was to be properly learned only from its practice and its exercise.

—Thomas Sydenham

We have the best government that money can buy.

—Mark Twain

You Have Got to Be Kidding Me

Here is the punch line: no one can get Emergency Use Authorization for a product if there are existing effective treatments for the condition. Read that line again and think of the implications. No one can get Emergency Use Authorization **for a vaccine** if there are existing effective treatments for the condition. Bringing a novel mRNA vaccine to market for universal use *could never have happened* if clinicians showed they could treat COVID effectively, which many did.

The War on Repurposed Drugs

When a new and concerning infectious illness sweeps through a population, the tried-and-true method of responding is to determine whether there are existing therapies that could be effective. Clearly this is much faster than new drug or vaccine development! Many scientists and clinicians worked on using existing medications empirically, studying the results, and reporting their experiences. Instead of the kudos and "thanks on behalf of many grateful nations" that they deserved, these clinicians mostly got denigrated, deplatformed, and/or fired. Let me thank those around the globe who contributed to this body of research. You are truly unsung heroes and deserve profuse thanks.

Many empirically trialed treatments did fail to provide reproducible and definitive proof of saving people infected with COVID. One notable exception was using corticosteroids in the treatment of moderate or severe COVID disease. An early study from Spain was published in June 2020; many other positive studies were published by autumn 2020.[98,99,100] Note the dates—many months before shots were unleashed. Another sterling exception was ivermectin. The developer of ivermectin got a Nobel prize in 2017 for the drug that emerged from the soil in Japan because of its near-miraculous safety and effectiveness in treating parasitic diseases, including river blindness.[101]

Evidence of Efficacy

Ivermectin emerged as a potential treatment strategy because it has numerous antiviral and anti-inflammatory effects. Under the aegis of the FLCCC (Front Line COVID Critical Care Alliance), Drs. Pierre Kory and Paul Marik looked at "published peer-reviewed studies, manuscripts posted to preprint servers, expert meta-analyses, and numerous epidemiological analyses of regions with ivermectin distribution campaigns."[102] What they found is worth quoting verbatim:

> A large majority of randomized and observational controlled trials of ivermectin are reporting *repeated, large magnitude improvements in clinical outcomes* [emphasis added]. Numerous prophylaxis trials demonstrate that regular ivermectin use leads to *large reductions in transmission* [emphasis added]. Multiple, large "natural experiments" occurred in regions that initiated *"ivermectin distribution" campaigns followed by tight, reproducible, temporally associated*

decreases in case counts and case fatality rates [emphasis added] compared with nearby regions without such campaigns.[103]

Ivermectin Decreased COVID Mortality

Their conclusions about controlled trials are also worth quoting verbatim:

> Meta-analyses based on 18 randomized controlled treatment trials of ivermectin in COVID-19 have found large, statistically significant reductions in mortality [emphasis added]., time to clinical recovery, and time to viral clearance. Furthermore, results from numerous controlled prophylaxis trials report significantly reduced risks of contracting COVID-19 [emphasis added].with the regular use of ivermectin. Finally, the many examples of ivermectin distribution campaigns leading to rapid population-wide decreases in morbidity and mortality [emphasis added]. indicate that an oral agent effective in all phases of COVID-19 has been identified.[104]

Wow! This spectacular review article came out in spring of 2021 when many of us were still licking our wounds and emotionally eating our way to an average twenty-pound COVID weight gain. I read it carefully, and it made so much sense to me! I was so excited that there was so much evidence of ivermectin's value that I could not wait to see if results in my pediatric patients compared to what had been reported elsewhere.

Many Pharmacists Refused to Fill
My Ivermectin Prescriptions

However, I could not get my prescriptions filled from chain pharmacies in my town. Walgreens enforced a complete ban in my town. Prescriptions at CVS *might* be filled, depending on who was working that day. One local pharmacy would dispense the pills at $6 per pill, another would compound an oily liquid for $80 per five-day course, and distant compounding pharmacies would ship ivermectin (but that caused a bit of a delay).

Based on the wealth of data publicized on the FLCCC website, I was eager to use ivermectin on my patients. I thought it was a crime to deny a cheap, repurposed medication with an extremely low side-effect profile and an excellent chance of preventing death in any patient. I wondered if the pharmacists would change their behavior if they were aware of the data, so I passed copies of the Kory/Marik paper through drive-in pharmacy windows and waited for the pharmacists to be wowed and quit blocking my prescriptions. It did not happen. Fortunately, there were two pharmacies in my town who honored my prescriptions: Timberlake Family Pharmacy and Forest Pharmacy.

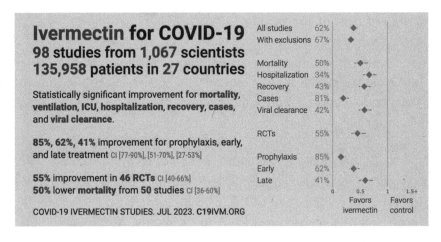

Courtesy of the Front Line COVID Critical Care Alliance.

A Decorated Marine Is Denied Requested Treatment

Several months later, a highly decorated Marine veteran named Jim got COVID when he was in my town to take his daughter to a local university. He and his wife requested he be treated for COVID with ivermectin and the doctors in my local Emergency Department refused. His wife began a search all over town to find a pharmacy to fill an ivermectin prescription and could not find one. I met Jim's wife, Bonnie, when I was testifying in front of the Virginia legislature on the issue of local pharmacists refusing to fill legitimate ivermectin prescriptions from licensed physicians for COVID patients.

This veteran's medical journey was long and tortuous. He was transferred to the University of Virginia (where I had trained and served as chief pediatric resident in the early 1980s). He was in the ICU for many months and developed bed sores and all sorts of complications from COVID and/or his treatments. He finally needed lung transplants but was prohibited from getting the surgery until he consented to a COVID vaccine. Remember, having COVID, with spike protein exposure, was what landed him in the ICU in the first place! In previous years, doctors did not typically vaccinate someone in the throes of the illness for which the vaccine could be used because of the concept of natural immunity. In the past, natural immunity, which develops after the person gets an infection, was recognized as robust and durable and almost always longer-lasting than vaccine-induced immunity. That concept went out the window in the era of dictates from on high about treatment protocols and strategies to get jabs in every arm worldwide. The official guidance was that, even when patients were recovering from COVID, they should get mRNA shots to tell their bodies to make spike proteins to generate protective antibodies. Huh? This recommendation was not supported by a century of data about the protection by the immune system after coronavirus infections. The power of natural immunity was acknowledged belatedly in the era of COVID.[105] In the meantime, many COVID infection survivors got unnecessary "vaccines." Some had bad side effects. First do no harm!

Eventually, the Marine obeyed orders, got a COVID vaccine, and had a temporary setback after the shot. He got his lung transplant, went through rehabilitation, and went home on Friday the 13th in May, more than nine months after his illness began. Thanks for your service, eh? My favorite part of the story is that his wife smuggled in ivermectin and hid it in his Bible. He swears he felt better, and she swears he had more improvements when she was able to get him his ivermectin stash. I met him at an event at our State Capitol and he looked great. Jim's story is another example of how the results of published trials, which may have errors of omission or commission or conflicts of interest, and which discuss mean and median results for a population, should not be extrapolated to deny cheap, safe, available treatments to individual patients.

Ivermectin Totality of Evidence

My friends and colleagues at the FLCCC have done a remarkable job excavating the research about how to treat COVID and the value of ivermectin. What follows is reported from their remarkable assembly of "the totality of evidence." Contrast their well-documented work below with the statement from the FDA website, "C'mon y'all, it's horse paste," which supplied endless fodder for late-night comedians to make fun of those of us who prescribed it. The efficacy of ivermectin in COVID-19 is supported by the following:

- Ivermectin inhibits the replication of SARS-CoV-2 in vitro (in a test tube). The authors wrote that ivermectin is widely available since it is on the WHO model list of essential medicines.[106]
- The nucleus is central to key stages of infectious cycles for many viruses. Ivermectin can dissociate the host importin α/β1 heterodimer/prevent reassociation. Translation: Ivermectin acts like a guardian at the gate of the nucleus, keeping bad visitors out.[107]
- Ivermectin docks to the SARS-CoV-2 spike receptor-binding domain attached to ACE-2. This docking may interfere with spike protein's ability to attach to cell membranes.[108]
- Avermectin (a close chemical relative of ivermectin) downregulates nuclear transcription factor kappa-B and mitogen-activated protein kinase activation pathway. Avermectin significantly regulates tumor necrosis factor alpha and interleukin 10. Translation: avermectin may inhibit pro-inflammatory cytokines (chemical messengers for inflammation) and have valuable anti-inflammatory effects.[109]
- Ivermectin inhibits inflammatory cytokines. Lipopolysaccharides are widely used to induce inflammation in animal models. Ivermectin significantly decreased the production of tumor necrosis factor alpha, interleukin 1B and interleukin 6 (often

elevated in COVID) both in an animal model and in test tubes.[110] Holy moly, this is great news!

- Effect of ivermectin on the cellular and humoral immune responses of rabbits: bottom line—it was good.[111]
- In Uttar Pradesh, early use of ivermectin (both preventatively and therapeutically) since the first wave of COVID helped achieve a low COVID positivity rate despite high population density.[112] I remember Pierre Kory systematically and passionately presenting this data on several FLCCC webinars. How might the COVID pandemic have played out if other countries followed India's lead?
- Ivermectin: partial monitoring results provided an extended use in positive COVID patients in Argentina, with significant documented improvements.
- "Regular Use of Ivermectin as Prophylaxis for COVID-19 Led Up to a 92% Reduction in COVID-19 Mortality Rate in a Dose-Response Manner: Results of a Prospective Observational Study of a Strictly Controlled Population of 88,012 Subjects." I think this might have been one of the studies that prompted Pierre Kory to exclaim, "Put a fork in it—it's done!" People in Argentina who used ivermectin preventatively had an 86 percent decrease in dying, decreasing death risk to 0.00024 percent. The risks decreased in a dose-dependent manner also dependent on compliance, which lends validity to the conclusion that ivermectin saved lives.[113]

A Great Book Straight from the Horse's Mouth

Dr. Kory has written a superb book about his experiences trying to bring a life-saving treatment to COVID-infected patients. Check out *The War on Ivermectin: The Medicine That Saved Millions and Could Have Ended the Pandemic*. It may not surprise you to learn that Dr. Kory lost or quit three jobs over the course of his in-depth research about ivermectin for speaking out in favor of using it acutely to treat COVID. As I write this, the Board of Internal Medicine has just withdrawn his internal medicine board certification. Shame on them! By tying the hands of experienced intensive-care and front-line physicians, millions of people were denied effective outpatient treatments. Many people died as a result. The advice to "stay at home until you turn blue then go to the Emergency Room" ignored basic principles of early treatment of viruses and fundamental tenets of good medicine. My analysis is that COVID countermeasures and the suppression of existing drugs was ethically reprehensible and responsible for immeasurable global grief. They lied; people died.

FOOD FOR THOUGHT

What did you think about ivermectin during COVID? Do you know people who wanted to use it and found doctors to prescribe it who could not get their prescriptions filled? Do you know people who took ivermectin and said it really helped them? Did you believe the hype that ivermectin was only for horses?

CHAPTER 7

How Did COVID Countermeasures Hurt Children?

What experience and history teach is this — that nations and govern-ments have never learned anything from history, or acted upon any lessons they might have drawn from it.

—Georg Wilhelm Friedrich Hegel

When at some future date the high court of history sits in judgment on each of us . . . our success or failure . . . will be measured by the answers to four questions:

> *First, were we truly men of courage . . . ?*
> *Secondly, were we truly men of judgment . . . ?*
> *Third, were we truly men of integrity . . . ?*
> *Finally, were we truly men of dedication . . . ?*

—John F. Kennedy

Making Decisions in a Crisis in Real Time

My husband is the best Monday morning quarterback I know. When I come home from work with a story about how I had a conversation with a commercial entity that frustrated me or did not go my way, he is full of spectacular ideas about what I could have/should have done. I'm sure many women can relate. We want to vent, be understood, and validated; men want to fix our problems.

It is understandable that, early in 2020, it was difficult to know what to recommend or what people should do. In writing this book, I have tried to look at the issues I critique in the context of what we knew when. While it does not seem fair to judge early governmental or agency responses based on everything we know now, I focused on what was known in 2020 and 2021 that could have changed the trajectory of the crisis for the better. By late 2020 and the first half of 2021, it was reasonable to expect better from our officials and agencies. Public health recommendations lagged far behind the published science. Messages from the CDC often changed and were sometimes internally contradictory. Many citizens were confused. I argue they were misled.

Warnings About Collateral Damage

Early in the pandemic, economists, epidemiologists, and Children's Health Defense warned about the collateral damage that would result from the extreme measures being adopted. See what John Ioannidis (Stanford epidemiologist), Anders Tegnell (epidemiologist from Sweden), JB Handley (autism activist), and Children's Health Defense said in 2020.[114,115,116,117,118,119,120]

They opined the measures were too severe and the damage to children would be huge.

Lockdowns and Job Loss Effects on Mental Health

During COVID, Dr. Fauci was focused on vaccine development and casting doubt on repurposed drugs that front line doctors were using to successfully treat patients. He was less concerned with the economic and social impacts of his proposals. In retrospect, he says that he knew the "draconian" COVID-19 policies he pushed for would lead to "collateral negative consequences" for the "economy" and "schoolchildren."[121] At the time, however, he tried to silence those who warned about dire consequences who turned out to be correct.[122,123]

Confining children and teens to their homes had disastrous effects. Lack of personal contact with peers has devastating mental health impacts on adolescents. Suicide attempts began rising as early as February 2020 and, by March, were double the same month of the previous year.[124] Comparing spring 2020 to spring 2019, obsessive-compulsive disorder (OCD) in adolescents also nearly doubled.[11] Depression or anxiety increased fourfold.[125] Suicide attempts in youth stayed elevated throughout 2020.[126]

Younger children had to deal with a wide variety of fears directly related to the coronavirus, despite their limited ability to understand infectious disease and medical complications. Children have vivid imaginations and often envision scenarios that are worse than reality. Children were forced to cope with difficult real-world situations such as being unable to visit grandparents who were sick in the hospital. The flip side of that coin is that grandparents were often confined alone in their homes, and many were unable to see their grandchildren for two to three years! I have met many elderly people whose spirits were crushed, who were widowed early in the pandemic or forbidden from seeing family members. For children who lost loved ones, the bereavement experience was very different from usual. Typical cultural grieving practices like wakes, burials, and visitation with friends and relatives were shut down. Sometimes bodies were burned on site at hospitals against the wishes of family members of the deceased due to unsubstantiated fears about contagion of the dead. Aaron Kheriaty, MD, schooled in ethics, spoke poignantly about the cultural importance of grieving rituals at a Senate committee meeting.[127] Dr. Kheriaty, a psychiatrist, ethicist, and all-around good guy with five children, advised appropriately that mandating experimental COVID vaccines to medical students and residents at his university was unethical. True, but he lost his job.

Domestic Violence and Child Abuse

During times of economic instability, especially when parents have lost their jobs, rates of domestic violence and child abuse rise.[128] The economic fallout from COVID-19 lockdown measures was no exception. When parents are under severe stress, it is hard to parent effectively and consistently. Some parents with addictive personalities turn to alcohol or drugs to cope, and the impact on their children can be harrowing. Removing children from imaginative play with their peers and from the normalizing experience of interacting at school with teachers and classmates also worsens children's stress. They have fewer people to talk with and their coping mechanisms are compromised.[129,130,131]

Teaching Children to Be Afraid of People and Their Environment

Kids were indoctrinated during COVID to wash their hands often to get rid of "bad germs." The "Happy Birthday" song was even repurposed as a timer for effective hand washing. Well-meaning parents deployed hand sanitizers widely to protect their children from what was perceived as a deadly virus. There is credible data that widespread use of hand disinfectants has detrimental consequences.[132]

Decades of evidence have proven that children who play in the dirt and are exposed to animal licking are healthier than kids who grow up in overly sterilized environments. This concept is known as the "hygiene hypothesis."[133] The value of exposure to nature and animals for modulating a healthy immune system in childhood has been confirmed by evidence from multiple cultures. I am delighted that my son's dog Rory and my daughter's dog Olaf have been licking my granddaughters since they were babies. Exposure to dirt and animals plays a particularly important role in preventing autoimmune disorders, which have risen dramatically in the past several decades. In fact, pathogenic viruses, bacteria, and parasites have strong protective properties against autoimmune diseases like type 1 diabetes, Hashimoto's thyroiditis, and celiac disease.[134]

Responding to this body of evidence, pediatricians are beginning to use probiotics to help children develop good gut flora, especially if the child has been treated with lots of antibiotics or was delivered by C-section (and not exposed to the mother's beneficial vaginal flora during childbirth). Dr. Yehuda Shoenfeld, the "grandfather of autoimmunity," told me a funny story about the birth of his grandchild. He was in the delivery room and was tasked by his daughter with providing a "perineal picnic" for the newborn. What is a perineal picnic, you ask? It's a catchy phrase coined by my mentor, Dr. Sid Baker. Yehuda was supposed to get some secretions from his daughter's nether regions on a piece of gauze and wipe the new baby's mouth with the gauze. Theoretically, this inoculates babies born by C-section with normal vaginal flora. He got so excited by the birth that he forgot to do his assigned task! He was glad that there was a family dog at home who was happy to lick the baby.

"COVID kids" have been taught to fear other people as a potential source of dangerous germs. Some kids are preoccupied with obsessive thoughts about contagion and handwashing. They cannot be expected to understand complex immunology which includes extensive data about the value of various microbiomes. What they hear is what the adults around them say; what they learn is what they see adults do. Many adults were understandably fearful of COVID infections and indoctrinated their kids to be overly afraid of germs. I tell patients and parents that childhood is a "journey through various infections." As the child gets sick with various viruses, their immune system modulates its ability to respond appropriately. That is the way nature works. We defy nature at our peril.

The Effect of Lockdowns on Diet and Exercise

Processed foods are pro-inflammatory and should have been avoided during the pandemic. Unfortunately, because the entire population was being fed a steady media diet of fear, people turned to "comfort foods" to cope. Lockdowns and lost jobs led many families to rely more on cheap processed foods. Even one fast-food meal is associated with significant elevations in pro-inflammatory cytokines—chemical messages that turn on inflammation.[135]

High inflammation is the opposite of what you want if you get an infection. Because SARS-COV-2 is marked by the potential for abnormal cytokine responses, it would make sense not to stoke the fires of inflammation with pro-inflammatory foods. Fast food also impairs the adaptive immune system, which lowers host defenses against viruses like COVID and influenza.[136]

In kids, the combination of 24/7 access to processed foods, decreased exercise, and lack of peer relationships was often associated with significant weight gain. Documenting increased consumption of soft drinks, simple carbohydrates, and sweet and salty snacks, German researchers found that body weight had increased significantly in ten- to twelve-year-olds, with 24 percent of boys affected and 13 percent of girls.[137]

The health benefits of exercise were also curtailed as children were kept inside by anxious parents during the height of the COVID pandemic. The German researchers found that physical activity declined in 38 percent of all children and in 60 percent of children aged ten years and older. It is unfortunate that so many parents were afraid to let their kids play outside, since the risk of transmitting or catching COVID in open air spaces was infinitesimally small.[138] Many parents took social distancing very seriously and limited kids from playing outside in parks and playgrounds unless they could maintain six feet of separation from other children.

Loss of Interactive Play

The essence of early childhood development is unstructured play. On the playground, children learn how to negotiate their role in various groups and nurture their imagination. The question that many have been asking since early 2020 is whether the trade-off of sacrificing interactive play in the name of avoiding a respiratory virus was worth it. If kids were a major source of infection to others, one might argue that the sacrifices of staying inside for a year and being told "not to share" were justified to protect vulnerable groups like the elderly or the chronically ill. But the data show that kids have largely not been responsible for transmitting COVID infections.[139] One of my favorite second cousins has three young boys, including a set of twins. When they were finally able to go back to day care, they were placed at desks six feet apart and given a plastic box full of toys to play with alone. Stories like that broke my heart!

Educational Fallout

As schools closed their doors to in-person learning, online methods were employed for age groups developmentally unable to master the technologies being used. Elementary school–aged children do not have the developmental skills to thrive with computerized distance learning. My daughter and her husband teach elementary school. "Shock and awe" are the words that describe my reaction to their mandate to teach kids as young as five via zoom. My son-in-law described the challenges of just getting kids to keep their videos on. He suspected that some kids turned off their videos so they could take a nap or play. I can't really blame those kids. We were asking the absurd. Many teachers now report that their virtual students have learned less than half of the expected academic content during virtual compared to in-class instruction.

The results of a study that polled teachers, students, parents, and policymakers from multiple countries revealed numerous adverse effects of lockdowns on education, including learning disruptions, decreased access to education, and increased student debt.[140] Poor infrastructure (such as limited network access, poor digital skills, and power outages) also hampered the switch from in-person learning to online classes. My daughter got very upset that some of her students who did not have Wi-Fi at home had to sit in the car at Walmart or McDonalds to connect to her zoom classroom.

Brown University released a horrifying study about the impact of COVID lockdowns on early childhood development. They were already in the middle of a large ongoing longitudinal study of child neurodevelopment. They compared childhood cognitive scores in 2020 and 2021 to the same measures in 2011 to 2019. They found that children born during the pandemic have significantly reduced verbal, motor, and overall cognitive performance compared to children born pre-pandemic. Not surprisingly, they found that males and children in lower socioeconomic families were most adversely affected. They wrote:

Even in the absence of direct SARS-CoV-2 infection and COVID-19 illness, the environmental changes associated with the COVID-19 pandemic are significantly and negatively affecting infant and child development. . . . we find mean ELC values from 2011 to 2019 ranging from 98.5 to 107.3, with standard deviations of 15.2 to 19.7 . . . Means and standard deviations for 2020 (March to December) and 2021 (January through Aug) were: 86.3+/-17.9 and 78.9+/-21.6, respectively.[141]

To oversimplify, most infants and toddlers prior to the COVID crisis scored around 100, like IQ studies in adults. Infants could be below the mean down to 85 or above the mean to 115 and still be considered cognitively normal. Now the averages were 86.3 from March to December 2020 and 78.9 from January to August 2021.[142,143] From a pediatrician's viewpoint, this is a disaster!

The Downside of Masks in Childhood

Children's Health Defense has reviewed the scientific literature about the effectiveness of masks in preventing transmission of SARS-CoV-2 and concluded, as did these authors in *The Journal of Infectious Diseases*, there is no consistent, unbiased, and credible evidence that masks work to prevent transmission in community settings. Children's Health Defense voiced concerns that mandating masks in school or community settings has adverse impacts on children's social and emotional development. Brownstone Institute published an article by another one of my COVID heroes, Dr. Paul Alexander, who tried to bring reasoned argument to the Trump administration early in the crisis. Paul compiled a list of 170 studies that showed face masks ineffective.[144]

During the COVID crisis, educators and psychiatrists raised concerns that "covering the lower half of the face reduces the ability to communicate, interpret, and mimic the expressions of those with whom we interact," noting that "Positive emotions become less recognizable, and negative emotions are amplified."[145] These experts also point out that masking reduces "emotional mimicry, contagion, and emotionality," affecting "bonding between teachers and learners, group cohesion, and learning."[146] The mask-related amplification of negative emotions comes on top of the 24/7 cycle of bad news that ramped up children's fears and activated the amygdala. When children are afraid, they are even more prone to misinterpret benign events as threatening, making it more likely they will fight or flee.

A whole series of social-emotional milestones are an integral part of childhood development. In infancy and early childhood, being able to recognize and interpret the facial expressions of others is one of the most crucial developmental tasks. Pediatricians watch for the development of a social smile between six and ten weeks of age as an early marker of normal bonding and social interaction. Two decades ago, a study of 120 school-aged children (ages five to ten years) sought to determine how visual spatial parameters of facial expressions develop in relationship to pathways of emotional language.[147] The researchers concluded that "emotion cognition is a variegated domain"[148] and is complex. We do not know the extent to which covering the lower half of the face is confusing for preschool and elementary school–age children.

We have seen data about the higher prevalence of language delays in "COVID babies." There are many potential causes for this—depression in caregivers leading to fewer vocal interactions, for example. But one must wonder how babies can learn to talk if they cannot see the mouths of those talking to them because they are covered in a mask. Adults wearing masks is an especially significant problem for children on the autism spectrum. Studies show that kids with autism look at mouths more than eyes, which is different from neurotypical children who look mostly at eyes.[149] A large body of research has confirmed that children are able to match expressions of joy, sadness, surprise, and anger with varying degrees of accuracy. However, child development experts have discovered that children match happy expressions with fewer errors than when they try to match surprised or angry faces.[150]

I noticed that when parents wore masks, I often had to ask them to repeat themselves. It made me realize how much I augment my hearing with lip reading day by day. We do not know the full extent to which exposure to less visual and auditory input from masked faces affects children's ability to "read" faces and decipher or remember emotional content. We do not know the full extent to which masking interferes with the child deciphering words.

There are also clear impacts on our environment from the production and widespread use of face masks.[151,152] James Roguski, one of the journalists who has maintained his ethics and integrity during the COVID crisis and is a regular attendee of our Thursday international COVID calls, wrote a nice Substack review of the issue of masks entitled, Mask Charade.[153]

Conclusion

From my perspective as a pediatrician trained in child development, there were no significant gains in limiting the spread of COVID by keeping kids out of school and away from their peers. And I fear the negative effects of being a "COVID kid" are wide reaching. Educators are worried that many children will not ever catch up despite our best efforts. I worry we have harmed some children irrevocably, and they will not "catch up." However, the long-term consequences are still unfolding, and we may find that surprising opportunities for root cause healing will emerge. The adults in the room need to learn the lessons from the COVID crisis and mobilize to prevent similar harms to children in the future.

FOOD FOR THOUGHT

Did you have a teen who got severe anxiety or depression during COVID lockdowns? Do you think your children have more fear of germs now compared to before COVID? Do you think masking made it difficult for your children to interpret emotions?

CHAPTER 8

What Happens to Good Doctors Who Voice Concerns About Medical Dogma?

I learned that courage was not the absence of fear, but the triumph over it. The brave man is not he who does not feel afraid, but he who conquers that fear.

—Nelson Mandela

Courage isn't having the strength to go on – it is going on when you don't have strength.

—Napoleon Bonaparte

Profiles in Courage

Many clinicians came forward during COVID to advocate for better early treatments and raise concerns about the quickly manufactured "vaccines." They often met resistance, criticism, or even defamation for their efforts on behalf of their patients. Here are the stories of two such physicians, which I tell in detail because I don't think the average patient knows that retaliation against such good doctors is so common.

Paul Thomas: Courage of Convictions

"Growing up in Africa seared my heart and soul. I dream in Shona," (an African language) the good doctor said. Paul Thomas grew up in Rhodesia which is now Zimbabwe. His parents were missionaries immersed in the African culture. When he was four years old, he moved to a village in the middle of land that was farmed by whites. Displaced Africans were given some land on which they built homes. "We lived in a three-room house with a living room and two bedrooms, a separate thatched roof kitchen and no running water, electricity, or indoor toilets. The village consisted of the church, a three-room school and 50 to 100 homes. When I was there Rhodesia was segregated like apartheid in South Africa. I was one of two white kids in a black school." Mai Rukunda lived next door. She became Paul's second mother; her children became a set of second siblings to Paul.

Paul's father set up a United Methodist Church in the village. Since there was no health care within five miles, Paul's mother started a clinic that eventually became a major maternity center. From an early age, Paul had role models who showed the importance of service to others. Paul's family moved to Saulsberry, now Harere, the capital of Rhodesia, where his father worked at Epworth, the theological college. For the first time, Paul was not in the racial minority, for now his classmates were almost exclusively white. Paul was an excellent student and a great athlete. He was the best field hockey player, the best high jumper, second in sprinting, and "just good" in soccer. His education followed the British system. In high school, every member of the senior class was a "prefect," expected to support law and order among the younger students. Paul was selected to be head boy, which was a huge honor. One of the ceremonial responsibilities for the head boy was to take down and put up flags during an annual patriotic ceremony. In his first act of civil disobedience, Paul—the esteemed head boy—refused to raise the flag of the oppressor. As he explains,

> I was already aware of the need to fight for the liberation of African people. Having grown up with so many black playmates, I did not think of myself as white. In fact, if I saw the group of whites coming toward me, I would cross the street to avoid them. I was completely comfortable when I saw a large group of black boys approaching me.

In another act of civil disobedience, twelve-year-old Paul distributed educational information about how to resist illegal acts. At the time, local trains were extremely racially segregated. First class with all the trimmings was for whites only. Second-class was also limited to whites. The native African Blacks were consigned to third class (sitting on a wooden bench) or fourth class (standing packed together in a cattle car). For his audacity in educating black Africans about their options, Paul was arrested and interrogated by the police. He explained "I kept refusing to give them my name. I don't know what I was thinking because I was a kid and eventually, I would need to go home."

Missionaries typically go back to the United States every three to five years to share stories of their work and raise funds. Paul wanted to be a minister like his father until adolescence, when he decided to become a doctor. Paul went to college in Kalamazoo, Michigan, where an aunt lived nearby. He had been told that every student the college recommended for medical school got in. What he did not know was that they only recommended one of twenty students. The atmosphere among pre-med students was brutally competitive, characterized by students stealing one another's textbooks and sabotaging one another's lab experiments.

After his freshman year, he transferred to the University of the Pacific in Stockton, California, where he graduated with a bachelor's degree in biology. During college he lived in a room in a *mortuary* down the hall from the embalming room. In exchange for being a corpse guard at the funeral home, he got the room for free and earned some tuition money. Paul stayed in Stockton for two more years to earn a master's degree in biology. During his master's education, he ran science labs and taught anatomy and physiology.

From 1981 to 1985, Paul attended medical school at Dartmouth in New Hampshire. Even though he got scholarships, tuition was still very expensive. During medical school, Paul lived in one room that had no running water and therefore no toilet, which was daunting during the winters when snow was five feet high. He sneaked into the med school dorm to shower. He went to Fresno, California, for the first two years of his residency training in pediatrics.

Dr. Thomas spent the last year of his pediatrics residency training at the University of California San Diego. He was chosen by his peers as Best Teaching Resident. He moved to Portland in 1988. His first job highlighted his teaching skills. He worked at Legacy Emmanuel Children's Hospital and taught students from Oregon Health Sciences University for about five years. He was an attending physician in the hospital and went to court multiple times to advocate on behalf of abused children. In 1993, he joined Westside Pediatrics, a group where he was in the minority again—the other four physicians were women. He worked as a busy pediatrician in that group setting for many years.

In 2002, Paul got a call that his African "sister" Tzitsi had been killed, leaving her children orphaned. Their father had died of colon cancer seven years before. "In the moment I felt a calling washing over me," Paul explained. "I turned to my wife and asked if we could take care of Tzitsi's kids." The African custom is for friends and family to take care of orphans. Instantaneously, Paul and his wife got four more children to add to their three biological children and two adopted children.

In 2004, Dr. Thomas was struck by his case in which a child who was developmentally normal at one year of age regressed into autism by age two. Paul wondered why that had happened. When he saw the same types of regression in patients in 2005, 2006, and 2007, he was already on the way to learning much more about the phenomenon of autistic regression. One day while trying to do a well child check on a two-year-old, Dr. Thomas could not get the child to interact with him at all. The child's development had been perfect at twelve months, okay at eighteen months (except for a slight language delay) and now he was unreachable at the age of two. This was the final straw that broke the proverbial camel's back and propelled Paul to start his own practice. Thus, in June 2008, Integrative Pediatrics was born.

During these years, he attended conferences organized by the Autism Research Institute. I was the medical director of ARI at that time, busy designing continuing medical education courses about the medical problems of children with autism. We had discovered that, by finding and managing underlying medical problems of children with neurodevelopmental disorders, some children got dramatically better and a few even "lost their diagnosis" of autism.

Dr. Thomas attended many clinical training sessions offered by the Autism Research Institute. He began treating children with autism and was struck by stories from the parents that symptoms often began after a set of vaccinations. Sometimes the child had been given lots of vaccines at once, sometimes the child was vaccinated while sick, and sometimes a regression followed closely on the heels of an MMR vaccine. Based on his clinical judgement, and experience, Dr. Thomas modified the CDC immunization to give fewer vaccines at once. He honored the parents' right to give informed consent. He wrote a book with Jennifer Margulis, the daughter of Lynn Margulis (the scientist who discovered mitochondria) who had been married to Carl Sagan (billions and billions of stars), entitled *The Vaccine-Friendly Plan*. This brought him to the attention of the Oregon Medical Board.

The board challenged Dr. Thomas to prove that his modified schedule was as safe as the one recommended by the CDC. Paul took on the challenge. A data miner was hired to comb through the records of the sixteen thousand patients in his practice to discover evidence of harm or benefit. An inconvenient truth emerged: the unvaccinated or under vaccinated had many fewer chronic illnesses. Dr. Thomas's peer-reviewed paper was published in November 2020.[154] Five days later, on Sunday, December 3, he received a call that he would not be allowed to work at his office the next day since he had been deemed "a threat to the safety of children." His medical license had been emergently suspended.

Thus began a multi-year legal battle against the Oregon Board of Medicine. At first, Paul had the energy to clear his name and get his license back. As time went on, his lawyer informed Paul that to play out the legal process to a final verdict would likely cost a million dollars. A successful outcome could not be assured due to the structure of the proceedings. Dr. Thomas eventually decided to surrender his license and become a health coach and child advocate, not because he did anything wrong, but because of the toll the legal fight had on his marriage, health, financial status, and sense of well-being.

I volunteered to be an expert witness on Paul's behalf. I reviewed the charts of patients the board alleged he had harmed. I did not find any evidence he had hurt them in any way. I did find evidence that he was a well-informed and careful doctor. If my grandchildren lived near Portland, I would be thrilled for Dr. Thomas to be their pediatrician.

One patient's case is illustrative of the ludicrous accusations made by the board—a patient who lived in rural Oregon. The boy had not been vaccinated against tetanus when he suffered a head wound from a rusty farm tool. He developed tetanus, was hospitalized at a university hospital, got treated in the ICU, and survived. The parents called Dr. Thomas's office for the discharge follow-up the hospital mandated. The parents reported other practices in town would not see their son because he had not been up to date on his tetanus vaccine when he was injured. Paul agreed to see him for the first time when he got out of the hospital. To be clear, *the boy had not been Paul's patient prior to the hospitalization.* During that first visit, Paul had to review copious hospital records, establish rapport with the family, and plan for the boy's ongoing care. The board's criticism of that outpatient encounter was that Dr. Thomas did not give the child a tetanus shot during that first visit. That criticism seemed absurd to me. First, the patient had just had tetanus and would have developed antibodies from the illness. Second, how many clinical tasks could the board realistically expect Dr. Thomas to achieve during one initial outpatient visit? Third, the boy had spent weeks at a university hospital, and none of the hospital's myriad interns, residents, or attendings had given him a tetanus shot. They were the academicians with weeks to think about his case. If a tetanus shot was indicated (which it was not, in my opinion), why not hold the university doctors responsible for the omission?

If you want to read more about Dr. Thomas's case, Jeremy Hammond has written a book called *The War on Informed Consent: The Persecution of Dr. Paul Thomas by the Oregon Medical Board.* In the author's words, "*The War on Informed Consent* exposes how the medical board suspended Dr. Thomas's license on false pretexts, illuminating how the true reason for the order was that, by practicing informed consent, he posed a threat to public vaccine policy, which is itself the true threat to public health."[155]

Dr. Thomas is a well-trained, energetic, ethical, and compassionate pediatrician. He is among the best and the brightest. If a medical board action can derail his career, it can happen to any of us. I grieve for the loss of another decade of his life's work as a practicing pediatrician. I think of the families he could have helped. You can see him now on a weekly Children's Health Defense show called *CHDTV – Pediatric Perspectives*.

Paul Marik: A Cautionary Tale of an Eminent ICU Specialist

Dr. Paul Marik is an astute clinician and ICU specialist from South Africa. One of the outstanding contributions of his career was his revolutionary work in developing a lifesaving protocol for sepsis (a bloodstream infection that causes more than 250,000 deaths each year in the US), which included the use of IV vitamin C. Because of the seminal work of Linus Pauling and Paul Marik, I became a fan of vitamin C. When our hospital held its first COVID meeting for the medical staff in March 2020, I advocated using vitamin C on our wards. It never happened.

Dr. Marik has specific training in Internal Medicine, Critical Care, Neurocritical Care, Pharmacology, Anesthesia, Nutrition, and Tropical Medicine, and Hygiene. He was a tenured Professor of Medicine and Chief of the Division of Pulmonary and Critical Care Medicine at Eastern Virginia Medical School (EVMS) in Norfolk, Virginia. Dr. Marik has written over five hundred peer-reviewed journal articles, eighty book chapters, and authored four critical care books. These publications make him the second most published critical care physician in the world! His papers have been cited over forty-three thousand times in peer-reviewed publications—forty-three thousand! He has delivered over 350 lectures at international conferences and visiting professorship and received numerous teaching awards, including the National Teacher of the Year award by the American College of Physicians in 2017.[156]

My brother, a family physician in Richmond, Virginia, turned me on to the protocols developed by the FLCCC. I used their treatments when I had COVID after treating COVID-positive patients. My oxygen levels never dropped, and I never needed to see a doctor. Aside from one day of muscle aches, a few days of fatigue, and one day of crying that could not be explained by my life circumstances, I was fine. In retrospect, I think the crying was due to inflammatory cytokines from the infection having a party in my amygdala (the part of the brain that regulates strong emotions).

Paul knew from his work with FLCCC that early treatment and repurposed medications could save lives. Sentara Hospital in Norfolk, Virginia, insisted he follow protocols that included remdesivir and ventilators but excluded ivermectin and high dose steroids Paul knew could help. In text messages to me from October 27, 2021, Paul was despondent, writing "I am rounding. It is a nightmare," "This is awful. Have seven patients dying from COVID that I cannot treat," and "This is murder. They are killing my patients." He testified under oath about that experience, and you can find it on YouTube labeled "false" and "misinformation."[157] Tell me, when an eminent ICU specialist testifies under oath about his personal clinical experiences, how can a journalist or fact-checker legitimately label that sworn testimony false? I beg you to watch his demeanor and see if you think he is lying.

When I heard that Dr. Marik was to be in court on November 18, 2021, for his conflict with Sentara, I felt compelled to attend his hearing to support him. He had filed suit against Sentara Healthcare for instituting a policy that prevented physicians from administering proven, safe, off-label, FDA-approved, life saving therapeutics for the treatment of COVID-19. Many supporters were waiting outside the Norfolk courthouse carrying signs that said things like "Ivermectin saved my husband's life," "Dr. Marik is a hero," and "God bless you, Dr. Marik." I watched in horror as his entire head turned bright red with anger when the Sentara administrators were on the witness stand. I was worried he was going to have a stroke.

In December 2021, Paul was called before hospital administrators for a peer review to decide whether he was incompetent or engaging in unprofessional conduct that would impact his clinical privileges. This action was taken under the auspices of the Healthcare Quality Improvement Act, passed by Congress in 1986. As Dr. Marik explains,

> This was, seemingly, a well-intentioned move. Recently, however, the Act has carried a number of unintended negative consequences, not least of which is the rise of sham peer reviews. These

are essentially "kangaroo courts" used to consign highly qualified and guiltless physicians into obscurity. . . . Sham peer review is an adverse action taken in bad faith by a hospital for purposes other than "the furtherance of quality health care." It is a process disguised to look like a legitimate peer review, but which in actual fact is not objectively reasonable precisely because it is not performed in the interests of patients.[158]

As a well-trained and well-published academician, Dr. Marik turned to the medical literature to illuminate his dilemma. He quoted an editorial by Dr. Lawrence Huntoon, editor in chief of the *Journal of American Physicians and Surgeons*: "The truth and the facts do not matter because the outcome is predetermined, and the process is rigged."[159] After Dr. Huntoon wrote his editorial, he gave dozens of talks on the subject. Many doctors attending his lectures said they recognized tactics that had been used against them also.

Dr. Marik tells his side of the story on the FLCCC website. I believe him.

Sadly, sham peer reviews have become fairly common. . . . Hospital officials are especially resistant to whistleblower physicians who bring patient safety or care quality concerns to the public's attention. After the court hearing as agreed by all parties, I was scheduled to work in the Intensive Care Unit at Sentara Norfolk General Hospital the weekend starting November 20. When I arrived at work that day, I found a hand-delivered envelope (it had no postmarks) on my desk. Inside was a letter from the Sentara Medical Staff Office, marked overnight delivery, which indicated it had also been sent by email, though no email was ever received. Signed by the President of Medical Staff, Sentara Hospitals, and the President, Sentara Norfolk General Hospital, the letter said:

> *The purpose of this letter is to inform you that a series of events have recently been reported to Hospital Administration and Medical Staff leadership at Sentara Norfolk General Hospital that have caused significant concern about your ability to conduct yourself in a*

professional and cooperative manner in the Hospital, which is essential for the provision of safe and competent patient care.[160]

The letter went on to list a number of outrageous claims, with no substantive evidence to support them, and with no patient details that would allow me to refute them. Based on the alleged incidents, the letter continued:

Medical Staff leadership has determined that your behavior causes such concern that there are grounds to impose a precautionary suspension of your Medical Staff appointment and clinical privileges.[161]

Although the hospital claimed this action was unrelated to my legal case and that the timing was purely coincidental, it is categorically clear that this sham peer review process was nothing more than corrupt retaliation.[162]

Dr. Marik continues with a review of the sham review playbook:

Although no court of law would permit depriving an accused person of files or records needed to defend themselves, as it is fundamentally unfair and in violation of due process, hospitals that employ sham peer review frequently refuse to provide records to the physician under review. Consider this: In the criminal justice system, accused serial murders, rapists, child molesters, drug dealers and thieves are entitled to due process and presumed innocent until proven guilty. But accused physicians in the hospital sham peer review process are presumed guilty and afforded limited (if any) due process. My very own "kangaroo court" happened on December 2, 2021. Not only was I not allowed legal representation, no other person could accompany me—not a colleague, friend or business representative. The proceedings were not recorded, and no transcripts were made available. I was confronted by about 25 angry people, none of whom introduced themselves.

During the course of my entire career, I have prided myself on my professional conduct, being courteous and polite to students, residents, nurses, patients, and families. In my more than 35 years on the job, during which time I have published more scientific articles in medical journals on critical care than any other physician in America, I have never been sued, never had a single patient complaint, and consistently received positive evaluations from those with whom I work. No one has ever lodged a complaint about my "unprofessional behavior."[163]

Long story short, Dr. Marik resigned from the hospital and medical school soon after. He now devotes his time to research and education in a new venue he co-founded: the Front Line COVID Critical Care Alliance. In March 2022, the Virginia House of Delegates unanimously awarded a commendation to Dr. Marik for "his courageous treatment of critically ill COVID-19 patients and his philanthropic efforts to share his effective treatment protocols with physicians around the world."[164] Well deserved, Paul.

Conclusion

These two Pauls are among the most caring, compassionate, and competent doctors I have even known. The world needs to know that organized medicine, now in an unholy alliance with Big Pharma, Big Tech, and Big Food, may go after exceptional doctors who see what others do not and put the well-being of their patients over their own incomes and careers. Such clinicians, and there were many during COVID, deserve our gratitude and support.

FOOD FOR THOUGHT

Did you wonder if your doctor was telling you the whole truth during COVID? Did your own instincts make you wonder if something nefarious was going on? Had you heard about Dr. Marik or Dr. Thomas before you read this chapter?

PART II
COVID "VACCINES"

CHAPTER 9

How Do COVID "Vaccines" Work?

We hang the petty thieves and appoint the great ones to public office.
—Aesop

Few things are more irritating than when someone who is wrong is also very effective in making his point.
—Mark Twain

How Do Traditional Vaccines Work?

Traditionally, vaccines against viruses were developed by weakening the virus so that it could no longer infect the recipient. When injected, the attenuated or inactivated virus would induce an immune response leading to antibodies against the virus. Viruses can be attenuated (weakened) by passing them through various cell cultures under suboptimal conditions to remain infectious or inactivated by exposure to temperatures outside the virus's comfort range.[165] However, in recent years, vaccine development has gone more and more high-tech with genetic recombination of viruses, use of "subunits" of viruses, and genetic manipulation of harmless adenoviruses to express the desired antigen (the technology used by the Johnson & Johnson COVID vaccine). From the time I heard the first descriptions of the novel mRNA technology COVID products, I was concerned about the unknown effects of an experimental technology.

Novel Gene Therapy Products

In my opinion, mRNA "vaccines" are potentially dangerous gene therapy products, unleashed on unsuspecting populations after a campaign that promoted fear. This new generation of vaccines uses genetic mechanisms. Synthetic (not natural) mRNA, which tells cells which proteins to produce, induces human cells to manufacture the spike protein of SARS-CoV-2. The spike protein is presented to the immune system as an entire protein or in fragments. ". . . it is essential to underline that every human cell that intakes the LNPs and translates the viral protein (in case of the mRNA vaccines), or that gets infected by the adenovirus and expresses and translates the viral protein (in case of the adenovirus-based vaccines), is inevitably recognized as a threat by the immune system and killed."[166] By the end of this book, you will understand why I am so worried about each cell that is exposed to lipid nanoparticles, replicating spike protein, and any contamination that occurred in manufacturing. Human cells are targeted for attack from within the body's intricate immune system.

Why Would We Think the Vaccine Stays in the Arm?

Remember the spiritual "Dem Bones" that we sang in elementary school, "the knee bone's connected to the thigh bone"? Human bodies are exquisite networks made of complex integrated systems. Cells talk to one another via complex biochemical messengers. This is a foundational concept of medical science. Biodistribution studies use biologic markers to tag vaccine ingredients and see where they go in the body. In my opinion, adequate biodistribution studies were not done (or had not been released) prior to injection of billions of people with mRNA vaccines.

One very concerning rodent study, done by Pfizer for the regulatory agency in Japan, demonstrated that the lipid nanoparticles accumulated in multiple organs, including the spleen, thyroid, liver, and pituitary gland, and were continuing to build up in the ovaries when the study ended.[167] Studies from the EMA (European Medicines Agency) replicated the distribution of LNPs to various organs in rodents.[168,169]

How Can Spike Proteins Hurt Us?

Dr. Bhakdi and Dr. Palmer enumerated the ways eloquently in their article, "Elementary, my dear Watson: why mRNA vaccines are a very bad idea." The consequences of widespread distribution of spike protein and the severity of potential damage depend on the type of tissue (ovary versus testes versus kidney, etc.) and the individual's immune response. Patients with preexisting autoimmunity, for example, might respond differently than someone with a robust normal immune response. Since myocarditis is a known adverse event from COVID injections, let's use the example of a heart cell (called a myocyte). If lipid nanoparticles deliver mRNA to the inside of heart cells, which then manufacture spike proteins, and the immune system recognizes those spike proteins as foreign and attempts to kill them, the heart cell may die also.[170] This is not even collateral damage. Since the spike protein is in the cell, it is targeted assassination. Scary stuff on a conceptual level, I think.

Immunology Basic Insights

Here goes a deeper dive into the cellular biology. Spike protein gets processed by the Golgi body, an apparatus inside the cell named after an Italian scientist. The immune system "sees" the spike in one of two ways: either as an entire protein or as fragments. In the entire protein scenario, B cells and helper T cells recognize the invader, so both arms of the immune system (humoral and cellular) are involved. In the fragments scenario, the major histocompatibility complex (MHC 1) presents the foreign fragments to CD8+ T lymphocytes. MHC 1 is found on cell membranes of all cells with a nucleus. When our immune system "sees" a foreign protein, a process is started to kill that protein. Make sense so far?

Major histocompatibility complexes 1 and 2 allow us to constantly screen cells for which proteins are being made. MHC 1 is on cells with nuclei; MHC 2 are on antigen-presenting cells (called APCs) like macrophages—the Pac Men of our immune systems prone to gobbling up foreign invaders. "When a CD8+ or CD4+ lymphocyte detects a cell expressing a viral gene (e.g., due to an infection), a mutant gene (e.g., due to cancer) or a foreign gene (e.g., due to a transplant), it binds to

MHC activating the immune response that leads to the destruction of the abnormal cell."[171,172] Bottom line: if foreign products from the novel injections make it inside our cells, our Pac Men macrophages might gobble up important cells.

Logical Concerns from a Small-Town Pediatrician

Knowing the above, I was quite concerned about the basic concept of turning on cellular machinery with synthetic mRNA (engineered by fallible people and not from nature) to instruct the body to make spike protein, which was a known driver of illness in those infected with COVID. What if the process ran amok, and instead of the body recognizing the spike as foreign immediately, making antibodies against it quickly and killing it with surgical precision, the body destroyed cells that make up our hearts and brains? What if the spike production facilities turned on by the COVID injections went berserk and kept making spike protein for days, weeks or months? My fears were well founded. We now know that spike protein can persist at least four months after injections.[173] Another study found spike protein persistence for at least ten months.[174]

Changing the Definition of Vaccines

During COVID, the CDC changed the definitions of vaccines to better fit what the novel experimental injections could achieve. Prior to 2015, vaccines "prevent disease." In 2015, the definition of vaccines was altered to say they "produce immunity" which did not require that they prevented disease. After the novel COVID injections were unleashed, and the inconvenient fact emerged that they did not necessarily "prevent disease" or "provide immunity," the definition was changed again to say that vaccines just "provide protection."[175,176] Thanks to the intrepid investigative reporter Sharyl Attkisson and Representative Thomas Massie for bringing attention to these evolving changes. Seems like a case of moving the goalposts.

Conclusion

Like Cassandra, with her ability to see into the future but being powerless to change it, I predicted negative results that were ignored. There is evidence that my concerns were justified.

FOOD FOR THOUGHT

If you got the COVID injections, did you realize it was not a traditional vaccine? If you chose not to get them, were concerns about them being a novel gene-based therapy part of your decision?

CHAPTER 10

What Are the Components in COVID "Vaccines"?

Fear not the path of Truth for the lack of People walking on it.
—Robert F. Kennedy

I got a bad feeling about this.
Lando Calrissian, *The Rise of Skywalker*

Trojan Horses, Spike Proteins, and the Goldilocks Dilemma

Synthetic messenger RNA is wrapped in a layer of fatty nanoparticles to trick human cells into making spike proteins and marketed to the world as "vaccines" under Emergency Use Authorization. What could possibly go wrong? Spike proteins are now known to be major pathologic drivers of bad outcomes in COVID. Can manufacturers be sure that their injectable product is "not too hot" (not so much spike protein that it causes damage to the recipient), "not too cold" (not so weak that an immune response is not generated), but just right? Does it seem to you that COVID vaccine porridges are "just right" and would please Goldilocks?

Spike Proteins

COVID spike proteins fuse into the receptors on the membranes of host cells, allowing the virus to gain entry to the innards of the cell to utilize cellular metabolism to replicate. Stephanie Seneff, an MIT scientist, explains: the "spike protein, which facilitates both viral binding to a receptor (in the case of SARS-CoV-2 this is the ACE 2 receptor) and virus fusion with the host cell membrane. The SARS-CoV-2 spike protein is the primary target for neutralizing antibodies."[177] We know that spike proteins have the ability to bind to receptors on multiple types of cells. We also know, from elegant studies by Vojdani and colleagues in 2020, that there is extensive immune cross-reactivity between SARS-CoV-2 antibodies and many different antigen groups. They report:

> We found that SARS-CoV-2 antibodies had reactions with 28 out of 55 tissue antigens, representing a diversity of tissue groups that included barrier proteins, gastrointestinal, thyroid and neural tissues, and more. We also did selective epitope mapping using BLAST and showed similarities and homology between spike, nucleoprotein, and many other SARS-CoV-2 proteins with the human tissue antigens mitochondria M2, F-actin and TPO. This extensive immune cross-reactivity between SARS-CoV-2 antibodies and different antigen groups may play a role in the multi-system disease process of COVID-19, influence the severity of the disease, precipitate the onset of autoimmunity in susceptible subgroups, and potentially exacerbate autoimmunity in subjects that have pre-existing autoimmune diseases[178]

In other words, antibodies made to neutralize the COVID spike protein in patients infected with SARS COV-2 cross-react with more than half of the human tissue proteins tested, raising concerns for future increases in autoimmune diseases, in which the body attacks its own tissues. You will read more about autoimmunity in part III.

Synthetic Messenger RNA

Synthetic mRNA gives instructions to our cells to make spike protein, which has been shown to be toxic and pathogenic, responsible for most of the bad effects of SARS-CoV-2 infections. What happens to those spike proteins after we make the hoped-for antibody response? In spring of 2021, a group of scientists at Harvard published work that showed that eleven of thirteen recipients of Moderna's mRNA injections demonstrated antigen from the shots circulating in the blood within one day. Rather than staying localized at the injection site as promised by the manufacturers, the CDC, and the media, the spike protein antigen goes into the blood and can then go everywhere in our bodies.[179]

Trojan Horse Lipid Nanoparticles

A scientist in Canada, Byron Bridle, obtained data through the Japanese equivalent of a Freedom of Information request showing that the lipid nanoparticles circulate in blood for at least two weeks after injection and accumulate in the spleen, bone marrow, liver, adrenals, and ovaries.[180] Many of us are particularly worried about the accumulation in the ovaries, because the levels were still rising when the study was terminated. Ideally such biodistribution studies would have been done in a variety of animals prior to unleashing the novel spike protein "vaccines" on healthy people on a global scale. The spike proteins carried in the lipid nanoparticle (LNP) Trojan horses bind to platelets and endothelial cells lining our blood vessels, which can lead to clotting or bleeding. This is being seen clinically after vaccination with COVID vaccines and reported to the Vaccine Adverse Event Reporting System (VAERS), which you can access at https://vaers.hhs.gov/.

Does it surprise you to learn that Bridle was soundly criticized and ostracized for bringing this information to our attention? Thanks, Byron, for doing the right thing and for suffering the consequences. Perhaps one day Justin Trudeau will realize you are a Canadian hero, though I'm not holding my breath.

Predictions Came True

One of the very credible scientists who raised timely concerns about unanticipated effects of artificially manufactured spike proteins was J. Patrick Whelan, MD, PhD. Whelan submitted a document to the FDA *prior to* the December 10, 2010, meeting when the committee would review the Pfizer/BioNTech SARS-CoV-2 vaccine data for emergency use authorization. Whelan's training at Harvard, Texas Children's Hospital, and Baylor College of Medicine and his degrees in biochemistry, medicine, and rheumatology made him well qualified to raise the concerns he articulated. I thank quintessential warrior mom and registered nurse Lyn Redwood for bringing Whelan's concern to attention. Whelan sought to alert regulatory authorities about the potential for vaccines that were designed to create immunity to the SARS-CoV-2 spike protein to cause injuries and explained biologically plausible mechanisms by which such harm could occur. Specifically, Whelan was concerned that the new mRNA vaccine technology used by Pfizer and Moderna has "the potential to cause microvascular injury (inflammation and small blood clots called microthrombi) to the brain, heart, liver and kidneys in ways that were not assessed in the safety trials."[181]

His point should have been well taken. He cautioned about the unforeseen consequences "if hundreds of millions of people were to suffer long-lasting or even permanent damage to their brain or heart microvasculature as a result of failing to appreciate in the short-term and unintended effect of full-length spike protein-based vaccines on other organs."[182] Remember, he took these concerns to the appropriate authorities (Vaccines and Related Biological Products Advisory Committee—VRBPAC) *before* the shots were approved for distribution. When his concerns manifested as serious adverse events after mRNA vaccination, he wrote about that, too. Another Cassandra.

Limited Data and Short Trials; FDA Endorsement December 2020

Unfortunately, the FDA relied on very limited clinical trial data collected in an unprecedentedly short amount of time, and endorsed both the Pfizer and Moderna mRNA vaccines for emergency use in December 2020. The FDA allowed the Pfizer-BioNTech COVID-19 vaccine to be widely distributed in individuals sixteen and older without calling for the additional studies that Whelan felt were critical to assure safety of the vaccine, especially in children (bearing in mind that the risks of long-term problems from Covid-19 infection itself to children and adolescents were low). Whelan was correct in insisting on a high bar for safety standards for use of the new vaccine technology on the pediatric population. I nominate him for a future decision-making position at the FDA.

Why was Whelan so worried about the mRNA vaccines causing blood clots and inflammation? One of the peculiar and often deadly findings with regard to SARS-CoV-2 infection is widespread damage occurring in numerous organs in addition to the lungs. Clinicians around the world have seen evidence that suggests the virus causes heart inflammation, acute kidney disease, neurological malfunction, blood clots, intestinal damage, and liver problems and that the spike protein is the factor responsible for cardiovascular complications. In fact, COVID-19 is now widely accepted as being primarily a disease of the microvasculature (small blood vessels) with systemic effects.

So, a salient question is this: Is the whole virus the main pathogen, or does the spike protein alone cause a lot of the damage? If the spike protein is the major villain, have we unleashed pathogenic protein production in cells that some people will not be able to neutralize without collateral damage? Hence the Goldilocks dilemma: Did we create spike proteins that are too "hot" (reactive causing damage) or too "cold" (unable to elicit an immune response)—not always "just right"? Remember there were billions of dollars on the line for the winners of the race. The odds seem stacked against achieving perfection and outwitting Mother Nature.

Crossing the Blood-Brain Barrier
Can Cause Brain Damage

Nature Neuroscience published a study a few days after Whelan's warning letter to the FDA that found commercially obtained COVID-19 spike protein (S1) injected into mice readily crossed the blood-brain barrier. Spike proteins were found in all eleven brain regions examined and entered the parenchymal brain space (the functional tissue in the brain).[183]

Researchers acknowledged that such widespread entry into the brain could explain the diverse neurological effects of spike protein such as respiratory difficulties, encephalitis (brain inflammation), and anosmia (the loss of smell). The injected spike protein was also found in the lung, spleen, kidney, and liver of the mice. Can spike protein generated from the vaccine have similar effects? Continuing reports of widespread adverse events are extremely concerning, as you will learn in part III of this book.

Buzhdygan and colleagues reported in December, 2020, that the SARS-CoV-2 spike proteins showed a direct negative impact on endothelial cells and provided "plausible explanations" for the neurological consequences observed in patients with COVID-19. The researchers demonstrated that the angiotensin-converting enzyme 2 (ACE 2), a known binding target for the SARS-CoV-2 spike protein, is "ubiquitously expressed throughout various vessel calibers in the frontal cortex" of the brain.[184] In another investigation, Nuovo studied brain tissues from thirteen fatal COVID-19 cases and found pseudovirions (spike, envelope, and membrane proteins *without* viral RNA) present in the endothelia of microvessels of all thirteen brains. They concluded that ACE 2+ endothelial damage is a central part of SARS-CoV-2 pathology and *may be induced by the spike protein alone* (emphasis added). Injection of the full-length S1 spike subunit in the tail vein of mice, as part of the same study, led to abnormal neurologic signs.[185]

Let's now circle back to the concerns voiced by Whelan in his prescient letter to the FDA: "I am concerned about the possibility that the new vaccines aimed at creating immunity against the SARS-CoV-2 spike protein have the potential to cause microvascular injury to the brain, heart, liver and kidneys in a way that does not currently appear to be assessed in safety trials of these potential drugs."[186] The original vaccine trials were very limited in the scope of laboratory investigations, and participants who had various chronic conditions were *specifically excluded* from the trials (but gleefully vaccinated during the rollout). Cha-Ching!

Summarizing the Mechanisms of Action

To recap, mRNA injections work by using the genetic blueprint for SARS-CoV-2's spike protein to instruct our own cells to make that spike protein. In theory, the body then will make antibodies against the spike protein to protect against SARS-CoV-2 infection. In what seems like a leap of faith, vaccine developers assumed that the cell would make just enough spike protein for humans to develop adequate antibodies, but not too many spike proteins to cause harm. In another leap of faith, vaccine developers hoped that the injected products would be broken down quickly and eliminated from the body.

The Law of Unintended Consequences from COVID Injections

Your teen could have a stroke from a blood clot or a cerebral hemorrhage after a COVID vaccine. Your child could have a liver and spleen full of toxic proteins known to attach to critical ACE 2 receptors for several weeks after the jab. Your preschooler girl's ovaries could contain spike proteins known to be at the heart of COVID pathology. How rare are these outcomes in the population at large? We do not know yet. But when the disease itself poses very little risk to a young, healthy person, the safety profile of any recommended COVID vaccine should be exemplary. In the next few years, we will find out how "rare" the side effects of these so-called vaccines are in young people. In the meantime, careful risk versus benefit calculations must be made. Based on the research conducted by the time of vaccine rollout, it was very likely that many recipients of the spike protein mRNA vaccines would experience the same or worse symptoms and injuries as they would have from the virus itself. Based on accumulated research on pediatric patients since the rollout, the risks of COVID injections seem worse than the risk of COVID infections.

MIT Scientist Tried to Warn Us Early in the Rollout

Stephanie Seneff of MIT published an important paper that is worth examining in detail. Seneff points out that the spike protein ". . . is a class I fusion glycoprotein, and it is analogous to haemagglutinin produced by influenza viruses and the fusion glycoprotein produced by syncytial viruses, as well as gp160 produced by human immunodeficiency virus (HIV)."[187] A cautious scientist would pause to reflect if there were lessons to be learned from influenza vaccine (low efficacy year after year) and HIV vaccine (four decades and counting without an effective vaccine) that might apply to our wrestling match with SARS CoV-2 vaccine development. A cautious scientist might worry about unintended consequences of turning on cellular machinery in a novel way to make a protein which has been demonstrated to cause endothelial damage and widespread inflammation.[188] Seneff goes on to discuss protein configurations in the pre-fusion and post-fusion state. Some COVID mRNA vaccine makers tweaked the spike protein by changing certain amino acids to encourage it to stay in the pre-cell-membrane fusion state. Seneff muses, "What might be the consequence of this? We don't know."[189] Other COVID vaccine makers changed some other amino acid sequences so that vaccinated people would make abundant spike proteins.

Many thoughtful scientists who were not encumbered by the bias of fortunes to be made have raised grave concerns about the unintended consequences of tricking Mother Nature to make spike proteins which are homologous (very similar) to so many human tissues. Such men and women of integrity did not reverse the momentum pushing populations to line up for "one size fits all" two-dose novel gene manipulation "vaccines," but the public's interest in boosters has declined dramatically.

Seneff continues with a series of concerns about potential unintended consequences and potential side effects. "We have reviewed the evidence here that the spike protein of SARS-CoV-2 has extensive sequence homology with multiple endogenous human proteins and could prime the immune system toward development of both auto-inflammatory and autoimmune disease. This is particularly concerning given that the protein has been redesigned with two extra proline residues to potentially impede its clearance from the circulation through membrane fusion. These diseases could present acutely and over relatively short time spans, such as with MIS-C (multisystem-inflammatory syndrome in children), or could potentially not manifest for months or years following exposure to the spike protein."[190] That secondary exposure could come via natural exposure to the virus or booster doses of a vaccine.[191] Bear in mind that these insights were published a mere *five months* after vaccine rollout, and were generated based on science available *before* the vaccines were distributed.

The Possibility of Prion Disease

Prions are proteins which induce misfolded proteins in the brain. Creutzfeldt-Jakob disease is the most common prion disease. Prions contain specific amino acid sequences called "motifs." The genetic analysis of the SARS-CoV-2 spike protein reveals five motifs, so it is plausible that it could behave like a prion. The spike proteins in Pfizer and Moderna "vaccines" have mutations that make it harder to fuse with the cell membrane region. It is reasonable to wonder if this would enhance the potential for prion-like activity.[192] Prion-like behavior could happen with COVID the disease as well as with COVID "vaccines."[193,194] Bart Classen proposed in 2021 that COVID spike protein could cause prion diseases through specific mRNA regions in the injections that bind to many known proteins, causing them to misfold.[195] Seneff and Nigh suggested the possibility of prion disease following vaccination.[196]

As early as May 2021 there had been more than 25,000 cases of noninfectious encephalopathy and/or delirium after COVID "vaccines." Correlation does not equal causation, but those numbers should make a careful clinician or scientist wonder about neurologic damage, especially since there is a scientifically plausible biologic mechanism.

What, Me Worry?

Many of my colleagues and I were worried about the incomplete mRNA fragments present in mRNA COVID injections. Pfizer officials dismissed those concerns, saying that mRNA would be quickly degraded by the cell. This is correct about the mRNA that humans produce, but it is not necessarily true about the synthetic mRNA found in the "vaccines," especially if the vials are not maintained in minus eighty-degree freezers.[197,198] I worry that these mRNA fragments will perpetuate the cell danger response you will read about in part IV of this book. As stated by mRNA scientists, quality checks are important, including high-volume mRNA targets for mRNA identification, mRNA sequence length, sequence accuracy, and integrity."[199,200]

Did you notice that many articles I cite above were from 2020 and 2021? While the injections were being developed, and in the months shortly after their release, well-credentialed scientists and clinicians were publishing their concerns, trying to warn the world what might happen with Warp Speed's novel technology. Remember, the initial widely quoted studies were performed in extremely healthy populations. However, the vaccine was rolled out to the population at large, many of whom had pre-existing conditions or were unhealthy. The vaccine trials did not include pregnant women, people who had already had COVID-19, or patients with multiple medical conditions. All those groups were jabbed in the real world, and some suffered severe effects.[201]

Concluding Questions

Have we thrown caution to the wind in a frenzy driven by fear and money? Have we created a Trojan horse composed of gene instructions enclosed in a fatty and potentially allergy-inducing vehicle to make a dangerous spike protein which travels to human beings' most vulnerable organs? Have we forfeited our common sense and critical thinking skills in exchange for permission to hug our friends, fly on a plane, or go to college? Have we planted the seeds of destruction in some unknown percentage of populations that did not meet criteria for study during COVID vaccine trials but are now being asked to roll up their sleeves for the greater good? Did it really serve the greater good?

FOOD FOR THOUGHT

How do you feel about scientific manipulation of genetic mechanisms in humans? Should all technologies that we are able to design be deployed? Did you understand the differences between COVID injections and traditional vaccines before you made decisions for yourself or your family?

CHAPTER 11

Did Anyone Foresee Mechanisms of Harm from the "Vaccines"?

Haste in every business brings failures.

—Herodotus

Good decisions come from experience.
Experience comes from making bad decisions.

—Mark Twain

The Narrative and the Truth

The answer is yes, people did warn about harms, but you may not have heard much about it. Government officials, and Anthony Fauci in particular, constructed a story in which vaccines would be the only answer to our woes. Remember Fauci telling you in November 2020 that the calvary was coming?[202] In the meantime, you should stay home and seek medical care only if you had respiratory distress?[203] Then we heard the repetitive and non-specific message that the vaccines were "safe and effective."[204,205,206] Cartoon videos reassured parents that the injections are safe and effective for children.[207]

The Cautionary Tale of Dengue Vaccines

People in my circle of clinicians and researchers raised concerns about prior attempts to develop coronavirus vaccines leading to infections that were more severe than would have happened without the vaccine. This phenomenon is called "antibody-dependent enhancement" or ADE. ADE can occur when someone is exposed to either a pathogen or vaccine that triggers the production of antibodies and is later exposed to a slightly different version of the pathogen. With rapidly mutating viruses (which includes coronaviruses), ADE is a very real likelihood. Even people like Paul Offit, a vaccine zealot and pediatric infectious disease specialist who developed a rotavirus vaccine, and Peter Hotez, the Dean of Tropical Medicine at Baylor, raised concerns about antibody-dependent enhancement. Both appear frequently on national news shows to pontificate about the value of vaccines. This is what Dr. Offit said in April of 2020 on the ZDoggMD podcast:

> So, you have the neutralizing antibodies, and you have the binding antibodies. You want to make sure that the quantity of the neutralizing antibodies that you have, and the persistence of those antibodies, is much greater than the binding antibodies. Because the binding antibodies could be dangerous and cause something called **antibody-dependent enhancement**. And we've seen that. We saw that with the dengue vaccine. . . . The dengue vaccine, in children who've never been exposed to dengue before, actually made them worse when they were then exposed to the natural virus. Much worse. Causing something called dengue hemorrhagic shock syndrome.[208]

Sanofi took more than twenty years to develop and license Dengvaxia. The vaccine against dengue fever cost more than 1.5 billion US dollars.[209] Sanofi tested Dengvaxia in more than thirty thousand kids around the globe, and the prestigious *New England Journal of Medicine* published the results.[210] The Philippines launched a very visible mass vaccination campaign in 2016, and *six hundred children died* due to ADE associated with the dengue vaccine. The Philippine government indicted fourteen officials, alleging they acted too hastily in launching that campaign. Dengvaxia has been permanently banned in the Philippines.[211,212] Compounding the tragedy is the fact that ADE was already a known factor in dengue fever by 2011.[213] In fact, one US scientist warned numerous times that the vaccine would end up killing a small percentage of the children who received it long before the Philippines debacle.[214]

The dengue vaccine Philippines story was all over the news in 2019. Vaccinologists and manufacturers involved in COVID vaccine development must have been aware of the danger of ADE. Perhaps the demands of "Operation Warp Speed" and the lure of billions of dollars to be made led them to discount the possibility that antibody-dependent enhancement could make people sicker if they got the so-called vaccines than if they declined.

Lessons from Prior Coronavirus Animal Trials

Dr. Peter Hotez worked on a vaccine for the SARS coronavirus outbreak in 2002–2003. Some animals given the experimental vaccine got more severe disease than the control animals. Here is what he said during government testimony:

> [W]hat happens with certain types of respiratory virus vaccines, you get immunized, and then, when you get actually exposed to the virus, you get this kind of paradoxical immune enhancement phenomenon. And what—how—and we don't entirely understand the basis of it, but we recognize that it's a real problem for certain respiratory virus vaccines. That killed the RSV program for decades. . . . When we started developing coronavirus vaccines . . . we noticed in laboratory animals that they started to show some of the same immune pathology that resembled what had happened 50 years earlier (referring to the RSV debacle), so we said, oh, my God, this is going to be problematic.[215]

Hotez was interviewed by Reuters on March 11, 2020. Here is what he said then:

"I understand the importance of accelerating timelines for vaccines in general, but from everything I know, this is not the vaccine to be doing it with," due to the risk of immune enhancement. "The way you reduce that risk is first you show it does not occur in laboratory animals."[216]

In spring of 2020, so much emphasis was placed on the development of a vaccine for COVID-19 that vaccine manufacturers were allowed to begin human trials without waiting for animal trial results. At the same time, my colleagues who were getting excellent treatment results with repurposed drugs as an alternative to waiting for a vaccine could not gain traction with policy makers. As time went on, those clinicians were discredited and derided for their efforts.

A landmark article reinforced the importance of proceeding with animal trials prior to human testing when they analyzed four SARS coronavirus vaccine trials. In mice, ferrets, and monkeys, all four vaccines induced antibodies and protected against infection. Unfortunately, when the animals were exposed to the same virus later, they got terrible lung disease. Hence, the authors urged caution in developing coronavirus vaccines due to this finding—antibody dependent enhancement—and its implications for those who are infected after receiving the injections.[217]

Those Who Fail to Learn from History
Are Doomed to Repeat It

An article published in October 2020, prior to the rollout in December 2020, summarized multiple examples of vaccine-associated disease enhancement, citing examples of dengue, SARS-CoV (the original), MERS (Middle East respiratory syndrome coronavirus) and respiratory syncytial virus (RSV). The authors argued for a robust safety and efficacy profile and cautioned against rushed approvals. Their conclusion makes so much sense to me.

"Vaccines for viruses with *high transmissibility but low case fatality*, such as SARS-CoV-2, should usually have a higher bar for safety than those for viruses with low transmissibility but high case fatality, such as Ebola virus, because many more healthy individuals will have to use them."[218] How I wish the powers that be had heeded that warning! How I wish proper attention had been given to repurposed drugs and the value of good immune health. How I wish Dr. Fauci had not said "hang on, the calvary is coming," referring to COVID vaccines as THE thing that would save us. If wishes were horses, then beggars would ride. . . .

FOOD FOR THOUGHT

Did you hear warnings that the COVID jabs might make subsequent infections more dangerous? Do you recall mainstream media covering antibody-dependent enhancement? Do you know people who got shots or boosters and got a bad case of COVID anyway? Do you know people who have had COVID three or four times despite taking several shots and boosters?

CHAPTER 12

Were There Shortcomings in the Original Warp Speed Adult Trials?

The difference between stupidity and genius is that genius has its limits.
—Albert Einstein

Always do right. This will gratify some people and astonish the rest.
—Mark Twain

Shortcomings, Side Effects, and Suspensions at Warp Speed

Were there shortcomings to Operation Warp Speed? Yes, yes, a thousand times yes! Now that COVID injections have been distributed around the planet, we have seen significant side effects that could have been predicted with a careful analysis of the original trials and their shortcomings. Now, because of problems recognized post-marketing, Germany, France, Finland, and Sweden have suspended the use of Moderna vaccines for people under thirty. Denmark no longer recommends vaccination for low-risk individuals under fifty and has suspended the Moderna vaccine for people under eighteen. Taiwan suspended the second Pfizer vaccine for adolescents twelve through seventeen years old.[219] If only we had listened to the sages who had valuable insights.

Critical Insights from Dr. Peter Doshi at the *BMJ*

One of my heroes in the COVID debacle is Peter Doshi, PhD, who works at the University of Maryland and for the *BMJ* as an editor. He wrote an amazingly prescient publication in November 2020. It is worth examining what he wrote in detail. Doshi argued persuasively that "full transparency and vigorous scrutiny" about the novel injections was necessary for clinicians, health departments, and populations to make truly informed decisions. In Dr. Doshi's words:

> The topline efficacy results from their experimental covid-19 vaccine trials are astounding at first glance. Pfizer says it recorded 170 covid-19 cases (in 44,000 volunteers), with a remarkable split: 162 in the placebo group versus 8 in the vaccine group. Meanwhile Moderna says 95 of 30,000 volunteers in its ongoing trial got covid-19: 90 on placebo versus 5 receiving the vaccine, leading both companies to claim around 95 percent efficacy.[220]

He quickly puts those numbers into context, however. He astutely points out that the astonishing 95% refers to relative risk reduction, not the absolute reduction in risk. In fact, the absolute risk reduction was only about 1%, which neither Pfizer nor Moderna advertised. In other words, there were so few people in the forty-four-thousand-participant trial who got COVID, that getting the vaccine only decreased an individual's *actual risk by about 1%*. Hmm. Doubtful governments would buy billions of dollars, pounds, euros, yen, franc, or rand worth of novel injections to reduce their citizens' absolute risk by a mere 1%. This can be a hard concept to grasp, so let me clarify. Absolute risk is the number of people experiencing an event, in this case getting COVID, compared to the total population at risk (in this case the people in the study, who either got the "vaccine" or did not). Absolute risk changes over time. Relative risk is the comparison between two groups, in this case those who got the vaccine compared to those who did not. The results for the number of people getting COVID in each group showed a reduction in the absolute risk of COVID illness of about 1% for the vaccine group. Such percentages do not sell drugs or vaccines or procedures. But relative risk reduction of 95%—that sounds like a lot!

Furthermore, Dr. Doshi points out that the trial's primary end points were COVID-19 of *any* severity. The trials did not address the "vaccine's ability to save lives, nor the ability to prevent infection, nor the efficacy in important subgroups (e.g. frail elderly)."[221] Frankly, when my pediatric patients get a mild case of COVID, it is like their experience with many other coronaviruses or rhinoviruses. I do not want a world in which we try to muck with nature to remove human experience with mild upper-respiratory infections. Experience with mild illnesses helps modulate children's immune systems. Doshi also reminds us that the results reflect a time soon after vaccination, so "we know nothing about vaccine performance at 3, 6 or 12 months."[222] Prior to the vaccine rollout, he hit the nail on the head. We now know that immunity from COVID injections is indeed short lived.[223,224,225]

Healthy People Results Do Not Translate to Ill People Results

One of my biggest concerns when dissecting the original Pfizer trial design was that only extremely healthy people were recruited for the trial—"the Avengers among us," as Bobby Kennedy Jr. has said in his Defender podcasts. Remember the original Pfizer and Moderna trials did not include pregnant women or many elderly people. Those with chronic diseases and immunosuppression were excluded. Yet the vaccines were rolled out worldwide to people with diabetes, heart conditions, chronic lung disease, and autoimmunity. This is an excellent example of recruiting bias in clinical trials, and a sad example of extrapolating results from a healthy population to populations with many co-existing conditions and genetic differences.

Financial Rewards Without Proof of Preventing Transmission to Others

Dr. Doshi argued early on that the mRNA trials were studying the wrong endpoint. Here he writes in *BMJ* in October 2020:

Peter Hotez, Dean of the National School of Tropical Medicine at Baylor College of Medicine in Houston, said, "Ideally, you want an antiviral vaccine to do two things . . . first, reduce the likelihood you will get severely ill and go to the hospital, and two, prevent infection and therefore interrupt disease transmission.[226]

Yet the phase III trials are not actually set up to prove either. None of the trials currently under way at this time are designed to detect a reduction in any serious outcome such as hospital admissions, use of intensive care, or deaths. Nor are the vaccines being studied to determine whether they can interrupt transmission of the virus.[227]

Read that quote again and compare it to what you were told in 2020 and 2021. *The trials did not study whether the "vaccines" stopped transmission.* Hence, all the rhetoric about getting vaccinated to prevent sickness in others was not based on science from the trials. It was as if altruism was being weaponized to motivate the public to take the shots!

Dr. Doshi continues: "In all the ongoing phase III trials for which details have been released, laboratory confirmed infections even with only mild symptoms qualify as meeting the primary endpoint definition."[228] Yikes! Pfizer made $103 billion in the first year by including prevention of mild upper-respiratory symptoms in their statistics! Moderna's CEO, who has an MBA and is not a doctor or scientist (despite what you might think from his appearances in ads wearing a white coat), got rich from bringing his company's first product to market during the COVID crisis.[229,230]

"In Pfizer and Moderna's trials, for example, people with only a cough and positive laboratory test would bring those trials one event closer to their completion."[231] Remember, Dr. Doshi wrote this in October 2020, before the rollout! Remember all the exhortations from on high to "get the vaccine to protect granny?" None of the trials were designed to show that to be true. Peter, you hit the nail on the head. I congratulate you for your foresight and sharing your perspective before the shots rolled out to an unsuspecting world.

How to Wipe Out Your Control Group and Sabotage Long-Term Follow-Up

As someone who was schooled in the importance of long-term placebo-controlled trials, I was shocked when Pfizer sent a letter to trial placebo recipients offering them the vaccine.[232] This move (copied by the other COVID vaccine manufacturers) essentially negated our ability to study the long-term side effects of the vaccine. This is bullshit!

How to Guess Which Group Your Subjects Are In?

As someone who was also schooled in the importance of blinding in clinical trials, where neither the recipient nor the investigator knows who receives the vaccine and who is in the placebo group, I was concerned that those conducting the trials could make educated guesses about the status of the volunteers. Dr. Doshi shared that concern in late 2020 before the roll out:

> [W]e need data-driven assurances that the studies were not inadvertently unblinded, by which I mean investigators or volunteers could make reasonable guesses as to which group they were in. Blinding is most important when measuring subjective endpoints like symptomatic covid-19, and differences in post-injection side-effects between vaccine and placebo might have allowed for educated guessing.[233]

If a vaccine has a high rate of side effects, as the COVID injections have proven to have, it becomes much easier to guess who got the novel injections versus who got the placebo.

Did They Look for COVID Equally in Both Shot Recipients and Placebo Cases?

As a clinician who constantly had to make judgments about which patients to test for which illnesses, I wondered about the criteria for deciding which patients to test for COVID. Doshi had the same concern:

One way the trial's raw data could facilitate an informed judgment as to whether any potential unblinding might have affected the results is by analyzing how often people with symptoms of covid-19 were referred for confirmatory SARS-CoV-2 testing. Without a referral for testing, a suspected covid-19 case could not become a confirmed covid-19 case, and thus is a crucial step in order to be counted as a primary event: lab-confirmed, symptomatic covid-19. Because some of the adverse reactions to the vaccine are themselves also symptoms of covid-19 (e.g. fever, muscle pain), one might expect a far larger proportion of people receiving vaccine to have been swabbed and tested for SARS-CoV-2 than those receiving placebo.[234]

Yikes again! Another way to screw up the true number of cases in controls versus vaccine recipients, which could inflate the relative risk percentages to seem more significant.

Did Taking Acetaminophen or Ibuprofen Skew the Results?

As someone who has spent over forty years observing potential vaccine reactions, I was concerned that with such a reactogenic vaccine, investigators would intuit which patients were in the placebo trials—the ones with less arm soreness, fever, or fatigue. Dr. Doshi has similar concerns and raises another crucial point:

Data on pain and fever reducing medicines also deserve scrutiny. Symptoms resulting from a SARS-CoV-2 infection (e.g. fever or body aches) can be suppressed by pain and fever reducing medicines. If people in the vaccine arm took such medicines prophylactically, more often, or for a longer duration of time than those in the placebo arm, this could have led to greater suppression of covid-19 symptoms following SARS-CoV-2 infection in the vaccine arm, translating into a reduced likelihood of being suspected for covid-19, reduced likelihood of testing, and therefore reduced likelihood of meeting the primary endpoint. But in such a scenario, the effect was driven by the medicines, not the vaccine.[235]

To my knowledge, Pfizer and Moderna did not release the information given to trial volunteers prior to the rollout, so we simply did not know yet.

Why Not Test Everyone to See the True Occurrence Rate of COVID?

I naively assumed that all people in either group who presented with symptoms that could be COVID would get lab tests for the virus. The symptoms were so broad, from respiratory to gastrointestinal to brain fog, and the presence of COVID in the two groups was a crucial bit of data. However, investigators were instructed to use their clinical judgment, which cannot be standardized for appropriate comparisons, to decide who got tested. From Moderna:

> It is important to note that some of the symptoms of COVID-19 overlap with solicited systemic Adverse Reactions that are expected after vaccination with mRNA-1273 (e.g. myalgia, headache, fever, and chills). During the first 7 days after vaccination, when these solicited ARs are common, Investigators should use their clinical judgement to decide if an NP swab should be collected.[236] [NP stands for nasopharyngeal—nose and throat—by the way.]

In my best medical judgment, a proper trial should have tested vigorously for COVID cases in all participants. In my rudimentary understanding of epidemiology, this would prevent differential measurement errors and ascertainment bias. When either of those errors is present, one cannot depend on the trial results. Those errors should not be allowed in trials for an intervention that will be given universally.

Short-Term Reactions Versus Long-Term Adverse Events

One of my pet peeves about vaccine-safety trials is that side effects are rarely studied more than seven to ten days. Honestly, I am not that worried about transient, immediate side effects like sore arms and low-grade fevers. I am much more concerned about long-term effects on children's cognition or health. In the mRNA trials, the first seven days were examined for side effects. For COVID, a phone app called V-Safe was used to collect data daily, then weekly, then at six and twelve months. We will learn in another chapter what that data revealed. Hint: it was not good.

FOOD FOR THOUGHT

Did you or those you love get the COVID injection because you believed it would protect those around you? Did you wonder if there were problems associated with completing trials in such a brief time? Did you know that the placebo recipients in the trial were offered COVID jabs months after the trial ended and that this would prevent long-term follow-up of side effects?

CHAPTER 13

Were There Shortcomings in the Pediatric Trials?

All men know their children mean more than life.

—Euripides

It is easier to build strong children than to repair broken men.

—Frederick Douglas

Children Deserve Better

When I dug into the details of the pediatric clinical trials for COVID jabs in kids, I was frankly shocked at the study designs and tepid results. First, the studies were extremely underpowered. Let's talk about the concept of powering studies for those of us (including me) who do not do power calculations with any regularity. To find effectiveness of drugs and occurrence of side effects, scientists figure out how many people must be enrolled in the trials. If there is an illness with infrequent but bad effects on the population being studied, it is even more imperative to enroll large numbers, so that the study answers questions like "Are there significant side effects that would make it unwise to give this to healthy children who are at low risk from the disease itself?"

Underpowered Studies

For the study published in the *New England Journal of Medicine* that was used as a basis for CDC recommendations for mRNA COVID injections in children, there were 1,517 kids five to eleven years old who got the first mRNA shot and 751 who got a placebo in the part 2/3 phase. Median follow-up was 2.3 months.[237] In the Pfizer study for vaccine efficacy in six month to four-year-old children there were 873 subjects in the mRNA shot group and 381 in the placebo group.[238] In another trial for COVID injections under age five, there were 1,178 babies six to twenty-three months who got the real injections and 598 who got the placebo. In the two-to-four-year range, 1,835 toddlers got the real material and 915 got the placebo.[239] In another trial for COVID injections in age five to eleven, there were 1,305 who got the real primary series and 663 who got placebo.[240] Can you see how such relatively small numbers could obscure serious side effects that occurred in 1 in 1000 or 1 in 2000 shot recipients?

Immunobridging

In Pfizer trials, a series of immunobridging studies were done. Measuring antibody responses to the "vaccine" served as a surrogate marker for protection. After the primary vaccine series, eighty-two children in the six- to twenty-four-month age range were studied. One month after dose three, eighty children six to twenty-four months were studied. In what was labeled a vaccine efficacy study in which cases of COVID that occurred more than seven days after dose three, 873 kids six months to four years were studied; there were 381 kids in the placebo arm.[241]

Shockingly, the pediatric clinical trials for boosters in the five- to eleven-year-old cohort only had 140 participants. No, I did not forget a zero or two; there were a total of 140 kids.[242]

This is unequivocally *much too small* to determine the prevalence of somewhat rare side effects. That is a monumental error because the injection was intended for *tens of millions* of kids; significant adverse effects occurring in 1 in 500 or 1 in 1,000 trial participants would affect many thousands of children.

What Happened to Most Children Who Did Not Complete the Trial?

Pfizer data presented to the FDA to obtain an Emergency Use Authorization to jab children six months to four years included very concerning data about safety and efficacy. One of the doctors who pointed out severe shortcomings in the pediatric trials was a pathologist in the UK named Clare Craig, who wrote a book released in June 2023 titled *Expired: COVID the Untold Story*. Pharma recruited 4,526 children from age six months to four years. About three thousand did not make it into the final analysis! Findings were reported on three percent of the children recruited.[243] When I was teaching in a residency program, if a resident presented such a high dropout rate in her research, she would fail the course and the results would not be considered valid. Did the babies and children drop out because they had side effects, and the parents were scared to continue? We do not know. Did the toddlers have behavioral changes or sleep disruptions that caused the parents to drop them out of the study? We do not know. This level of attrition is highly disturbing.

Case Definitions Matter

"Severe COVID" was defined as an increased heart or respiratory rate. As a pediatrician who has treated children for more than forty years, I know that rapid heart rates and breathing can be caused by a multitude of factors, including fevers, and may be anything from quite transient to very prolonged. There were six cases considered severe in the vaccinated group and just one in the unvaccinated group. There was only one child in the trial who had to be hospitalized. They had fever and a seizure. Guess which group that child was in? Yep, the vaccinated group. When one does the math, one in about two hundred vaccinated children got severe COVID or was hospitalized. It would be rational to compare that to the rates of severe COVID or hospitalization seen in pediatric patients before the vaccine was available (which was also when COVID strains were more virulent).

One Side Effect of COVID Jabs Is COVID Illness in the First Three Weeks

There were three weeks between the first and second doses in the trial. During that time, thirty-four vaccinated kids got COVID compared to thirteen kids in the unvaccinated group.[244] We now know that COVID vaccines adversely affect the immune system during the first few weeks after receipt.[245,246] During that time, "vaccine" recipients are more likely to get various infections (which could include COVID) and be impaired in their ability to kill circulating cancer cells. You may have heard of "breakthrough" infections with COVID shortly after the injection. These changes in immune function are one mechanism that increases the likelihood of such infections. There were eight weeks between the second and third doses during which time more kids in the vaccinated group got sick. In the first few weeks after the third dose, more kids in the vaccinated group got COVID.

Cherry Picking Is for Farms, Not Pharma Trials

Later, after the third dose, three children in the vaccine group got COVID compared with seven children in the unvaccinated group. That particular data point was the one extracted to make efficacy claims. This is what is called *cherry picking*; 97 percent of the COVID cases from earlier time points were omitted from the analysis. Furthermore, later data would show a significant decline in efficacy after two to three months, suggesting even the cherry-picked "protection" was short lived.

There were eleven children in the trial who got COVID twice. Only one was unvaccinated. Many of the vaccinated children who got COVID twice had received all three doses. This finding foreshadowed the finding of negative efficacy (increased risk of COVID with multiple doses, discussed in another chapter.)

Lowering Efficacy Standards for Shots in Kids

Pfizer clinical trials for kids two through four years old did not meet FDA-specified requirements for vaccines of 50 percent efficacy or a 30 percent lower limit with a 95 percent confidence limit (meaning that 95 out of 100 times, the efficacy would fall in the defined range) Why, why, why would we relax our standards so egregiously for our toddlers? Why, why, why would our CDC recommend putting these vaccines on the Guidelines for Childhood Immunization Schedules with such a dismal showing in these underpowered trials? It must be for reasons other than concern for the health and well-being of our children.

How Many Deaths or Disabilities Are Enough to Stop a Clinical Trial?

Maddie de Garay is a twelve-year-old girl who developed symptoms within twenty-four hours of her Pfizer COVID injection and is now confined to a wheelchair because she cannot walk and has a feeding tube because she cannot eat. Her mom testified before a Senate committee that she faced significant roadblocks when she tried to bring her daughter's paralysis to the attention of the trial center (Cincinnati Children's) and the "vaccine" manufacturer (Pfizer).[247] Maddie was one of about one thousand children in the pediatric Pfizer trial. If we took her case to a high school statistics class, they would conclude that the risk of severe neurologic disability was about 1:1,000 in the pediatric trial. Many parents would conclude that is an unacceptable risk. But Pfizer did not truthfully report her adverse events. In the final report, her illness was described as "functional abdominal pain. "

The system is not supposed to work in such a callous and non-scientific way. According to the World Medical Association Declaration of Helsinki,

> Physicians may not be involved in a research study involving human subjects unless they are confident that the risks have been adequately assessed and can be satisfactorily managed. When the risks are found to outweigh the potential benefits or when there is conclusive proof of definitive outcomes, physicians must assess whether to continue, modify, or immediately stop the study.[248]

12 -15 ADOLESCENT TRIAL FAILURE TO REPORT SERIOUS ADVERSE EVENTS

- Maddie is now **paralyzed** from the waist down.

- She has **gastroparesis** with great difficulty swallowing food and water.

- Maddie needs a **wheelchair** or walker to get around, and a **feeding tube** for nourishment.

- At one point, Maddie was having **20 or more blackout/fainting episodes per day.** POTS

Reported as functional abdominal pain

Maddie de Garay is a 12 year old trial participant who developed a serious reaction after her second dose and was hospitalized within 24 hours.

Maddie developed gastroparesis, nausea and vomiting, erratic blood pressure, memory loss, brain fog, headaches, dizziness, fainting, seizures, verbal and motor tics, menstrual cycle issues, lost feeling from the waist down, lost bowel and bladder control and had an nasogastric tube placed because she lost her ability to eat. She has been hospitalized many times, and for the past **10 months she has been wheelchair bound and fed via tube.**

In their report to the FDA, **Pfizer described her injuries as "functional abdominal pain."**

- One participant experienced an SAE reported as generalized neuralgia, and also reported 3 concurrent non-serious AEs (abdominal pain, abscess, gastritis) and 1 concurrent SAE (constipation) within the same week. The participant was eventually diagnosed with functional abdominal pain. The event was reported as ongoing at the time of the cutoff date.

When Maddie was 12 years old, she heard about the Pfizer vaccine trials at Cincinnati's Children's Hospital and told her parents she wanted to sign up as a test subject. Her brother, Lucas, also volunteered.

Mumper PowerPoint presentation 2022.

Risk-to-Benefit Ratios Do Not Justify Universal Childhood COVID Injections

During the first ninety days of the COVID jab rollout, thirty-four children were given the injections erroneously since pediatric studies were not concluded and recommendations for use in kids had not been made. There were 134 recorded adverse events in that group of thirty-four children. Pfizer knew.[249]

In my view, the "one size fits all" recommendation for COVID vaccination is egregious. Clearly there are variable risks depending on factors such as age and health status. Pushing vaccines on an entire population takes away two important basic rights: bodily sovereignty and the right to confidential physician-patient relationships. By disguising serious side effects like paralysis and the need for a feeding tube and wheelchair as "functional abdominal pain," Pfizer misled the FDA and committed fraud, in my opinion.

FOOD FOR THOUGHT

Had you heard of Maddie de Garay's paralysis or her mom's fight to get help for her daughter? Are you surprised by the small number of pediatric research subjects studied prior to recommending COVID shots to babies, children, and teens? Did the revelations about the shortcomings in pediatric COVID shot trials surprise you? What should pharmaceutical companies provide to their research subjects who are harmed by vaccine trials?

CHAPTER 14

What Are the Principles and Challenges in Vaccine Research?

There are known knowns . . .
We also know there are known unknowns . . . we know there are
some things we do not know. But there are also unknown unknowns—
the ones we don't know we don't know.

—Donald Rumsfeld

If we knew what we were doing, it would not be called research, would it?

—Albert Einstein

Holes in the Vaccine Approval Process

When one looks behind the curtain, the vaccine approval process is not as airtight as one would hope for products that are so widely distributed to healthy people for the purpose of avoiding sickness. For the interested reader, there are websites from the Food and Drug Administration so you can learn about the process in detail.[250,251]

The Use of "Placebos" in Vaccine Trials

Placebos are meant to be inert substances like saltwater so that investigators can compare outcomes in groups of people who get an intervention (like a medicine or vaccine) to those who do not. Many people are surprised to learn that when there is an existing medication or vaccine, new medications or vaccines are usually evaluated against the *existing* treatment, not against a true placebo. ". . . there is uniformity on the use of placebos, i.e. that if a proven effective intervention exists, the trial intervention should generally be tested against it. Failure to do so deprives participants in the 'control' arm of an intervention that is likely to benefit them."[252]

Why Are There [Virtually] No True Placebo-Controlled Vaccine Clinical Trials?

The essence of the argument is that it is unethical to have a true placebo-controlled trial of vaccines because you will deprive the control group of the benefit of a vaccine. Even though ". . . RCTs [randomized control trials] are considered the 'gold standard,' where participants are randomly allocated to receive either the investigational or the control vaccine (placebo, different vaccine, or nothing)," in practice, vaccines are usually compared to the previous generation of the existing vaccine, not a true inert placebo.[253]

Here are several examples of vaccine "placebos" that are not inert. In a children's trial for Hepatitis A vaccine, "Like the vaccine, each dose of the placebo—aluminum hydroxide diluent—contained 300 µg of aluminum and thimerosal at a 1:20,000 dilution."[254] Both aluminum and thimerosal (which is half mercury by weight) are known neurotoxins and inducers of autoimmunity.[255,256,257,258,259,260,261,262,263,264,265,266,267,268,269] In fact, Yehuda Shoenfeld, a leading world authority on autoimmunity, wrote a whole book about how vaccines can induce autoimmunity.[270]

Regarding a trial for a rotavirus vaccine, the brainchild of Dr. Paul Offit, a leading vaccine spokesperson: "The placebo consisted of all components of Rotarix, but without any RV particles."[271] "The placebo was identical to the vaccine except that it did not contain the rotavirus reassortants or trace trypsin."[272] To borrow from Shakespeare, a placebo by any other name would smell as sweet (or stink as much).

Choosing Study Participants: Inclusion/Exclusion Criteria

One problem with the studies evaluating COVID vaccines was that participants were chosen because they were healthy and not pregnant. Then the shots were given to the population at large, including pregnant women and very unhealthy people. From the FDA's guidance document on choosing control groups:

> It is always difficult, and in many cases impossible, to establish comparability of the treatment and control groups and thus to fulfill the major purpose of a control group. . . . The groups can be dissimilar with respect to a wide range of factors, other than use of the study treatment, that could affect outcome, including demographic characteristics, diagnostic criteria, stage or severity of disease, concomitant treatments, and observational conditions. . . .[273]

Study in One Group, Give to Other Groups—Not Cool

My concerns about safety and efficacy issues with COVID vaccines might be less dramatic if the vaccines were offered to populations stratified for risk factors, such as age and comorbidities. Instead, COVID injections were marketed aggressively on a global scale to groups in which the vaccines had not been studied. The risk-to-benefit ratios change depending on multiple health factors including age, lifestyle, current medications, and existing chronic illnesses. This is not a difficult concept.

For the Good of the Child, or Children as Human Shields?

Cline and Nelson, authors of a 2015 paper on ethical considerations for research using children, explain that "A fundamental pillar of pediatric research is the ethical principle of 'scientific necessity.' This principle holds that children should not be enrolled in a clinical investigation unless necessary to achieve an important scientific and/or public health objective concerning the health and welfare of children."[274] During COVID, a new rationale for vaccines for children was put forth: that children were vectors for illness that could be transmitted to *others*. Children were asked to take the shot to protect the elderly. One of the most poignant interactions I ever had with a child during my career of more than forty years happened during COVID. After I diagnosed a six-year-old child with COVID, I told his mom about extra fluids, vitamins D and C, and anti-inflammatories. I assured my little patient that his body knew how to manage the disease. Then he looked at me with sad eyes and asked, "Am I gonna kill my granny?"

COVID Trials Never Showed Childhood Shots Would Protect Granny

Even as early as 2020, evidence indicated that children were not significant vectors for disease transmission to the vulnerable. Sweden, which never shut schools down, did not show any significant increase in disease or death rates among schoolteachers.[275,276,277,278] As stories emerged about young children apologizing to their grandparents on their death beds for giving them the disease that was killing them, my heart broke as I thought about the lifelong impact of guilt and trauma we were causing this generation of children by laying this false burden on them.

The Helsinki Agreement and Children

The Helsinki Final Act was an agreement signed by thirty-five nations in 1975. Here is the language from the Helsinki agreement about pediatric patients: "For a potential research subject who is incapable of giving informed consent, the physician must seek informed consent from the legally authorized representative. These individuals must not be included in a research study that has no likelihood of benefit for them unless it is intended to promote the health of the group represented by the potential subject."[279] So, the Helsinki Declaration does not endorse the use of a vaccine trial in children if the main outcome is to protect people *other* than children. Throughout human history, most cultures endorse that adults act to protect and preserve their offspring; putting themselves at risk to potentially protect elders is not part of a child's job.

The Institute of Medicine Examines
Causality for Vaccine Side Effects

In 2009, the IOM was asked to "convene a committee of experts to . . . review the epidemiological, clinical, and biological evidence regarding adverse health events associated with specific vaccines covered by the VICP."[280] The committee was charged with developing a consensus report with conclusions on the evidence bearing on causality and evidence regarding the biological mechanisms that underlie specific theories for how a specific vaccine is related to a specific adverse event.

The committee was asked to analyze the scientific data applicable to a potential causal link between eight different vaccines and 158 adverse events. This was the first comprehensive review since 1994. "The committee concluded the evidence convincingly supports 14 specific vaccine–adverse event relationships. In all but one of these relationships, the conclusion was based on strong mechanistic evidence with the epidemiologic evidence rated as either limited confidence or insufficient."[281]

When Evidence Is Sparse, Absence of Evidence
Does Not Mean Evidence of Absence

There are two important messages embedded in the above sentence from the IOM: 1) Adverse events clearly occur from vaccination, and 2) epidemiological evidence is inadequate to identify causal relationships. The second point is particularly irritating to me because the CDC uses epidemiologic evidence to "prove" to parents "vaccines don't cause autism." Even if there was no connection between vaccines and autism, the correct epidemiological assertion would be "the evidence favors the rejection of the hypothesis that vaccines cause autism." I would argue that, even if the epidemiologic evidence argued against the autism and vaccine hypothesis (which is only true if you ignore a lot of the circumstances surrounding that evidence), clinical cases that satisfy Bradford Hill criteria of causation would prove the connection for individual cases.

This is the "black sheep" phenomenon. If you do a study in a certain place and find only white sheep, you might conclude that there are no black sheep. But if I see and photograph a black sheep, I have disproven your hypothesis. If the CDC cites epidemiological evidence that vaccines do not cause autism, but I have patients who were healthy the day of vaccination, deteriorated within twenty-four hours without an alternative explanation, have underlying genetic predispositions that put them at higher risk of a reaction, have lab data that supports the kind of inflammation and mitochondrial dysfunction that vaccines can cause, improved with various treatments until they got the next vaccine and deteriorated again, I have proven a vaccine/autism connection in that individual.

In medspeak: "It is important to note that mechanistic evidence can only support causation. Epidemiologic evidence, by contrast, can support ("favors acceptance of") a causal association or can support the absence of ("favors rejection of") a causal association in the general population and in various subgroups that can be identified and investigated . . ."[282] The IOM committee accepted fourteen specific vaccine adverse event relationships. They rejected others, not because there was clear evidence to disprove the connection but because "the vast majority of causality conclusions in the report are that the *evidence was inadequate* to accept or reject a causal relationship."[283] Another way to phrase that is "absence of evidence does not mean evidence of absence."

It Is Hard to Find What You Do Not Look For

Bernadette Healy was a cardiologist and a heroine in the field of women's and children's health. She served as the first female director of the National Institutes of Health. I wonder if things would have been different if she were still in charge of NIH during the COVID crisis. In 2008, Healy told CBS news that public health officials are hesitant to research whether "subsets of children are 'susceptible' to vaccine side effects. . . . There is a completely expressed concern that they don't want to pursue a hypothesis because that hypothesis could be damaging to the public health community at large by scaring people."[284] Sadly, Dr. Healy died in 2011 before she could wield her influence to mobilize any such studies.

It Is Hard to Be Objective About Your Own Babies

Those of you who are parents or grandparents may realize that you perceive your children or grandchildren to be—fill in the blank—cuter, smarter, or friendlier than other babies. A similar phenomenon happens with vaccine developers. In this case, the "babies" are the vaccines they create. Despite the obvious inherent conflicts of interest, investigators who are being paid to do vaccine trials are the ones tasked with deciding whether side effects observed in trial participants are due to the vaccine or unrelated. Not surprisingly, most of the time they conclude the adverse events are unrelated. Here are some examples:

- DTAP: "Whether an adverse event was attributable to vaccination was judged by the investigators considering temporality, biologic plausibility, and identification of alternative etiologies for each event. Serious adverse events were reported by 22 participants. . . . *None* were judged to be attributable to Tdap vaccine."[285]
- HPV vaccine: "*No* SAE [serious adverse events] in the HPV-16/18 vaccine group was considered related to vaccination or led to withdrawal."[286]
- Meningitis vaccine: "1755 (63%) subjects had adverse events that were considered *not to have any causal relation* with the vaccine."[287] [emphases added]

Calculating Power

I am not talking here about the power of the WHO, the CDC, the NIH, or various governments to mandate vaccines or shut down economies. I am talking about power calculations when doing research. To calculate statistical power, one needs to apply appropriate formulas depending on the type of study. Now there's an app for that! Larger trials increase statistical power. Vaccine trials are expensive, especially if adequate long-term follow-up is done. Therefore, many vaccine trials are not powered to detect rare side effects. Rare is a concept, like beauty, that is in the eye of the beholder. When a vaccine is deployed globally and taken by billions of people, even rare side effects can add up to thousands or even millions of people being adversely affected. "Clinical trials of new vaccines have typically involved a relatively small number of individuals (usually fewer than 10,000) and thus cannot usually detect uncommon adverse events."[288]

Conclusions

A careful analysis of the protocols and results from pediatric COVID shot trials suggests to me that the studies were underpowered to find serious side effects that could affect a generation of children who received these injections. Even bad side effects that did occur in the trial were covered up and/or misrepresented by the manufacturers. A careful analysis of vaccine science to date uncovers significant gaps which the public rarely knows about. Our children deserve better.

FOOD FOR THOUGHT

Had you heard about children who suffered serious side effects from vaccine trials? Were you surprised at the small number of children who were studied? What sources did you use to decide about giving COVID injections to your kids? Were you surprised that the Institute of Medicine found that vaccine science was inadequate to answer important safety questions?

CHAPTER 15

Why Did the ACIP Recommend Adding COVID "Vaccines" to the Childhood Immunization Schedule?

If we don't stand up for children, then we don't stand for much.
— Marian Wright Edelman

History will judge us by the difference we make in the everyday lives of children.
— Nelson Mandela

The CDC and ACIP

My answer to the question I pose in this chapter is "I do not know; it does not make sense to me." But let me give you some background so you can decide whether you disagree. The CDC's Advisory Committee on Immunization Practices (ACIP) is composed of people with expertise in vaccine development and deployment. ACIP can make recommendations about vaccines, including whether one should be added to the Childhood Immunization Schedule, but the Centers for Disease Control makes that ultimate decision. Often the committee members work at academic institutions or have developed a vaccine themselves.

Dr. Paul Offit, whom you have probably seen on TV talking about the benefits of vaccines, was one member of the ACIP. His team developed the rotavirus vaccine. He is the Maurice R. Hilleman professor of vaccinology at the University of Pennsylvania, which means his position is funded by Merck. He directs the Vaccine Information Center at CHOP – Children's Hospital of Philadelphia. He is currently a member of the NIH working group on "Accelerating COVID 19 Therapeutics and Vaccines." He is also a member of the FDA's Vaccines and Related Biological Products Advisory Committee (VRBPAC).[289]

What Is the National Childhood Vaccine Injury Act of 1986?

Once a vaccine is added to the childhood schedule, it is covered under the National Childhood Vaccine Injury Act of 1986,[290] which means pharmaceutical companies have extremely limited liability for vaccine recipients who are injured or die as a result of the vaccine. Without any liability to worry about, vaccine manufacturers have become less and less concerned with vaccine safety. So, quite ironically, this bill guaranteed that devastating vaccine injuries would happen more frequently. Even if your child dies within hours of a vaccine and there was no other potential cause, you could not sue the pharmaceutical industry directly. The bill set up the National Vaccine Injury Compensation Program to be an alternative to parents taking vaccine manufacturers to court for alleged vaccine injuries. It was intended to be more efficient and non-adversarial. Yet, parents who have been through the process have told me it drags on for years and is highly adversarial. Parents must file a case in "vaccine court" within three years of the symptom onset.[291] Employees of "vaccine court" called "special masters," who are lawyers—not judges or doctors—decide the cases. A table of known vaccine injuries lists side effects that are eligible for compensation.[292] Your taxpayer dollars fund the court. $5,134,788,935.74 (yes, the total is more than five billion dollars) has been awarded in compensation so far.[293] The 1986 bill also establishes a "National Vaccine Program" in the Department of Health and Human Services to:

1. direct vaccine research and development within the Federal Government
2. ensure the production and procurement of safe and effective vaccines
3. direct the distribution and use of vaccines
4. coordinate governmental and nongovernmental activities.

It also requires the director of the program to report to specified congressional committees. See H.R.5546–99th Congress (1985–1986): National Childhood Vaccine Injury Act of 1986 | Congress.gov | Library of Congress.[294]

Testifying in the Omnibus Autism Proceeding

I was an expert witness in vaccine court in DC for the Omnibus Autism Proceeding (OAP). It was a harrowing experience. The government is the payor of any damages, and government lawyers from the Justice Department cross-examine witnesses for the families of children who have been hurt by vaccines. I testified on behalf of approximately five thousand children. My testimony included arguing that mercury in vaccines could cause neurologic and autoimmune damage in susceptible children. I think the science was on our side, but we lost the case. I still feel horrible about it. Children's Health Defense recently filed a motion alleging (and demonstrating) fraud on the part of the OAP special masters. An expert witness for the government thought that vaccines could cause autism in susceptible subsets and was dropped from the list to testify.

ACIP Meets to Discuss Myocarditis After COVID Vaccines

On June 23, 2021, the CDC's Advisory Committee on Immunization Practices met to review reports of post-COVID-vaccination myocarditis (inflammation of the heart muscle) and pericarditis (inflammation of the sac around the heart). Pediatricians and intensive care physicians around the world had reported a striking uptick in number of cases since the rollout of COVID vaccines, and the Vaccine Adverse

Reporting System had received an increased number of reports of heart damage after the novel shots. VAERS captures only a small fraction of adverse events, perhaps less than 1 percent according to a 2010 Harvard report by Lazarus and colleagues.[295] The Harvard project was funded to improve the quality of adverse event detection and reporting. Despite finding that as many as 99 percent of adverse events were not reported, the CDC has not proceeded with the proposed improvements in the decade and a half since the Lazarus report.

ACIP's slide presentation for that meeting was available on the CDC website, so readers could look at the information themselves.[296] I was going to encourage you to look at the slides and evaluate how your tax dollars are being used, but as we go to publication the information has been archived and not able to be found. About eighty people at the CDC were working on this issue as part of the COVID-19 Vaccine Safety Technical Work Group (VaST).[297]

ACIP meetings are televised on the internet. Interested citizens are invited to submit comments or apply for the chance to make a three-minute oral presentation. Here are the comments I sent ahead of the meeting.

As a pediatrician with over 40 years of experience both in the private and academic sectors, I strongly oppose recommending COVID vaccines for children or teenagers. Once again, the pharmaceutical industry has conflated relative risk reduction with absolute risk reduction. For example, from Pfizer's data, we see eight cases in a vaccine group of 20,000 and 86 cases in the placebo group of 20,000. This is an attack rate of 0.0004 in the vaccine group and 0.0043 in the placebo group. By emphasizing relative risk, they generated their 90 to 95% effectiveness marketing mantra. However, the absolute risk reduction for the individual is only 0.0039. Is a less than 1% absolute risk decline in [having severe] COVID symptoms worth risking as yet unknown side effects or adverse events that could last for decades in young people? I emphatically think not. Furthermore, the original trials were not designed to demonstrate effects on transmission, hospitalizations, or deaths. COVID vaccines are being marketed to young people as a way to allow them to see their friends again, travel, or go to college. Since vaccines in my state started being marketed to young adults, I have seen significant adverse events. One 19-year-old college student experienced 27 days and counting of such overwhelming fatigue that he thinks it is not safe for him to drive several miles to the grocery store to get one item for his mother. Several others have had significant side effects, such as not being able to raise their arms to brush their own hair or having such fatigue that they were not able to work for four days and had to stay in bed. I am at a loss as to how to recover these patients. I am worried that they are suffering extreme oxidative stress, mitochondrial impairment, or microvascular pathology that I do not yet understand. Notably, public health officials seem to discount such side effects as coincidental. Regrettably, I have not seen guidelines from federal agencies about how to help these young people regain their health. We have already done grave harm to our young people in terms of their mental health and education. Suicides, opioid overdoses, anxiety, and depression have essentially doubled during lockdown compared to prior years. Elementary school students have lost [as much as]an estimated 70% of their academic progress while trying to learn virtually. I beg you to remember our oath: "first do no harm." The world is watching.

Results of the ACIP Meeting

I wonder how many of the submitted remarks the committee members read. I noticed something interesting when listening to the daylong meeting. Most presenters talked in terms of "benefit-to-risk" ratios rather than the more customary "risk-to-benefit" ratio. Was this shift in language and emphasis intentionally chosen to suggest that benefits were paramount, and the risks were less? Language matters, and it is an interesting question from a marketing/propaganda standpoint. The group did not limit itself to reviewing vaccine-induced myocarditis but discussed other causes of heart damage, including viruses, parasites, protozoa, bacteria, fungi, toxins, hypersensitivity reactions, and immune-mediated problems.

From ACIP's discussion, we learned that gender seems to play a significant role, with myocarditis being reported in males 76 percent of the time. Among the medical literature cited, the committee discussed a case series in *Pediatrics* describing seven healthy males who developed myocarditis within four days of receiving their second mRNA COVID vaccine.[298] Four required intravenous immune globulin (IVIG) and steroids, while three improved with supportive treatment. Small numbers, but the data was just beginning to be published, since the "vaccines" had only been out for six months and not yet added to the childhood schedule.

The group also discussed studies in adult patients—but the crucial point is that there is still no long-term data. According to the pediatric cardiologist presenting at the ACIP session, the risk of dying from "garden-variety" myocarditis was, before COVID, about 4 percent to 9 percent. Several committee members stated their impression that the post-vaccine myocarditis seemed "milder," voicing optimism that outcomes would be good. However, because ejection fractions and other objective markers of cardiac function were not presented—and because the patients were still in a phase of illness characterized by cardiac medications and severe limitations on activity, there were no data to confirm those positive impressions. However, I do want to respect their clinical experiences, just as I would want them to respect the clinical acumen of clinicians in the trenches, many of whom have voiced concerns that fatality rates post mRNA injection are quite high.

One speaker at the ACIP meeting presented a working document with a chart of recommendations for second doses of vaccine if the patient developed pericarditis or myocarditis after the first dose. For pericarditis (heart sac damage) beginning after the first shot, the working draft recommendations were to give the second dose of vaccine if the patient had recovered. I was shocked! For myocarditis (heart muscle damage) beginning after the first shot, the language included a recommended discussion between patient and caregivers and administration of the second dose if the patient had recovered. In my opinion, the risk of further heart damage being incurred by a second dose far outweighed the meager benefits of a shot that provided, at best, only temporary protection from an illness that was usually mild in that age group.

For pediatricians and family doctors who are trying to provide the best care for the patients before them, it is difficult and counterintuitive to order a second shot when the first dose caused a life-threatening heart condition. CDC officials (who may not provide direct patient care) have a different perspective from doctors who are directly responsible for and to their patients. At the meeting, Dr. Pablo Sanchez expressed his personal reluctance to give a second dose of vaccine, pointing out appropriately that "whatever triggered myocarditis the first time could cause a problem with the second dose." He is also to be commended for emphasizing the need to inform patients and parents—*before* COVID vaccines are given—of the risk of myocarditis. Thank you, Dr. Sanchez, for standing up for children.

Risk of Death Needs to Be Considered

The committee did not mention death as a potential outcome, even though at that time we knew of at least two deaths in young people that seemed directly related to the COVID vaccines—one in a thirteen-year-old male and one in a nineteen-year-old female. According to VAERS data as of August 13, 2021 (excluding reports from outside the US), there were over 2,000 cases of myocarditis and pericarditis, with 1,335 cases attributed to Pfizer's, 703 cases to Moderna's, and seventy-eight to Johnson & Johnson's COVID vaccines.[299] Remember those numbers may be undercounted by a factor of up to 100, according to the Harvard study commissioned to improve reporting to VAERS.[300] So, the real numbers could be as high as 200,000 even before the vaccines were made available to children, which is horrifying. The CDC tends to downplay adverse event reports, saying that some of the cases are not confirmed. But adverse event reporting is not fast, easy, or intuitive, so I think VAERS reports should be given the benefit of the doubt, not dismissed as maliciously generated.

Were There Fundamental Flaws in the Analysis?

The ACIP discussions about benefits versus risks included data about background rates of hospitalization and death that seem fundamentally flawed, in my opinion. As already discussed, several studies have concluded that COVID deaths in pediatric patients include a significant percentage of patients who died of other proximate causes but also had a positive COVID test. In many hospitals, patients were tested every day or two for infection-control reasons. As also discussed, PCR testing reliability is highly dependent on the number of cycles run. During the first eighteen months of the crisis, most PCR tests were run at cycle numbers that generated lots of false positives. Of the numbers used in ACIP analysis, it is difficult to know how many of the COVID cases were false positives, or false negatives for that matter. Yet, ACIP had to make their recommendation based on the information known at the time. We all have to make imperfect decisions in real time and judge them in retrospect.

Safe, Effective, and Inexpensive Treatments Were Available

Another factor muddying the waters is that safe, inexpensive, and effective treatments that could have thwarted significant illness in pediatric COVID patients were not widely publicized or made available. Some pharmacists refused to fill prescriptions for ivermectin due to guidance from their regulatory bodies, despite a meta-analysis showing the drug's tremendous impact on diminishing serious illness and death.[301] As discussed in the section on ivermectin, this analysis determined with moderate certainty that use of ivermectin can achieve large reductions in COVID-19 deaths and that the drug's early use can limit progression to severe disease. Yet the FDA did not recommend ivermectin outside of clinical trials. Even optimizing the health of as many pediatric patients as possible by promoting nutrient-dense foods, good sleep patterns, exercise, excellent vitamin D levels, and nurturing relationships—or using other supplements like zinc, vitamin C, and quercetin as appropriate—was not endorsed by public health officials. I wonder how much children's baseline SARS-CoV-2 risk would have diminished below the already low level of risk if such commonsense strategies had prevailed. Instead, clinicians like me who suggested traditional antiviral and anti-inflammatory strategies were ridiculed.

In the autumn of 2023, I attended the World Vaccine Congress. I had the opportunity to ask Dr. Ashish Jah, who was COVID coordinator for President Joe Biden, why no one from the White House team talked about the benefits of vitamin D (which decreased mortality) or vitamin C (excellent antiviral and anti-inflammatory) or the importance of nutrition and exercise for immune health. He replied that the White House did not want to make medical recommendations, and that they deferred to the American Academy of Pediatrics or the American Academy of Family Physicians to give medical advice. Huh? I could have sworn that recommending vaccines, which the White House did repeatedly, was giving medical advice. In my practice we used informed consent, discussed potential side effects, and felt compelled to answer parent questions *before* recommending vaccines. Do you remember President Biden endorsing COVID vaccines, clearly a medical procedure? Not only did they recommend getting COVID shots, they recommended getting flu shots at the same time! According to Jah, the reason God gave us two arms was to have "one for the flu shot and one for the COVID shot."[302] By the way, there were no studies at that time showing that giving those two vaccines together was safe.

Comments from the Public Were Insightful

At the ACIP meeting, individuals making public comments delivered a lot of valuable information that should be highlighted. One speaker pointed out the flawed premise for vaccinating young people against COVID, given a fatality rate on par with influenza and the seriousness of emerging heart complications. He also mentioned the numerous treatment alternatives available and their suppression. He said that vaccine adverse event surveillance mechanisms are a "disaster" and voiced little confidence in the data being captured during the vaccine rollout in young people. Finally, he reminded the committee of the unsuitability of PCR testing in this clinical situation.

Dr. Tom Perry, a retired cardiologist, pointed out that natural immunity is good and durable, while objecting to wasting doses and risking side effects in pediatric patients who have already had COVID. He also pointed out the limitations of testing immunity with antibody titers, which measures only one small part of the immune response to pathogens, stating correctly that T-cell immunity is more important.

A clinician trained in Chinese medicine reported her clinical experience with multiple patients presenting with a concerning constellation of symptoms after COVID injections, describing "frightening" changes in radial pulse characteristics and involvement of the heart in many patients—perhaps even more than are captured as myocarditis cases in our flawed surveillance.

Dr. Leslie Moore, a practicing physician, expressed incredulity that investigational COVID vaccines are being promoted to younger and younger children. She noted that frontline physicians are being left to deal with the consequences of vaccine adverse events without emerging recommendations for what might be helpful. She correctly pointed out that the biodistribution studies from the Pfizer vaccine showed accumulation of the lipid nanoparticles in the ovaries, which should ring alarm bells for anyone making recommendations for pediatric patients who may want to be parents someday.

Another citizen reminded the ACIP that Dr. Cody Meissner, an attending physician at Tufts, pointed out in a prior ACIP meeting that hospitalization rates of four per million in pediatric patients do not constitute an emergency. Logically, there should have been no call for emergency use vaccines. I wrote to Dr. Meissner after that meeting to thank him for his insights. The commenter said it was not ethical to ask children and adolescents to risk their health to protect other people, and she asked why ACIP was ignoring natural immunity. Remember, these comments were made in June of 2021, only six months after the "vaccine" rollout.

One part of the ACIP meeting that struck me as bizarre was the discussion of booster doses. Dr. Oliver explicitly acknowledged the problem of waning vaccine immunity and raised the specter of fear of variants. She did not concede, however, that the 30 percent to 50 percent of children the CDC estimated had already had COVID should have had durable immunity that was likely to be better than vaccine-induced immunity, which is highly specific and limited to one part of one arm of the immune system. She made it clear that boosters for children were on the horizon.[303] As we go to press, these important slides are no longer posted on the web.

It seems to me that the ACIP rarely meets a vaccine they do not like. Many decisions are unanimous in favor of the vaccine. One exception was when Paul Offit, MD, who, remember, is among the most zealous proponents of children's vaccination, was asked to endorse booster doses of COVID vaccine on June 28, 2022. He said he voted "no" only because "hell, no!" wasn't an option. I wrote to him to thank him for that vote. He was concerned that data did not support giving boosters, even if theoretically it seemed like a promising idea.

Conflicts of Interest

I am genuinely concerned about financial conflicts of interest in those who make up the CDC's ACIP committee and the FDA's Vaccines and Related Biological Products Advisory Committee (VRBPAC). One concern I have is that the CDC routinely gives waivers from conflict-of-interest rules. Even when advisory committee members are excluded from voting due to financial conflicts of interest, they can still take part in deliberations and lobby their colleagues for approval or recommendation. As we saw during COVID, peer pressure is a powerful force!

As the daughter of a history professor schooled in historic precedents of exploitation by people in power, I was interested to learn that in the past, a chair of an advisory committee owned six hundred shares of Merck stock. You may remember that Dr. Julie Geberding, former director of the CDC, who was instrumental in dismissing the MMR/autism link and fast-tracking the HPV vaccine Gardasil, left the CDC to earn millions working for Merck, the manufacturer of MMR and Gardasil. To use one example of conflicts that could influence voting on approval of the rotavirus vaccine, four of eight CDC committee members and three of five FDA committee members "had financial ties to pharmaceutical companies who were developing different versions of the vaccine," which you can read about in the House Committee on Government Reform's Majority Staff Report from June 2000 on Conflicts of Interest in Vaccine Policy Making.[304]

Conclusion

I think our children deserve better. People who vote to add vaccines to a schedule that will affect millions of children should not have worked on that vaccine or have financial conflicts of interest. Academicians who rarely see healthy general pediatric patients should listen to concerns about side effects when they are voiced by clinicians in the trenches.

FOOD FOR THOUGHT

Did you know about the risk of heart damage in young people after COVID shots? Were you surprised by the inner workings of these CDC committees? Did you know that the pharmaceutical industry has no liability for illness or death caused by childhood vaccines? Did you suspect conflicts of interest between government agencies and the pharmaceutical industry during the COVID crisis?

CHAPTER 16

Why Did I Lose Faith in the American Academy of Pediatrics?

There comes a time when one must take a position that is neither safe, nor politic, nor popular, but he must take it because conscience tells him it is right.

—Martin Luther King

Humans do have an amazing capacity for believing what they choose— and excluding that which is painful.

—Spock

Fellowship in the American Academy of Pediatrics

For many years I had the letters FAAP after my name, meaning I was a Fellow of the American Academy of Pediatrics. The dues were not cheap but gave me access to updates on conditions affecting children and various documents intended to provide education for patients. Over the years, I became disillusioned with the AAP's actions in areas in which I had some expertise. It seemed to me that the AAP was an agency affected by corporate capture from Big Food and Big Pharma.

The First Strike: Sugar and Processed Foods

More than a decade ago, I became impressed with the work of Rob Lustig, a pediatric endocrinologist at the University of California at San Francisco. I read his book *Fat Chance* and made an appointment to talk with him when I was at UCSF giving a lecture. He told me about his frustration trying to hold the food industry accountable for the damage processed foods were doing to children's health. Specifically, he demonstrated that sugar and processed foods were responsible for fatty liver disease in a shocking 11 percent of children at that time. The numbers are worse now, with nonalcoholic fatty liver disease, a condition traditionally associated with adults, noted in as many as 34 percent of obese children.[305]

He was due to give a lecture about this crucially important finding at an AAP convention but had to edit his talk to avoid offending one of the sponsors, the Coca-Cola corporation. Pediatricians were denied hearing the full story about something as fundamental as how they should counsel parents about the importance of limiting sugars when feeding their children. The reality is that sugared and artificially sweetened soft drinks are very bad for everyone. But Dr. Lustig was blocked from telling the whole story since a corporation whose advertising theme was "I'd like to buy the world a Coke" was a financial sponsor of the AAP meeting.

The Second Strike: Autism

My next disappointment came when I was medical director of the Autism Research Institute. ARI scientists have made tremendous progress in research on the medical problems of children with autism. For example, Dr. Jill James and her team published evidence that many children with autism had abnormalities in their methylation and sulfation biochemistry, and that correcting this crucial biochemical crossroad led to improvements in development and health. Dr. Judy van der Water and Dr. Paul Ashwood mapped out the impact of maternal immune activation in pregnancy on the neurologic development of infants. Gastrointestinal specialists like Tim Buie, Sydney Finegold, and Harland Winter showed abnormalities in digestive enzymes and gut flora in children with autism. Other researchers found that GI symptoms and autism symptoms improved when those deficits were corrected. International researchers like Karl Reichelt from Norway and Paul Shattock from England demonstrated opioid metabolites in the urine of children with autism who ingested casein in milk and gluten in bread. Alesso Fasano did groundbreaking work on zonulin and the immunological effects when the tight junctions of the intestine break down. Bob Naviaux, Dan Rossignol, and Richard Frye documented the devastating effects of mitochondrial dysfunction on the neurodevelopment of children with autism. Sid Baker and Jon Pangborn tied the diverse medical problems in autism together physiologically and developed treatment guidelines.

Lagging Behind the Science

We invited officials from the AAP to come to our conferences to hear the science presented and question the researchers themselves. Yet the AAP's approach remains focused on teaching pediatricians to identify autism early and refer the children for various therapies, like speech and applied behavioral analysis. As the numbers climbed from one in five thousand to one in thirty-six,[306] the AAP did not do enough, in my opinion, to investigate the causes of the exponential increase and teach pediatricians how to handle the medical problems of children with autism. Even in the latest edition of the AAP-endorsed book on autism (*Autism Spectrum Disorder*, pages 143–168, "The Role of Integrative, Complementary and Alternative Medicine"), pediatricians are discouraged from recommending diet changes. In fact, the AAP emphasizes counseling about the nutritional deficiencies that can occur if cow's milk and gluten are removed from the diet. I fear that the subset of children with autism who have opioid effects on their brains from maldigestion of gluten and casein, or have cerebral folate deficiency due to blocking or binding antibodies, are being advised by well-meaning pediatricians that the AAP says there is no science behind the idea of taking wheat or dairy out of the diet. Just because taking wheat and cow's milk dairy out does not help *every* child with autism does not mean it does not help those with opioid peptide problems[307,308,309,310] or cerebral folate deficiency.[311,312,313] Children deserve individualized care.

The Third Strike: PANS and PANDAS – Autoimmune Encephalitis

You may have heard of this condition, either as it was originally known, PANDAS (pediatric acute neuropsychiatric disorder associated with strep), or its more general name, PANS (pediatric acute neuropsychiatric syndrome). Whatever you call it, it is now clear that infections can cause acute behavioral changes in children.

One of my dear friends and colleagues, Nancy O'Hara, MD, who co-taught physician training with me for decades, is an international expert in basal ganglion autoimmune encephalitis, which is inflammation of the basal ganglia in the brain due to antibodies produced by the immune system. It is crucial to recognize that acute behavioral changes, such as sudden onset of obsessive-compulsive behaviors or severe separation anxiety, or sudden onset of new medical issues like urinary incontinence or tics and twitches, may be due to an autoimmune encephalitis that is treatable with anti-inflammatories, antibiotics, and antivirals. Dr. O'Hara's book, *Demystifying PANS/PANDAS*, lays out the science and uses anonymized case histories to guide parents and clinicians to appropriate treatments. Many of my patients have benefited from what Dr. O'Hara has taught me.[314]

Dr. Swedo's Work

Dr. Sue Swedo was a career scientist at the National Institutes of Health. In the early 1990s, she and her colleagues discovered a connection between strep infections in children and acute changes in their neuropsychiatric status. She has since written or cowritten more than fifty papers on the subject.[315,316,317,318] Her research has been instrumental in my ability to help many dozens of patients who got a strep infection that triggered production of strep antibodies that instead of clearing the infection attacked the basal ganglia in their brains, where it caused them to have sudden horrific psychiatric symptoms like uncontrolled weeping, suicidal ideation, urinary incontinence, tics, twitches, and obsessive behaviors that could make school attendance difficult, if not impossible. My patients improved with the treatment approaches pioneered by Dr. Swedo, and she is one of my medical heroines. I know that my prescribing combinations of antibiotics and anti-inflammatories prevented some children from being misdiagnosed as having a primary psychiatric diagnosis and ending up in psychiatric wards.

At the Red Book panel during the 2023 AAP convention, Dr. Swedo very respectfully asked during Q & A if she or others with expertise in pediatric acute neuropsychiatric syndrome could contribute to the 33rd edition of the Red Book that was published in 2024. She quoted some research from other scientists and then pointed out that she thought there were significant errors in the prior two publications.[319,320] She made a strong case that children were being denied medically appropriate treatment and being sent to psychiatric hospitals for psychotropic meds because the AAP guidelines were outdated. Instead of addressing her concerns, the lecturer offered to talk to her after the session. I saw her in line to speak to him. I wonder if she made any progress.

I empathize with her perspective because I have had similar challenges in my career. She spent her career trying to help children with PANS or PANDAS and still many physicians "do not believe in it," as if it were a religion instead of an easily verified medical condition. I spent my career trying to teach others that some cases of autism could be prevented, and other cases could be treated to the point where some kids no longer met diagnostic criteria for autism.[321,322] Yet many pediatricians and family physicians still think of autism as a mysterious behavioral disorder (because that is what they were taught) and "do not believe" that autism includes medical problems that urgently need to be addressed (because they have not been trained).

The Fourth Strike: Vaccine Safety

The Complementary and Alternative Medicine (CAM) committee of the AAP reached out to me during the COVID pandemic to ask if I would lecture about COVID vaccines in kids. I eagerly accepted but made the mistake of mentioning that I did not recommend COVID vaccines for children, which led to me being disinvited. Someone higher up in the organization than the CAM committee felt the AAP should not highlight someone who disagreed with official AAP policy. When the subject is vaccines, many pediatricians get the message that questioning safety is taboo, since vaccinations are such an integral part of pediatric practice. I would argue that scientific progress falters in the absence of vigorous debate.

The AAP Convention

I decided to attend the 2023 AAP convention to get up to date on pediatric issues I address in this book. I was shocked to discover that the AAP required attendees at the autumn 2023 National Convention and Exhibition to be up to date on COVID vaccines. I obtained a medical exemption from the COVID vaccine by sending a letter giving more information about my medical history than the AAP had a right to know and cited references that the vaccine could cause a worsening of my health conditions. To their credit, the AAP granted the exemption.

About fifteen thousand pediatricians descended on the Washington, DC, convention center. I went to multiple lectures over four and a half days. The last half day, I went to my hotel room lamenting what I perceived as a willful blindness among the smart and clearly well-meaning lecturers. From my perspective, giving vaccines is such an integral part of the average pediatrician's practice and the typical academic pediatrician's career that the AAP leaders seem truly unaware of the damage that vaccines can do to a subset of people. I was surprised that the lecturers I heard were so enthusiastic about mRNA injections, considering the novel technology and the issues outlined elsewhere in this book.

I listened to a two-hour update about Red Book recommendations. The Red Book is a "bible" of vaccine recommendations and updates on infectious diseases that is revised every three years. None of the four panelists on that update session said they knew about the reports from around the world about excess mortality in 2021 and 2022 disproportionally affecting working age adults.[323,324] All of them recommended COVID vaccines for children, some stating they would give it to their own children or grandchildren. The AAP recommendations are to give COVID vaccines to every child starting at six months of age and any boosters the CDC recommends, stating that COVID-19 vaccination is safe for children six months and older. They reassured parents that children cannot get COVID-19 from the vaccine, relying on semantics that kids would not directly get COVID from what was in the shots. Remember that in the pediatric clinical trials, more kids in the "vaccine" group got COVID in the month after the shot than in the placebo group, and that immune suppression in the weeks after the shot is well documented.

Paul Offit, MD, Vaccinologist

I was eager to hear Paul Offit lecture. Dr. Offit wrote a book many years ago entitled *Autism's False Prophets*. If I had been more famous, I might have made the cut for the way I treat children with autism with gut health strategies, nutraceutical biochemistry, and mitochondrial support. His lecture topic was "COVID Vaccines: Lessons from the Pandemic." I thought his lecture was a great analysis of the chronology of the crisis. He was affable, well-liked by the crowd, and I understood why many pediatricians turn to him for vaccine advice. Before the lecture, I thanked him for voting "no" (when "hell, no" was not an option, as he explained on video)[325] at a meeting of an advisory committee to the FDA about COVID boosters. Offit thought there was not enough data for approval, so he voted his conscience despite the pressure for approvals that result in frequent unanimous yes votes from his colleagues. When I asked in the Q & A session about the documented DNA plasmid contamination of mRNA vaccines, he reassured me that there was no way the DNA from the shots could enter the nucleus. When I pressed back about the new data, he agreed to discuss it more with me one-on-one. I told him I was really worried about his decision to try to talk his daughter-in-law into giving his granddaughter the COVID mRNA vaccine and reiterated the research about contamination. To his credit, he told me he thinks babies should get an initial series (two or three doses, depending on the manufacturer) but did not need to get yearly shots, which the CDC and AAP were considering. I thought that was a big step from him, given his career in vaccine development. I emailed him after the conference with some references about the alleged contamination which had been replicated by four other sets of scientists by that time. I have to give him credit for answering me within twelve hours, but he didn't change his mind. Thank you, Paul, for having a cordial conversation with a vaccine-risk-aware pediatrician. Many others at the conference seemed to want to shut me up. Thank you, Paul, for saying that not everyone needs yearly doses at a time when the CDC was vigorously promoting boosters.

Full Speed Ahead on RSV Vaccines in Pregnancy

The Red Book committee recommends RSV "vaccine" (which is really a monoclonal antibody product that does not trigger antibody production) before or during RSV season at thirty-two to thirty-six weeks of pregnancy. The timing was moved to later than the usual twenty-eight-week gestation time when flu and pertussis vaccines are given because of better efficacy later in pregnancy and a safety signal of increased risk of prematurity. By giving RSV shots at thirty-two to thirty-six weeks of pregnancy, the baby will be bigger and further along if born prematurely. A spokesperson for the manufacturers stated to me that the prematurity signal was only seen in South Africa, not in the United States. They also told me in November 2023 that the European Medical Agency had just approved RSV vaccine down to twenty-four weeks in pregnancy. That decision gives me chills.

The Red Book lecturers discussed a study in which 7,358 pregnant women were given the RSV vaccine between twenty-four and thirty-six weeks of gestation. The researchers defined their end point as "medically attended severe lower respiratory infection." They calculated vaccine efficacy as 81.8 percent in the infant's first three months and 69.4 percent at six months of life. However, they acknowledged that the confidence intervals on the study were very wide, which means we cannot have much confidence that the results are correct. Remember, narrow confidence intervals increase confidence that the results are not due to chance.

The lecturers reported vaccine hesitancy in Ob-Gyn doctors leading to a "lack of strong recommendations for vaccines" stemming from logistical issues and lack of inclusion of pregnant women in most vaccine trials. I want more data from sources other than the pharmaceutical industry and their colleagues before even considering another vaccine in pregnancy. I am eager to see what the post-marketing results in preventing hospitalizations and deaths show during upcoming RSV seasons.

I cannot recommend this new RSV vaccine for pregnant women during the initial rollout for many reasons. Historically, we have been cautious about recommending medications in pregnancy, due to experiences with drugs like diethylstilbestrol (DES, which increased cancer risks in offspring) and thalidomide (which caused developmental defects in arms and legs). The end point of "medically attended lower respiratory infections" would include outpatient visits to pediatricians or family doctors who would provide supportive treatment and see the infant recover in a few days. I would like to see more documentation about how many deaths and hospitalizations are prevented when moms are vaccinated during pregnancy. Part of my hesitation comes from my knowledge about the adverse neurodevelopmental effects of maternal activation syndrome, which happens when Mother Nature's modulation of the immune system in pregnancy is perturbed. Doctors have traditionally been very careful about using medical interventions that interfere with the normal course of pregnancy. Now we have recommendations to add two new vaccines— COVID and RSV—in the same season. Remember, vaccines are often not tested in combination when the clinical trials are done. I am inclined to trust Mother Nature and be skeptical of new interventions with profit motives as a major impetus.

AAP Promotes COVID Injections for All Children Starting at Six Months

I have tried hard to see this recommendation from the perspective of the many pediatricians who have jumped on board, but I just do not understand their thinking. Clearly, I am in the minority. Perhaps I am jaded by my experience of listening to hundreds of parents tell similar stories about how their children regressed after vaccines. Perhaps my attendance at so many think tanks and international conferences where mechanisms of harm were debated has given me a perspective that most practicing pediatricians never have the opportunity to consider. Perhaps my debate training prior to my medical training honed my analytic skills to always see the other side of an argument. No matter what the reasons, I remain convinced that COVID injections are a bad idea for children.

FOOD FOR THOUGHT

Have you had experiences at organizations that led you to question their official positions or decisions? Were you aware of Big Food's efforts to suppress information about how sugar and soft drinks are bad for health? Have you known children who may have had PANS or PANDAS and could not get help? Do you trust the AAP and CDC for vaccine advice?

CHAPTER 17

How Well Do Vaccine Adverse Event Monitoring Systems Work?

Reject the evidence of your eyes and ears.
—The Party in Orwell's book *1984*

Data is like garbage. You better know what you are going to do with it before you collect it.
—Mark Twain

Background: Strategies for Safety Monitoring

A picture is worth a thousand words. Check this out. Have you seen this reported on traditional news channels during COVID times?

Post-vaccination deaths reported to the US VAERS system since 1990:

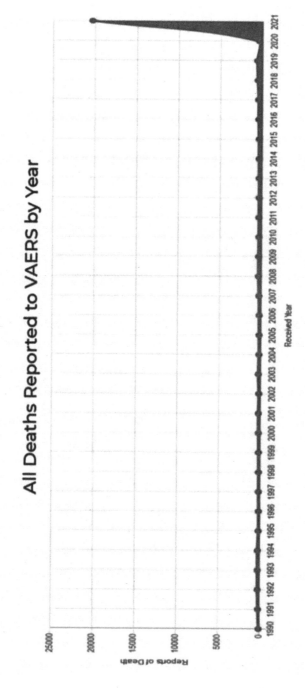

All Deaths Reported to VAERS by Year

Post-vaccination deaths reported to the US VAERS system (OpenVAERS)

What is VAERS?

VAERS (Vaccine Adverse Event Reporting System) was established to keep track of side effects thought to be related to vaccines reported after they come to market. Ideally, analysis of VAERS would pick up red flags of concerning adverse events to be investigated. Vaccine recommendations could change, or certain vaccines could be pulled from the market if they proved dangerous. Do you see any red flags in the chart reproduced above?

There are risks associated with any medical product. Patients should be aware of potential side effects from medications, surgical procedures, and immunizations. The Supreme Court has declared that vaccines are unavoidably unsafe, acknowledging that when the same product is given to a large population, some individuals will be allergic or have adverse events, both expected and unforeseen. However, pharmaceutical companies have unique protection from lawsuits arising from injury: "No vaccine manufacturer shall be liable in a civil action for damages arising from a vaccine-related injury or death associated with the administration of a vaccine after October 1, 1988, if the injury or death resulted from side effects that were unavoidable even though the vaccine was properly prepared and was accompanied by proper directions and warnings."[326]

Underreporting in VAERS

A study by Lazarus and colleagues was commissioned to determine how to make VAERS more effective. They determined that only about 1 percent of adverse events were picked up by the current system of passive reporting. They recommended that VAERS be automated so that reporting side effects would be easier.[327] In more than a decade since that recommendation, proper automation of VAERS has not happened.

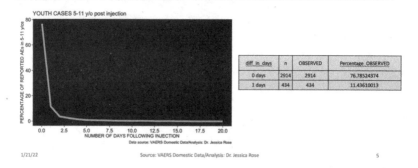

Sourced from Jessica Rose Substack.[328]

Health-Care Practitioners Often Do Not
Fill Out the Forms

I have been filling out VAERS reports for decades and have never been contacted for more information about the patient's history or clinical outcomes. In retrospect, it may have been a big waste of time if red flags like the one I showed at the beginning of this chapter are not acknowledged. Many physicians do not even know about VAERS or think it is too time consuming to submit a report. I am always shocked when I am teaching residents (doctors who have finished medical school who are getting specialized training in pediatrics or family medicine) who do not remember learning about VAERS during medical school or residency. During COVID, clinicians reported that it took about thirty minutes to fill out VAERS reports online, and that if they got interrupted, they had to start all over again.[329] In my experience, it is rare for a clinician to have thirty minutes of uninterrupted time during a workday. It seems like a system designed to fail (or at least be suboptimal) to detect many adverse events.

Early Warning Signs in January–February 2021

During COVID vaccine rollouts, reports of adverse events began accumulating in early 2021, soon after the mid-December launch. The baseline of about four hundred deaths per year from all vaccines combined spiked dramatically in lockstep with the launch of the novel injections made at Warp Speed and approved without the usual regulatory oversight. Public health officials and governments continued to encourage people to get vaccinated to stop other people from getting ill and to decrease the severity of their own infection with COVID. The trials were never able to show that the vaccines decreased transmission to others. As a practicing doctor in a crisis, I read the trials and learned about the shortcomings. One would hope that public health officials would read the documents for themselves. If they did read the trials and continued to say the injections stopped transmission, is that lying? If they saw the chart above showing more deaths reported after COVID injections than all other vaccines previously, was it ethical to continue to use the phrase "safe and effective" instead of more nuanced language? If they were overly fearful of the virus itself or worried about retaliation in the workplace for speaking out, were they making good judgements for the overall good of the people? We know that fear interferes with critical thinking by activating the amygdala, which is the seat of primitive emotions, and shutting down functions of the cerebral cortex.

Here is what Dr. Rochelle Walensky said during a video interview with Wolf Blitzer in August 2021. "Our vaccines are working exceptionally well. . . . They continue to work well for Delta, with regard to severe illness and death—they prevent it. But what they can't do anymore is prevent transmission."[330] Is that a fair representation of the facts? Her last sentence implies that they did prevent transmission previously, which was never proven in the trials.

Sudden Adverse Events

Jessica Rose, PhD found that 80 percent of the individuals who reported sudden adverse events had no current illness when injected with COVID "vaccines." Seventy-seven percent who died after COVID injections had no illness at the time of their shot.

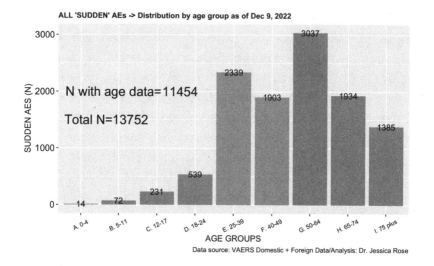

Sourced from Jessica Rose Substack.[331]

V-Safe for COVID – Just Check the Box

The CDC's V-safe smartphone monitoring system was designed to collect data about COVID vaccine side effects such as fevers, local inflammation, and muscle aches in the weeks after vaccination.[332] The purpose of the app was to rapidly detect clinically important adverse events. Early users included health-care workers, many of whom were required to get the jab for work and others who were enthusiastic about getting the vaccine. Having been told by Dr. Fauci that COVID "vaccines" were the equivalent of the calvary, they rushed to sign up. It was like standing in long lines to get Beanie Babies in the 1990s. The CDC promised to make the data available to the public and use the information to guide vaccine decisions.[333]

Three FOIA Requests and Two Lawsuits

As reports of vaccine injury after COVID vaccines surfaced in Spring 2021, ICAN (Informed Consent Action Network) submitted a FOIA request to get information in V-safe so that independent scientists could study the data. The CDC pushed back against releasing the data, arguing in June

2021 that ICAN could not have protected data that included patient identities and that they were not able to locate any de-identified data. Aaron Siri is a civil rights attorney who has researched vaccine injury extensively. His team knew, from the V-safe website, that the de-identified data had already been made available to Oracle for storage. So ICAN appealed the decision in August 2021, pointing out that the CDC's own literature said the de-identified data existed. In September 2021, ICAN submitted another FOIA, explicitly asking for the de-identified data that had been given to Oracle. That should have been easy to do. CDC did not comply. ICAN had to sue to try to get this information, which clearly was of value to public health. ICAN vs. CDC was filed on December 28, 2021, as Siri's law employees worked over the holidays. Remember, V-safe data was part of the post-marketing surveillance that had been promised by the manufacturers and government. By this time, the Federal government had mandated vaccines for some populations, so the analysis of the data was even more important.

Fighting for Transparency

Then ensued months of dialogue between Siri's firm and the Department of Justice, who serve as lawyers for the CDC. Siri maintains that the Justice department said that the CDC got it wrong, that the data had not been de-identified. Hmm . . . so when it was confirmed that de-identified data did exist, the DOJ was in the awkward position of arguing that either that the CDC got it wrong, or they had lied. Siri pushed back, arguing that this was such a crucial issue they would file another lawsuit if needed. He pointed out that more lawsuits would be a waste of taxpayer money and waste of ICAN resources, which is a non-profit. The DOJ was unmoved. In April 2022, ICAN filed a third FOIA, asking to see *all* the data. In May 2022, ICAN filed a second lawsuit, this time asking the CDC "to provide transparency to claim COVID 19 vaccines are safe and effective." The lawsuit argued that it could be valuable for independent scientists to analyze the data.[334] In the midst of a global emergency, an "all hands-on-deck" approach could help. Let's count. It took three FOIA requests and two lawsuits to get to the data that was intended to provide early warning signals about side effects of a novel vaccines technology that was being deployed around the world. Three FOIA requests over the course of a year! Two lawsuits!

A judicial order compelled the CDC to provide the first batch of data on or before September 30, 2022, nearly two years after vaccine rollout. How do those of you who had bad reactions to the COVID injections during that time frame feel about that? The judge determined that on or before October 14, 2022, both sides would meet and confer to evaluate the adequacy of what had been provided.

CDC Could Not Work Out the Technology, but ICAN Did in a Weekend

One of the initial arguments (after the failed argument that they did not have the data) was that CDC did not have the resources to provide the requested data. Aaron Siri reports he got a letter on September 30, 2022, saying that the CDC has not yet completed the technical and administrative processes required, so they released the data sets directly. Siri's team got the data by 6 p.m. on a Friday, and by Monday ICAN tech geniuses had created an interactive database dashboard that enabled anyone to navigate around the platform and examine the data in detail.[335] ICAN had thought about how they would look at the data when it emerged. One could argue that the CDC should have built such a data analysis tool when they launched the app.

The Data Was Shocking

Of the 10 million V-safe users, 7.7 percent of them had to seek medical care for adverse events after the shots.[336] That is 770,000 people! Twenty-five percent missed work or school due to the side effects.[337] That is 2,500,000 people! A shocking 40 percent (4,000,000 people) reported joint pain,[338] raising concerns about autoimmunity or inflammation of bones induced by the injections. Females in every ethnic group fared worse with more side effects than men. Shocker. The gender disparity is consistent with the reports from the FDA. In science, when the same finding is affirmed by different groups, the chances of it being true increase.

Several anticipated side effects were included in checkboxes for users to choose. Of the initial ten million, 1.1 million reported fatigue, 700,000 muscle aches, 700,000 headaches, and 500,000 other pains.[339] These side effects were pre-populated by the CDC. Noticeably missing from the choices were anaphylaxis, abdominal pain, chest pain, and heart problems, all of which happened.

At the ACIP meeting that gathered concerns about myocarditis and pericarditis after COVID vaccines in June 2021, a participant who suggested adding symptoms of chest pain and shortness of breath to V-safe received the response that the app's questions had already been programmed, with no plans to add the cardiac symptoms! Now that we know, and even the CDC has acknowledged that myocarditis is a risk, how hard would it be for a software programmer to add that option? Legal action to get information about the free text entries in V-safe was required. At the time V-safe was designed, perhaps there was no reason to list chest pain or shortness of breath to look for myocarditis. By not adding it once it was reported, it sure looks like they do not want to find it. No heart issues to see here. V-safe has not been looking for heart issues. Why should it be up to patients, parents, and physicians to detect myocarditis in vaccine recipients? Only some of the myocarditis patients even have signs or symptoms of myocarditis. Sometimes the heart damage is asymptomatic, so unless we look prospectively, we might miss it.

I was shocked to discover that 13,963 infants and toddlers under the age of three had gotten the COVID jabs and reported to V-safe during a time when the vaccine had not yet been approved for children. Of 13,963 children less than three years old, about a thousand had to seek medical care after the shot.[340] This works out to be 7 percent, more than other pediatric vaccines by an order of magnitude and far greater than the risk of serious complications from COVID itself. Most babies and toddlers with COVID have mild or absent symptoms, as you learned early in this book. Another shocker was that most of the V-safe reports were filed from December 2020 through May 2021. So, the CDC had most of the data prior to approvals for pediatric patients in June 2021.

V-Safe data provided insight into the differences between brands. More people got Pfizer than Moderna shots, but more people had side effects from Moderna than Pfizer in V-safe.

V-Safe Shuts Down Data Collection for COVID Shots

In the chapter about shortcomings in the COVID injection trials, we showed how there is no long-term control group to provide a comparison to chronic conditions that develop in those who got the COVID jabs versus those who did not. Now we have no long-term followup of the injection group of 10 million participants in the V-safe tracking system. The website states that data collection concluded on June 30, 2023.[341] The website directs people to VAERS, which you now know is passive and prone to underreporting by as much as a hundred times. Look at a screenshot of data from V-safe. Go to the ICAN V-safe dashboard at https://icandecide.org/v-safe-data/ and navigate around the data yourself.

You will see that there were about 10.1 million users and about 1.2 million of them could not go about their usual activities. Novel mRNA technology has never been so widely deployed in so-called vaccines. Long-term follow-up would be prudent, especially considering the unprecedented numbers of injuries and deaths reported after the injections.

Safety Reports Continue for Old Cars, but Not Young People

Dr. David Gortler is a pharmacologist, research scientist and Doctor of Pharmacy. He is a former member of the FDA Senior Executive Leadership Team, where he advised the FA commissioner about drug safety, science policy, and regulatory affairs. In an article for the Brownstone Institute[342] he drew a great analogy between vaccine safety regulation and car safety standards. He pointed out that he can still file a safety report on his thirty-year-old Ford Bronco II, which was handed down to him during his student years, as a resident and fellow, while a professor at Yale and while a senior medical analyst for the FDA.[343] Dr. Gortler wrote that "the CDC isn't accepting new safety reports on two-year-old novel mRNA vaccines."[344] The irony is obvious. He draws attention to the following data: 1,585,094 reports of adverse events through August 11, 2023.[345]

Data in the chart below shows more than twenty thousand heart attacks with the timing slope consistent with a causation pattern—deaths peaking at day zero when the shot was given.[346] If the shots were not implicated, scientists would expect a more random pattern with the deaths scattered relatively evenly throughout the month.

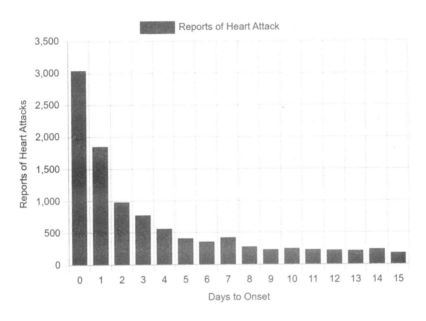

Source: OpenVAERS.com.

There were 28,769 cases of myocarditis or pericarditis when this book went to the publisher.[347]

Hope CDC Pays Attention to Its Own Data

I sincerely hope the CDC renders an opinion about the V-safe data they have collected so far. Is it too much to hope that it will inform decision making? Otherwise, what value is it to spend time and resources gathering the data? Remember, the US government explicitly granted immunity from liability to COVID vaccine manufacturers in the PREP Act. If you die or are permanently disabled from COVID jabs, you cannot sue your doctor or the vaccine makers who have made billions from these injections.

Data from Other Parts of the Government

Once I started looking at how the sausage was made, my faith in post-marketing vaccine safety surveillance was shaken to the core. As I listened to multiple ACIP committee meetings, I continued to be surprised that the committee saw the same slides and heard the same presentations that I did and came to opposite conclusions. Again, a matter of perspective. Committee members are biased because they are part of the culture of vaccine development. I am biased because I have seen a lot of children who have deteriorated after vaccines. My clinical research associates higher vaccine load with higher rates of chronic illness.[348] If there were interactions between practicing clinicians and bureaucratic officials, perhaps we could learn from each other and find some middle ground for the good of the populace.

Absence of Evidence Does Not Mean Evidence of Absence

Let's look at some more history. The Institute of Medicine, as part of the National Academy of Sciences, has looked at vaccine safety throughout the years. In 1994, they confronted the issue that vaccine science was not robust enough to make definitive determinations about causation for many side effect/vaccine relationships. Let's look at their words: "With the majority of vaccine-adverse event pairs the evidence was considered *inadequate to accept or reject causality*. . . . Concern about this unfortunate condition of uncertainty has led the committee to urge that more definitive research be done on possible adverse events during the development of new vaccines or vaccine combinations and to urge that efforts to sharpen current post marketing surveillance systems be accelerated."[349] Meaningful reform to address those concerns has not happened.

In 2004, I attended the Institute of Medicine meetings to investigate the possibility that thimerosal, which is 49.9 percent mercury, could be contributing to the startling rise in neurodevelopmental problems of children with autism. Many of my colleagues presented data at that meeting. Representative Dave Weldon presented reports of denigration of scientists who investigated concerns about vaccine safety. The committee found that:

"The evidence favors rejection of a causal relationship between thimerosal containing vaccines and autism." There was no change in licensure. The committee recommended: 1) Better surveillance; 2) Quantify other sources of mercury exposure; and 3) Restrict chelation to IRB-approved studies. Thimerosal was already being phased out, but more aluminum was added to vaccines. Flu shots were phased in for pregnant women and yearly flu shots were recommended for babies and toddlers. Many of those shots contained mercury, a known neurotoxin that also induces autoimmunity.[350,351,352,353]

It was a catch-22. There were not enough studies to prove causation, but it was hard to get scientists to do studies because of the taboo of questioning vaccine science. We had evidence that mercury is a known neurotoxin, that it induces autoimmunity, and that many children with autism had problems eliminating heavy metals like mercury and aluminum from their bodies. What we did not know at the time was the mindset of the committee prior to the meeting. Before the meeting, Marie McCormack, MD stated: "we are never going to come down that it [vaccines and autism] is a true side effect."[354] A full discussion of conflicts of interest of the IOM committee and suspect sources of private funding is beyond the scope of this book, but Mr. Handley's book, *How to End the Autism Epidemic*, is a rich resource.

A Rose Among the Thorns

Jessica Rose, PhD is an intrepid biostatistician with a wicked sense of humor who has been doing in-depth COVID side effect research since 2020. The COVID panic derailed her plan to tour as a professional surfer and planted her in front of a computer in Israel where she conducted extraordinary analyses of vaccine safety data from myriad sources. I encourage you to check out *Unacceptable Jessica,* her Substack series of in-depth articles.[355] She is my go-to expert for all things VAERS and the source of many of the charts in this book. Thanks, Jessica, for your dogged pursuit of the truths behind the numbers.

Conclusion

Courageous researchers are publishing data that demonstrate how adverse events to mRNA COVID products remain underreported "due to numerous clinical, systemic, political and media factors . . . these injections are not as safe as widely purported."[356] People who report life-altering adverse events from COVID injections continue to be gaslighted. Pharmacies continue to advertise free COVID shots. V-safe data is still not reported by mainstream media as we go to publication.

FOOD FOR THOUGHT

Did you know about VAERS before this book? Were you surprised by the spike in deaths seen in the VAERS chart at the beginning of this chapter? Were you surprised that the CDC and DOJ fought against the release of the V-safe data? Did navigating around the V-safe dashboard show you information that will help you make decisions about COVID vaccines for your child?

PART III

RISKS AND BENEFITS OF COVID INJECTIONS

CHAPTER 18

Can COVID Shots Cause Heart Problems?

The hottest places in hell are reserved for those who in periods of moral crisis maintain their neutrality.

—Dante, *Inferno*

The stories of past courage . . . can teach, they can offer hope, they can provide inspiration. But they cannot supply courage itself. For this, each man must look into his own soul.

—John F. Kennedy

Answering the Call to Action

As I sorted through articles about heart problems from COVID the virus vs. COVID injections during the dark years of 2020 through 2022, one doctor whose work made the most sense to me was Peter McCullough. Peter is an American cardiologist who was vice chief of internal medicine at Baylor University and a professor at Texas A & M University. His training includes a BS from Baylor and an MD from University of Texas Southwestern. He did a residency in internal medicine at the University of Washington in Seattle followed by a cardiology fellowship. After two years of practicing clinical cardiology, he earned a Master of Public Health at the University of Michigan. He was a co-editor-in-chief of *Cardiorenal Medicine* and editor of *Reviews in Cardiovascular Medicine*.

As of March 13, 2023, as I write this, he has published 1,212 articles in his career. His work has been cited by other scientists 68,339 times, with 23,529 citations since 2018. In 2022, he cowrote a book called *The Courage to Face COVID.*[357]

When one asks him a question about COVID, as I have done, he is likely to answer with some version of the following: "(Name) and colleagues published a paper in (Journal) on (Date) that showed (Results). More work needs to be done, but clinically we can use that information to consider (treatment strategies)." A medical academician must have an impressive command of the medical literature to be able to quote so many published articles off the top of his head! As the editor or co-editor of two cardiology journals, he read a great deal of cardiology research in a detailed and critical way so that he could identify the strengths and weaknesses of the papers and accept or reject the articles for publication. In the early days of COVID, he shared his clinical and research experience in the medical literature. He is an author on more than seventy-five scientific papers plus the whole book about COVID. In my judgment, he is absolutely credible! Let's look at his output during 2020, when most doctors were scratching their heads and following hospital protocols. Let's follow the paper trail.

Early Research Insights

An early contribution was to examine patients with COVID and see if there were any identifying factors that put patients at higher risk for death. Here is what Peter's colleagues found early in the pandemic: "In a cohort of 2,215 adults with COVID-19 who were admitted to intensive care units at 65 sites, 784 (35.4%) died within 28 days, with wide variation among hospitals. Factors associated with death included older age, male sex, morbid obesity, coronary artery disease, cancer, acute organ dysfunction, and admission to a hospital with fewer intensive care unit beds."[358] Valuable contribution to identify risk factors which could lead to targeted treatments, don't you think?

Peter used his public health training when he was first author on a paper that identified this problem: "Despite near hourly reporting on this crisis, there has been no regular, updated, or accurate reporting of hospitalizations for COVID-19 . . . many test-positive individuals may not develop symptoms or have a mild self-limited viral syndrome consisting of fever, malaise, dry cough, and constitutional symptoms. However, some individuals develop a more fulminant syndrome . . . leading to death attributable to COVID-19."[359] He articulated the wide variation in effects from COVID, setting the stage for targeted, individualized treatments.

One concern from cardiologists was that patients having heart attacks might delay seeking medical care for fear of getting COVID. In a study of nearly two hundred patients having acute myocardial infarctions, there was evidence of "significant delays in hospital presentation for patients under study during the pandemic compared to those during the same five weeks of the year before the pandemic."[360] Of great concern was that patients with ST-elevation heart attacks, which are more serious and have a higher risk of death, wait(ed) nearly three times as long during the pandemic to seek care compared to their pre-pandemic counterparts."[361] The authors reviewed data about the decline in ER visits around the world and the increased occurrences of DOA—dead on arrival. The authors concluded that "Every effort should be taken to increase awareness of the consequences in delaying treatment, perhaps through utilizing telemedicine and outreach programs."[362]

ACES Wild

Dr. McCullough and his colleagues heard the clarion call for action and began systematically studying the issues. One clinical dilemma doctors faced early in the pandemic was to decide: "Shall we continue or discontinue ACE inhibitors?" We found out that SARS-CoV-2 viruses binded to ACE 2 receptors to facilitate entry into the cell. We were not sure whether ACE 2 receptors were upregulated in patients taking angiotensin converting enzymes inhibitors (ACEIs) or angiotensin receptor blockers (ARBs) and therefore provide more opportunities for the virus to bind to cells and cause mischief. After a nice review of published papers on the subject, McCullough and his coauthors stated,

We caution against indiscriminate discontinuation of ACEIs and ARBs in patients who rely on these drugs for treatment of heart failure and who, additionally, might benefit from the postulated positive effects during overwhelming infection with SARS-CoV-2. Discontinuation of ACEIs or ARBs is associated with readmission to hospital and mortality among patients with heart failure.[363]

That is an important clinical tidbit for those in the trenches trying to figure out how to help patients with a new disease.

In another paper on the controversy about using or discontinuing ACE inhibitors or blockers in patients with COVID, the authors searched PubMed and CINAHL databases and pre-print servers for studies about these blood pressure medications in COVID. The meta-analysis (study of multiple studies) included twenty-one publications. This is what they found: "In combining both mortality and severe disease outcomes, the pooled odds ratio was 1.09 [0.80–1.48] p = 0.58 but with heterogeneity of 92%."[5] To translate those statistics, there was a wide variety of patients and no significant difference in COVID severity or death rates comparing ACE-inhibitor and ACE-blocker patients versus patients without those medications. "The results hold true that ACEI/ARB use is not associated with COVID-19 disease severity or mortality. To look for any potential beneficial effects, randomized controlled trials are needed."[364]

They looked at a lot of data, reported their findings and called for further research to see if other scientists duplicated their findings. Seems very responsible to me.

The Clot Shot

Early in the pandemic, it became clear that COVID-19 was associated with increased blood clotting, causing complications especially in critically ill patients.[365] In a comprehensive review of the current literature that outlined the rationale for early treatment of SARS-CoV-2,. McCullough and his coauthors highlight the pathogenesis, clinical features, and management.[366]Again, very helpful for the clinicians in the trenches.

Heart Rhythms Go Haywire

In the beginning of new infectious disease outbreaks, clinicians can provide a valuable service by reporting what they see in their clinics and hospitals. Peter and colleagues did just that by reporting the "clinical course of an otherwise healthy patient who experienced persistent ventricular tachycardia and fibrillation."[367]

While hospitalists and ER doctors were struggling to figure out the details of how to manage critically ill patients, Dr. McCullough was the last author of a Baylor University publication about intravascular volume management strategies in severely ill patients with COVID. Being the last author is not like getting picked last for kickball—it's more like being the quarterback of the article, which is an honor. The article gave clinicians valuable information on how to use IV fluids in ICU patients.[368]

In the six papers I have cited so far to which Peter contributed, we have not even gotten to the end of 2020 yet. Dr. McCullough is one of the most productive clinicians I have ever known. He deserves our gratitude.

What About Kidney Function?

In 2020, a group of more than two hundred clinicians and researchers from around the world collaborated to study COVID-19. Talk about herding cats! Just the administrative coordination involved in this study impresses me. In a retrospective cohort study, the "STOP COVID" researchers examined the records of 4,264 patients with COVID 19 with pre-existing kidney failure admitted to intensive care units at sixty-eight hospitals across the United States. Some needed maintenance dialysis, others did not. They compared kidney failure patients with 3,600 patients without preexisting kidney disease. This is what they found: "Dialysis patients had a shorter time from symptom onset to ICU admission compared to other groups (median of 4 [IQR, 2-9] days for maintenance dialysis patients; 7 [IQR, 3-10] days for non-dialysis-dependent CKD [chronic kidney disease] patients; and 7 [IQR, 4-10] days for patients without pre-existing CKD). More dialysis patients (25%) reported altered mental status than those with non-dialysis-dependent CKD (20%; standardized difference=0.12) and those without pre-existing CKD (12%; standardized difference=0.36)."[369] They also reported data on comparison of mortality between the three groups. They acknowledged their study shortcomings in the paper, as good scientists should do, noting potential residual confounding. They called for "identifying safe and effective COVID-19 therapies in this vulnerable population,"[370] which seems like an important contribution to me. One take-away message is that if your dialysis patient has a change in mental status, COVID infection should be in the differential diagnosis. Thanks to all of you out there that worked on this research.

How Should We Treat Patients?

Peter McCullough and his colleagues, along with Drs. Kory and Marik at the Front Line COVID Critical Care Alliance and many other unsung heroes around the globe who faced the challenge and treated outpatients, were pioneers in early ambulatory multidrug therapy for high-risk COVID patients. Their work was innovative, important, and well-reasoned. Dr. McCullough quarterbacked another paper that stated:

> We evaluated a total of 922 outpatients from March to September 2020. All patients underwent contemporary real-time polymerase chain reaction (PCR) assay tests from anterior nasal swab samples. Patients aged 50.5 ± 13.7 years (range 12 to 89), 61.6% women, at moderate or high risk for COVID-19 received empiric management via telemedicine. At least two agents with antiviral activity against SARS-CoV-2 (zinc, hydroxychloroquine, ivermectin) and one antibiotic (azithromycin, doxycycline, ceftriaxone) were used along with inhaled budesonide and/or intramuscular dexamethasone consistent with the emergent science on early COVID-19 treatment. For patients with high severity of symptoms, urgent in-clinic administration of albuterol nebulizer, inhaled budesonide, and intravenous volume expansion with supplemental parenteral thiamine 500 mg, magnesium sulfate 4 grams, folic acid 1 gram, vitamin B12 1 mg. A total of 320/922 (34.7%) were treated resulting in 6/320 (1.9%) and 1/320 (0.3%) patients that were hospitalized and died, respectively.[371]

Read the last sentence of that paragraph again and ask yourself how things might have been different if Dr. McCullough had been the quarterback giving White House briefings. Only six patients of 922 were hospitalized!

This approach made so much sense to me. A lot of the science was familiar to me as useful for children I had treated with multiple medical problems, including chronic inflammation, oxidative stress, and mitochondrial dysfunction. I loved the recommendations for both nutritional support and prescription medications.[372] Yet, as late as November 2023, I heard Dr. Mandy Cohen on NPR urging people to get their flu and COVID shots at the same time; she did not mention taking vitamins C or D for the winter.

Using Repurposed Drugs

Three doctors I admire, Joe Ladapo, Harvey Risch, and Peter McCullough (and one I just do not know) collaborated on a systemic review and meta-analysis of randomized clinical trials about hydroxychloroquine's role in preventing new infections or decreasing hospitalization and deaths in COVID patients. They used sophisticated statistical methods: calculating random effects, using Cochran Q and I parameters, an Egger funnel plot, and a fixed effect meta-analysis.[373] Honestly, it sounds like Greek to me, but that is why I am glad we have people like Harvey Risch, who is an epidemiologist from Yale School of Public Health, doing such calculations. Here are the results:

Five randomized controlled clinical trials enrolling 5,577 patients were included. HCQ was associated with a 24% reduction in COVID-19 infection, hospitalization or death, P=.025 (RR, 0.76 [95% CI, 0.59 to 0.97]). No serious adverse cardiac events were reported. The most common side effects were gastrointestinal.[374]

Hmm, an existing medication with a statistically significant improvement in outcomes, and a low side-effect profile. What's not to like?

Kill the Messenger

I felt compelled to establish Dr. McCullough's credibility because if you Google him, what pops up first are accusations that he peddled COVID misinformation. Despite his timely response to a crisis and a prolific body of research that was published *tout de suite*, he was criticized. And marginalized. And fired from editing cardiology journals. And dismissed from the faculty at Baylor. Hard to believe, but it happened.

Dr. McCullough's crime was to point out inconvenient truths. First, he pointed out that cheap repurposed drugs could prevent hospitalization and death, potentially derailing the lucrative approval of remdesivir and the rushed COVID vaccines. Then he warned about the risks of myocarditis (inflammation of the heart) and pericarditis (inflammation of the sac around the heart), both of which can be serious and deadly, reported after COVID injections. He wondered why so many young, healthy athletes were dying on soccer fields around the globe. He shared his opinions, always backed by his clinical observations and the medical literature, in podcasts and web interviews. So how did the pharmaceutical powers deal with him? They used the time-honored strategy of "kill the messenger."

CDC Messages

Let's look critically at the data and see what other researchers have found. Then let's look at the messaging from the CDC. As I write this in the summer of 2023, they acknowledge a safety "signal" from myocarditis but do not see red flags to back away from their campaign to inject babies, toddlers, elementary school kids, and adolescents. See what you think.

Two Myocarditis Deaths, Two Autopsy Reports, and Two Letters

In 2022, pathologists published autopsy results from two boys who died in their beds three and four days after a COVID jab. Such deaths in young people are not normal! The pathologists described their findings from examination of various tissues under the microscope. They note that myocarditis is rarely diagnosed at autopsy in deaths due to severe SARS CoV2 *infection*.[375,376] They cite seven references reporting myocarditis, particularly in teenage boys and young men, most often following the second injection.[377,378,379,380,381,382,383]

The first week after the second dose of Pfizer-BioNTech COVID-19 vaccine is the main window of risk. The clinical presentation may be mild or even asymptomatic,[384] which raises concerns for undiagnosed cases. Let's read that again: **the clinical presentation may be mild or even asymptomatic.** Think about people you know who were sick in bed for up to a week after their COVID vaccines. Is it possible they had subclinical myocarditis? I think it is a legitimate question.

The pathology analysis of histologic findings (examining cell structures under the microscope) suggested an excessive inflammatory response and a role for "cytokine storm"—a massive outpouring of inflammatory messengers from the immune system. Notably, forensic toxicologic testing showed no medications or drugs of abuse. I quote from their methods here:

> Standard medicolegal autopsies were performed including gross, microscopic, and toxicologic testing. SARS-CoV-2 nasal swab testing was performed by reverse transcriptase–polymerase chain reaction assay. Tissues were sent to the National Center for Emerging and Zoonotic Infectious Diseases branch of the Centers for Disease Control and Prevention for molecular studies. . . . The myocardial injury seen in these postvaccine hearts is different from typical myocarditis and has an appearance most closely resembling a catecholamine-mediated stress (toxic) cardiomyopathy.[385]

The first boy was previously healthy. Analysis of his heart showed "global myocardial injury" and "coagulative myocytolysis" (breakdown of heart muscles and cells). Inflammation was present around the vessels, and the types of white blood cells seen were mostly neutrophils (infection-fighting cells). There were no acute blood clots. PCR test for SARS-CoV2 infection was negative. Findings demonstrated myocardial fibrosis (abnormal scarring). The pathologists concluded that the death was compatible with "a stress cardiomyopathy with contraction bands and a neutrophilic/histiocytic infiltrate."[386]

The second boy was obese but had no other pre-existing medical conditions. His histologic findings showed more ischemic (low oxygen), widespread transmural (through the heart wall) changes and more interstitial (between the cells) inflammation. He also had a predominance of neutrophils with histiocytes, no acute clots, and a negative PCR test for COVID. He had cardiomyopathy (swelling of the heart which leads to poor heart muscle function).[387]

The clinical histories of these boys prior to their sudden deaths are striking. "Neither boy complained of fever, chest pain, palpitations, or dyspnea."[388] The first boy had a headache and some gastric upset after the vaccine but felt better three days after the shot. Their parents found them dead and cold in bed three and four days after the COVID jab. Can you imagine that experience?

The authors investigated potential mechanisms for these deaths. First, they looked at the differential diagnoses, which is medical-speak for various possibilities doctors should consider when patients present with symptoms: autoimmune conditions, toxic exposures, radiation damage, drug side effects, etc. Then the authors propose catecholamine toxicity as a possible cause and give some background and reasoning supporting that hypothesis. Catecholamine toxicity on the heart was first described in patients with pheochromocytoma. In these two cases, the teenage boys did not have obvious cardiac symptoms. Their histopathology did not demonstrate a typical myocarditis. This injury pattern is instead like what is seen in the myocardium of patients who are clinically diagnosed with Takotsubo, toxic, or stress cardiomyopathy, which is a temporary myocardial injury that can develop in patients with extreme physical, chemical, or sometimes emotional stressors.

Then the authors describe potential mechanisms which can be associated with catecholamine mediated myocardial damage. Catecholamines like epinephrine are a potential source of free radicals and oxidative stress. Catecholamines can interfere with sodium and calcium transporters across cell membranes, potentially leading to an overdose of calcium inside the cell. Epinephrine or other catecholamines may be involved in postvaccine reactions in which there is an overactive immune response, similar to what is seen in some cases of SARS-CoV-2 that led to multisystem inflammatory syndrome with cytokine storms.[389] The authors thought that an allergic hypersensitivity response was unlikely, due to the lack of eosinophils (allergy cells) seen on microscopic exam.[390]

After publication of this article, two groups of scientists wrote letters sharing their perspectives and offering more possibilities. This is the way scientific publishing is designed to work. One group of scientists develops a hypothesis and other scientists chime in to say if they can replicate the findings in their work or if there are other explanations for what is discovered. Nicholas Kounis and colleagues wrote that relatively few autopsies were being performed in myocarditis cases after COVID vaccination because many cases are seemingly mild. They reported one "fatal fulminant" (the opposite of mild) case of "necrotizing eosinophilic myocarditis . . . in a female patient following the initial dose of the Pfizer-BioNTech mRNA COVID-19 vaccine . . . abundant eosinophils and focal myocyte necrosis were found at autopsy." The authors characterized this case as an extremely rare idiosyncratic hypersensitivity reaction.[391]

Then they raised the possibility that the excipient polyethylene glycol in Pfizer-BioNTech mRNA COVID-19 vaccines could potentially cause hypersensitivity reactions and cause some cases of myocarditis. They note that a large percentage of the population is sensitized to PEG due to its presence in cosmetics, creams, and lotions.[392] In a video prepared with Children's Health Defense in August 2020 when COVID vaccines were being developed at "Warp Speed," we raised concerns about possible hypersensitivity reactions and the potential for anaphylaxis.[393] The authors argue for different excipients.[394] I agree.

Another set of scientists raised the possibility that underlying infection could have been the reason for the deaths and pointed out results from the CDC's immunohistochemical and molecular testing that were not included in the paper by Gill. They contend that Patient A had evidence of parvovirus B19, which is a known cause of myocarditis (true dat) and has been found in normal heart tissue (correct, we are all hosts to thousands of viruses). Regarding Patient B, they wrote that an immunohistochemical stain specific for *Clostridium* species demonstrated extensive staining of clostridia in each of these tissues, as well as in the microvasculature of the adrenal glands, liver, pancreas, kidneys, heart, lungs, and spleen. Based on a PCR assay and composite histologic, immunohistochemical, and molecular findings, they thought death was attributable directly to C septicum sepsis. This conclusion is supported by well-recognized clinico-pathologic characteristics of this disease. Clostridium septicum sepsis often presents with nonspecific signs and symptoms but is fatal in approximately 60 percent to 70 percent of cases.[395]

Fair enough, good to know. But these teenagers are still dead. One must wonder, if the second patient had a fulminant Clostridia infection which spread all over his body and got into his bloodstream—did the COVID injection play a role? (We will discuss immune-suppressing mechanisms associated with COVID jabs in another chapter.) If the parents of Patients A and B are reading this, I apologize for the dehumanizing language that is generated by laws requiring anonymity in medical reporting. I know that your sons were real people with friends, relatives, and teachers who love and miss them. I know that losing a child is an agonizing grief experi-ence. I cannot imagine the shock of finding them dead in bed. I hope my readers can imagine the real lives behind the cold hard numbers that are reported, tallied, and charted as if they were nothing more than pork sales or stock prices.

Myocarditis 2024: Summer Prior to Fall School Entry

Myocarditis has not been common in pediatrics, so many pediatricians and family physicians would not think of inflammation of the heart immediately as a diagnosis when a young patient presents with chest pain or shortness of breath. Doctors in training are taught that when kids pres-ent with chest pain, it is much more likely to be gastrointestinal or muscu-loskeletal in origin. Shortness of breath is more likely to be a problem with the lungs than with the heart. In the age of COVID shots, that mindset needs to change. Now, physicians who care for children and adolescents need to have a much lower threshold for suspecting inflammation of the sac around the heart (pericarditis) or heart itself (myocarditis). Further complicating the correct diagnosis is that new studies have confirmed that heart damage from COVID injections can occur in young people without dramatic symptoms. When scientists look prospectively for heart damage after the injections, they find very concerning numbers of patients affected.

Let's look at the data, keeping in mind that mortality from COVID the disease in those up to nineteen years old is 0.0027%.[396] A 2023 study in Hong Kong followed young people after COVID injections who had a second MRI after the first showed cardiac damage. Fifty-eight percent had concerning abnormalities on the second scan suggesting scars could be forming in heart muscle. Forty patients aged twelve to eighteen years were tested, using EKG, cardiac echocardiogram, and cardiac magnetic resonance imaging. Thirty-three were males. 73% had no symptoms, meaning that parents and doctors might not suspect heart problems. 18% had chest pain, 8% palpitations, and 3% fatigue. Approximately 18% had reduced left ventricular ejection fraction, meaning that the heart muscle was not as strong as it should be to push blood out to the rest of the body. A low LVEF is a risk for developing heart failure.[397]

In another study in Israel, where the Pfizer injections were deployed widely in the population in what some have criticized as a "national experiment without informed consent" researchers examined Clalit databases (the largest Health Care Organization in Israel) for a diagnosis of myocarditis up to forty-two days after first injection, using the case definition used by the CDC. The highest incidence of myocarditis was 10.69 cases per 100,000 male patients between the ages of sixteen and twenty-nine years, which works out to be 1 in 9,354. Of the fifty-four patients older than sixteen identified with myocarditis and followed for a median of eighty-three days, one died of "unknown cause" and one was re-admitted to the hospital.[398]

Pfizer and Moderna vaccines were rolled out starting December 17, 2020, in the US; the Janssen product followed on April 1, 2021. Hundreds of thousands of adverse events have been reported to the Vaccine Adverse Events Reports System (VAERS). Rose and McCullough analyzed the data to examine cardiac side effects, primarily myocarditis, after the first or second doses of COVID "vaccines." Myocarditis rates were significantly higher in youths between the ages of 13-23 (p value 0.0001) with ~80% occurring in males. [That p-value means that the odds that those results are due to chance are one in ten thousand]. Within eight weeks of the public offering of COVID-19 products to the 12–15-year-old age group, we found 19 times the expected number of myocarditis cases in the vaccination volunteers over background. . .[399]

There was a five-fold increase in myocarditis rates after the second dose compared to the first dose in fifteen-year-old males. 1.1 percent of the patients in the myocarditis group died. They point out that the temporal relationship between the dose and adverse event, biological plausibility of cause and effect, and internal and external consistency with emerging clinical data all support a cause-and-effect relationship. This was published less than one year after rollout, yet as I write this nearly three years after rollout, the injections are still being recommended or mandated for college entry in young people for whom the risk of COVID itself is very small. As this book goes to the publisher, many colleges have dropped those requirements.

Critical Thinking When Interpreting Scientific Papers

Let's examine a study from the FDA and use critical thinking skills to determine the real message. Under a mandate from the US FDA, a public health surveillance cohort study was conducted on patients aged 5–17 years who had received a BNT 162b2 COVID-19 vaccine. The study ended in May or July 2022 depending on the database used. There were 3,017,352 patients (50.1% male, 49.9% female and 95% urban dwellers) who were studied for twenty different health outcomes of concern. A safety signal emerged for myocarditis and pericarditis for 12–15-year-olds and 16–17-year-olds in all three commercial databases. Medical records were reviewed, and some (but not all) identified heart inflammation cases. Acute cases were identified, with a mean of 6.8 days and a median of three days from time of shot to time of presentation. 39.4 cases presenting within seven days of the shot were found per 1 million doses. That works out to be around 1 in 25,000, meeting criteria for a safety signal. In an odd interpretation, the authors concluded that because only myocarditis and pericarditis were identified as a safety signal, "these results provide additional evidence that COVID-19 vaccines are safe in children."[400] Huh? Is that really the take-home message?

I have multiple concerns about the interpretation of this research. First, it is likely that cases of myocarditis and pericarditis are undercounted because only cases where the diagnosis was made were counted. We know that many post-vaccine patients with myocarditis in this age group do not present with enough symptoms to seek medical help. In fact, two teenagers with autopsy-proven vaccine-induced myocarditis within a few days of their COVID injections died in their sleep without overt symptoms of heart problems.[401] Second, rather than concluding that the vaccine is safe because the only identified problems were with the heart, isn't the real issue how often do such heart problems occur at baseline or with COVID itself at this age? The authors acknowledge potential shortcomings of their study. Critics note that the time windows were very short, thereby potentially missing patients who developed heart issues later. Signal thresholds for other adverse events were set very high, not triggering signals unless the adverse event occurred twice as often in the vaccinated versus the historical controls.

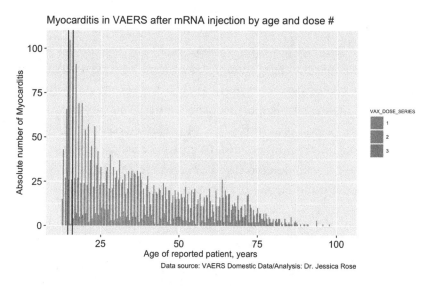

Sourced from Jessica Rose Substack.[402]

When You Look for Damage, You Find It

A troponin test looks for damage to cardiac muscle. When hospital employees were scheduled to receive an mRNA booster, researchers followed lab tests prospectively to look for evidence of heart damage with high sensitivity cardiac troponin levels. Yikes! They found that heart damage was more common than previously thought. The damage seemed to be mild and transient, and was more common in women than men, which was different from prior findings of more myocarditis and pericarditis in males who developed symptoms. But *one in twenty* people showed elevations in this common marker for heart damage on *day three* after the injection. Those who had elevations in troponin had similar side-effect profiles to those whose cardiac troponin remained normal.[403] The researchers looked at other markers that were different in those with and without vaccine-associated myocardial injury. Seems to me like this is a concerning finding, given that the vaccine does not prevent transmission of COVID illness. Let's see if other scientists reproduce this finding.

Conclusion

The high prevalence of heart damage in young people after COVID injections is particularly worrisome because of the pressure to mandate COVID injections for college attendance. Even in the fall of 2023, when most students had recovered from COVID infection, about one hundred colleges still mandated the shots for school entry. You can see which colleges still have mandates at https://nocollegemandates.com/. Medical and nursing schools with mandates may be creating a subset of future healthcare professionals who develop chronic heart problems, which will impact their productivity and quality of life.

Dr. McCullough's Plea

Since Dr. McCullough has been on the forefront of this battle from the beginning, I thought it only fitting that I give him the last word. Below is his open letter to clinicians around the world to re-think their promotion of COVID injections. He has since called for a complete moratorium on COVID "vaccines."

Note of Concern to Colleagues

As of August 25, 2023, the CDC has recorded 18,015 deaths reported to them in VAERS by healthcare professionals or pharmaceutical companies who believe the vaccine is related to the death. Approximately ~1100 deaths have occurred on the same day of vaccination. The largest autopsy study published to date indicates 73.9% of deaths after vaccination are a direct cause or significantly contributed to by COVID-19 vaccination. There are >3400 peer reviewed manuscripts in the medical literature concerning fatal and nonfatal COVID-19 vaccine injuries including those recognized by regulatory agencies around the world such as myocarditis, neurologic injury, thrombosis, and immunologic syndromes. The World Council for Health, June 11, 2022, has produced a pharmacovigilance report which is factual, scientifically grounded, and consensus driven calling for global market withdrawal of COVID-19 vaccines based on lack of safety. I am a widely regarded expert on COVID-19 and vaccine safety, (and) on December 7, 2022, in the US Senate, and on September 13, 2023, in the European Parliament, have called for with assent of an expert panels, removal of all COVID-19 vaccines from the US and EU markets for excess risk of death. On March 21, the Association of American Physicians and Surgeons issued a factual, scientifically grounded, and consensus driven statement calling for all COVID-19 vaccines to be removed from the market based on lack of safety and efficacy. The National Citizens Inquiry, a Canadian citizen-led and citizen-funded organization chartered to investigate governments' COVID-19 policies, on September 14, 2023, called for market removal of all COVID-19 vaccines. These observations and sources of evidence indicate the COVID-19 vaccines are not safe for human use. No large-scale, conclusive, randomized, double-blind, placebo-controlled trials have demonstrated reduction in infection transmission, hospitalization, or death as primary endpoints. Thus, the COVID-19 vaccines are not proven to be effective in reducing important clinical outcomes.

Please consider these developments if you have taken or supported COVID-19 vaccination.

Sincerely yours,
Peter A. McCullough, MD, MPH

Calling for an Immediate Moratorium on COVID Vaccines

Joseph Fraiman, lead author of peer-reviewed research analyzing Pfizer and Moderna trials of mRNA vaccines has called for an immediate halt to COVID vaccines due to serious harms. He states: "We have **conclusive evidence** that the vaccines are inducing sudden cardiac death."[404]

FOOD FOR THOUGHT

If you got a COVID vaccine, were you told about risks to your heart? Do you know of children who had "heart attacks" since the vaccines rolled out? Do you know people who had heart attacks or arrythmias they thought might have been related to their COVID vaccine? Do you know young people who got myocarditis after COVID vaccines or adults who died an unexpected cardiac death?

CHAPTER 19

What Are the Autoimmune Risks of COVID Injections?

Medicine is the restoration of discordant elements; sickness is the discord of the elements infused into the living body.

—Leonardo da Vinci

Self-conceit may lead to self-destruction.

—Aesop

Early Warning Signs

As early as summer of 2020, astute scientists were already aware of the potential for COVID-19 to cause autoimmunity. When I saw the Vojdani paper I discuss below, I worried about COVID vaccines in development causing autoimmune conditions, since the mRNA stimulated the vaccinee to produce spike protein, the very substance linked to cross-reactivity with multiple human tissues. This is important—the COVID mRNA jabs provoke the recipient to produce spike protein, which is what causes a lot of damage associated with COVID itself.

Ari Vojdani is a friend and colleague whom I met during my tenure as Medical Director of the Autism Research Institute. We published a paper together about the immune function of children with autism.[405] He directs Cyrex labs, which is a haven for tests of autoimmunity. He and Datis Kharrazian, one of my fellow faculty members at the Institute for

Functional Medicine, published an important paper, "Potential antigenic cross-reactivity between SARS-CoV-2 and human tissue with a possible link to an increase in autoimmune diseases," that demonstrated that

> "21 out of 50 tissue antigens had moderate to strong reactions with the SARS-CoV-2 antibodies and are a sufficiently strong indication of cross-reaction between SARS-CoV-2 proteins and a variety of tissue antigens beyond just pulmonary tissue, which could lead to autoimmunity against connective tissue and the cardiovascular, gastrointestinal, and nervous systems."[406]

Whoa! This was terrifying to me at the time. Their prediction turned out to be true.

Autoimmunity 101

Allow me to explain some background and why this is so significant. In autoimmune conditions, the patient's immune system does not distinguish self from non-self, and accidentally starts mounting an immune reaction against parts of her body, resulting in antibodies against various tissues. Autoimmunity has increased dramatically in the past several decades and disproportionately affects women. If you know a woman like me who is hypothyroid, it is likely she has autoantibodies against her thyroid gland, a condition called Hashimoto's thyroiditis. Other autoimmune conditions include rheumatoid arthritis, lupus erythematosus, autoimmune hepatitis, and type one diabetes.

HPV Vaccine: A Cautionary Tale

Aluminum, which is used in many vaccines to induce an immune response from the recipient, is known to cause autoimmunity. In fact, aluminum is used in laboratory animals to elicit autoimmunity so that the resulting disease can be studied. I think, based on the available evidence, that aluminum in vaccines contributes to increasing prevalence of autoimmunity in humans. The most egregious example seems to be the HPV vaccine, but that is a topic for another book. I recommend *The HPV Vaccine on Trial*, by Holland, Rosenberg, and Iorio, for those who are interested in learning about how an aluminum-containing adjuvant was used as the placebo in Gardasil trials.[407] Since aluminum is known to induce autoimmunity, the effect was to "wash out" evidence of the vaccine leading to autoimmunity, because 2.6 percent of girls in each group developed an autoimmune disease.[408] The package insert for Gardasil quotes 2.3 percent in each group.[409]

Markers of Autoimmunity Found

Early in the COVID pandemic, Dr. Vojdani and colleagues tested five blood specimens positive for SARS-CoV2 antibodies. They were surprised to find that three of the five specimens had significant elevations in antinuclear antibodies (ANA), anti-extractible nuclear antigen (ENA), anti-actin antibodies, and antibodies against mitochondria, the powerhouses of the cell. In other words, 60 percent of specimens showed important markers of autoimmunity. Trust me when I say that nuclei, actin, and mitochondria are crucial for optimal health. These red flags prompted them to study the relationship between spike and nuclear proteins in SARS-CoV-2 and target proteins in human tissue. Some of the strongest reactions were with transglutaminase 2 (tTG2) and 3 (tTG3), which are important for metabolizing gluten, myelin basic protein (MBP), which covers our nerves and speeds up motor messages, mitochondria, the sources of cellular energy, thyroid peroxidase (TPO), an enzyme in the thyroid, and collagen, the glue that holds tissues together.[410] Scary findings, don't you think?

Vojdani and Kharrazian asked, "Is it possible that some of the extensive organ, tissue, and cellular damage done by SARS-CoV-2 is due to viral antigenic mimicry with human tissue?"[411] By George, I think they've got it! At the time of their publication, clinicians were noticing that the virus caused damage in body parts way beyond the lungs, including the cardiovascular, nervous, digestive, and immune systems.

My colleagues noted that vaccine-induced autoimmunity from autoimmune cross-reactivity is associated with lots of clinical conditions, including Guillain-Barré syndrome, multiple sclerosis, demyelinating neuropathies, systemic lupus erythematosus, narcolepsy, and postural orthostatic tachycardia syndrome in susceptible subgroups. The "Grandfather of Autoimmunity," Dr. Yehuda Shoenfeld, did the seminal work in this area with his colleagues.[412,413,414,415]

The Grandfather of Autoimmunity Was Right

Shoenfeld, an eminent scientist from Israel, has published multiple books about autoimmunity for decades and written a whole book about its connection to vaccines, titled *Vaccines and Autoimmunity*, which I read and recommend.[416] In person, he seems like a kindly Jewish grandfather and is so gracious that one would love to spend a day with him. He and his colleagues noted that SARS-CoV-2 spike glycoprotein proteins shared thirteen out of twenty-four pentapeptides with lung surfactant proteins. By my math this means that spike protein has molecular similarity to more than half of the proteins tested.

This made Dr. Shoenfeld worry that the immune response following infection with SARS-CoV-2 may lead to cross-reactions with pulmonary surfactant proteins.[417] Surfactant is a "surface active agent" that has wetting and spreading properties. In the lungs of premature babies with respiratory distress syndrome, pediatricians give surfactant to help babies breathe more effectively by preventing the collapse of the small air spaces in the lungs. Shoenfeld warned that the use of the entire SARS-CoV-2 spike protein antigen sequence in the vaccines would be dangerous. I wonder how things might be playing out differently now if the executives at Moderna and Pfizer had listened to Shoenfeld.

Autoimmunity During C Diff Vaccine Trials

Another set of scientists made similar suggestions based on their experience trying to make a vaccine against *Clostridium difficile*. You may have heard of "C diff" in people who have been treated with long-term antibiotics, which can lead to depletion of the normal friendly bacteria in the gut and an overgrowth of harmful bacteria including *Clostridia difficile*. It can be quite serious; there are several hundred thousand cases in the United

States each year. Agnieszka Razim and colleagues showed that two peptide (protein) sequences they worked with were recognized by the blood of patients with autoimmune disease, as well as by the C difficile patients. They warned that "special care in analyzing the sequence of new epitopes should be taken to avoid side effects prior to considering it as a vaccine antigen."[418] No kidding! Their recommendation to analyze the sequence of tissue cross-reactive epitopes was exactly what Dr. Vojdani did.

Haste versus caution

My colleagues Ari and Datis wrote in 2020:

> At the moment, scientists are frantically trying to develop either a definitive cure, neutralizing antibodies, or a vaccine to protect us from contracting the disease in the first place, and they want it right now. We must consider that finding a vaccine for a disease may normally take years. There are reasons for all the precautions involved in developing a vaccine, not the least of which are unwanted side-effects. In light of the information discussed above about the cross-reactivity of the SARS-CoV-2 proteins with human tissues and the possibility of either inducing autoimmunity, exacerbating already unhealthy conditions, or otherwise resulting in unforeseen consequences, it would only be prudent to do more extensive research regarding the autoimmune-inducing capacity of the SARS-CoV-2 antigens. The promotion and implementation of such an aggressive "immune passport" program worldwide in the absence of thorough and meticulous safety studies may exact a monumental cost on humanity in the form of another epidemic, this time a rising tide of increased autoimmune diseases and the years of suffering that come with them.[419]

Prescient words. "It's tough to make predictions, especially about the future," as Yogi Berra said. Scientists are cautious when they do so. But Ari and Datis saw the writing on the wall and were courageous enough to stake their reputations on their findings. Did they hit the nail on the head or what? In the manic rush to put all our eggs in the vaccine basket, decision makers ignored (or actively suppressed) the strategies classically used to fight viral epidemics—early treatment with repurposed drugs and measures to enhance the immune function of the population. (More about this in another chapter.) Let's see how my colleagues' predictions played out.

Awakening a Sleeping Dragon?

"Awakening the sleeping dragon" was the intriguing subtitle of a letter to the editor of *Clinical Immunology* in May of 2021 by a group of Greek scientists. While acknowledging that reports of COVID vaccines were "welcomed with great joy," the authors soberly noted that "Nonetheless, COVID-19 vaccines have raised a number of concerns regarding their reactogenicity, while a large proportion of the field remains unexplored, especially among patients with autoimmune conditions."[420] They quoted the literature, including the prescient work by my colleague Ari Vojdani, discussed above, which showed that antibodies against spike glycoproteins cross-react with host protein sequences. The authors echo concerns raised by Talotta, who argued that giving a nucleic acid vaccine to young females who already have autoimmune or autoinflammatory disorders may increase their risk of unprecedented side effects.[421] After COVID injections were widely distributed (and remember how many people felt forced to take them to keep their jobs or travel to see family) reports began to appear concerning autoimmunity associated with the injections. Let's look at a few.

Autoimmune Liver Disease

"Autoimmune phenomena and clinically apparent autoimmune diseases, including autoimmune hepatitis, are increasingly been reported not only after natural infection with the SARS-CoV-2 virus, but also after vaccination against it," Michelle Ghielmetti wrote, reporting a case of a "man without a history of autoimmunity or SARS-CoV-2 natural infection who experienced acute severe autoimmune-like hepatitis seven days after the first dose of the mRNA-1273 SARS-CoV-2 vaccine."[422] Intriguingly, his lab tests showed anti-mitochondrial antibodies (AMA). Surprisingly, this pattern was different from classical anti-mitochondrial antibodies. Another surprise—further testing showed his anti-nuclear antibodies were different and unique. He had a genetic variation, an HLA DRB1*11:01 allele, which protects against the more typical autoimmune liver disease, primary biliary cholangitis.[423]

Are we seeing something new and horrible here? I think so. Ting Zhang reported a series of cases of autoimmune hepatitis after various COVID-19 vaccines. Susceptible patients included liver transplant recipients and patients with hepatitis C virus and primary sclerosing cholangitis.[424] Other scientists who reported cases of autoimmune hepatitis after COVID vaccines include Isabel Garrido,[425] Cathy McShane,[426] and Kenneth Chow,[427] noting reports of autoimmune hepatitis began increasing in the summer of 2021.

Autoimmune Clotting and Bleeding

Scientists from Egypt, Northern Ireland, India, Austria, Iran, Sweden, Jordan, Brazil, Italy, Saudi Arabia, and the United States collaborated to produce a paper published in 2021 outlining the autoimmune roots of clotting events after COVID-19 vaccination. This paper illustrates the ideal of international scientists working together toward common goals. Their work showed that SARS-CoV-2 vaccines can cause severe thrombotic events. The vaccine-induced clotting with low platelets is analogous to an autoimmune condition in which the patient makes antibodies against heparin, a substance that keeps blood from clotting excessively. They concluded more work needs to be done to discover the mechanisms involved.[428]

Yue Chen and colleagues include the obligatory kudos to the benefits of vaccination but note that COVID vaccines were approved without the benefit of extensive or long-term studies of side effects (or efficacy, for that matter). "Recently, new-onset autoimmune phenomena after COVID-19 vaccination have been reported increasingly (e.g. immune thrombotic thrombocytopenia, autoimmune liver diseases, Guillain–Barré syndrome, IgA nephropathy, rheumatoid arthritis and systemic lupus erythematosus)."[429] They acknowledge that information about autoimmune disease emerging after vaccination is controversial, "we merely propose our current understanding of autoimmune manifestations associated with COVID-19 vaccine."[430]

Will it be easy to sort out the autoimmunity that results from SARS-CoV-2 virus itself from that induced by COVID injections? No. Do we know the extent of autoimmunity that could result from COVID vaccines? Not yet. Autoimmune diseases usually take years to present and are surging in prevalence already, especially in women. "Increasing evidence shows a continuous rise in the incidence of autoimmune diseases. The incidence of rheumatic, endocrinological, gastrointestinal, and neurological autoimmune diseases has increased annually, at rates ranging from 3.7 percent to 7.1 percent, between 1985 and 2015.[431,432]

No Long-Term Studies from COVID Trials on Autoimmunity

The initial COVID-19 injection trials only lasted two months. After the blinding was broken to determine who had received the real mRNA injection and who got the placebo, placebo recipients were offered the injection. Most of them got the jab, effectively wiping out the original placebo group for ongoing safety and efficacy studies. Read that again: We do not have an ongoing control group from the original mRNA trials to compare to vaccine recipients for autoimmune side effects that could emerge in the three to twenty years that many autoimmune diseases take to be diagnosed.

Conclusions

Credible scientists warned us in peer-reviewed, well-documented published science that autoimmunity was likely from the Warp Speed jabs. They were right.

FOOD FOR THOUGHT

Do you know someone in your circle of friends and family who developed autoimmunity after either COVID the illness or COVID the jab? Are you surprised that credible scientists published concerns about autoimmunity even before the COVID injection trials were completed?

CHAPTER 20

Are There Neurologic Side Effects from COVID Vaccines?

There is no more difficult art to acquire than the art of observation, and for some men it is quite as difficult to record an observation in brief and plain language.

—William Osler

You only see what you know.

—Goethe

Early Warning Signs

You may have seen posts and videos on social media of teenagers developing partial paralysis, facial palsies, or acute changes in their mental status after getting COVID jabs. The medical literature has confirmed that a wide variety of neurologic complications can follow COVID vaccines. These articles started pouring into the medical literature during 2021, within the first year the jabs were released on the world. It often takes many months for medical papers to be peer-reviewed and published. These early post-vaccine papers about side effects imply that scientists and health professionals started noticing these adverse events soon after the roll out. We knew from previous vaccines that neurologic symptoms are possible.[433] The "safe and effective" mantra did not include the reality of what victims were experiencing and clinicians were observing.[434]

Common, Mild, and Serious

Some post-vaccine side effects are described as mild and transient. Such symptoms as fever, chills, fatigue, muscle aches, joint pain, and headaches are often written off as "normal" reactions. What was unusual during this vaccine rollout is that more severe reactions, like symptoms that confined recipients to bed for a week, were interpreted by health officials and reported in the press as a "sign that the vaccine is working."[435] That infuriated me. After reading this book, I hope you will ask yourself what might be going on in the body on a cellular level that causes these post-vaccine symptoms and whether those symptoms were trying to alert us to some very important concerns.

Inflammation in the Nervous System

Reported complications post-vaccine include acute transverse myelitis (inflammation of both sides of a section of the spinal cord), encephalomyelitis (inflammation of the brain and spinal cord), polyneuropathy (damage to nerves throughout the body), and Bell's palsy (muscle weakness on one side of the face giving the patient a lopsided look). Cerebral venous sinus thrombosis—a blood clot in the sinuses which drain blood from the brain—is one of the most devastating complications after COVID injection.

Correlation Does Not Equal Causation

Critics of vaccine safety advocates are quick to point out that "correlation is not causation." Just because a person gets a novel technology product that was developed at unprecedented speed does not mean that new symptoms that appear soon after the vaccine are caused by that product. Large collaborative research projects and the application of epidemiological laws such as the Bradford Hill criteria could settle the issues of true causation or whether the vaccines unmasked underlying pathology. My analysis of the totality of evidence about COVID vaccine side effects is that Bradford Hill criteria are met for multiple side effects that patients have reported. Bradford Hill criteria include things like timing association, specificity, consistency, and plausible biologic mechanisms that are used to establish epidemiologic probability of a causal relationship. We should take patient symptoms seriously and try to figure out mechanisms by which such adverse events could occur.

Warning Signs Published Early

Ravindra Kumar Garg, from the Department of Neurology at King George Medical University in Uttar Pradesh, India, and his colleagues did searches in the medical literature (including PubMed and Google Scholar) to identify published neurologic complications of COVID-19 vaccines. You may remember Utter Pradesh from my chapter on the wonders of ivermectin. Utter Pradesh made ivermectin widely available to its population and the state had astonishing low mortality from COVID. Garg and colleagues did their last search on August 1, 2021, meaning it captured only those events that were known and already published in the first six months of the rollout. The review was published online in October 2021, about ten months after the products were launched. Thank you, Ravindra and Vimal, for your important article about the spectrum of neurologic complications post-COVID injection.[436]

Garg reported that feelings of syncope and dizziness were often characterized as anxiety-related events. It reminded me of patients who reported what I would have interpreted as neurologic symptoms but were told they were just suffering from anxiety. If I were an ER doctor who sent people home with a diagnosis of anxiety instead of recognizing more serious neurologic issues, I would be berating myself, since I tend to be self-critical. Forty lashes with a wet noodle, as Ann Landers and my dad used to say.

The Data

Let's dive into the data. Pre-print data from Mexico showed 6,536 adverse events in 704,003 subjects after the first dose of Pfizer-BioNTech mRNA COVID-19 vaccines. Astonishingly, 65 percent of the subjects who reported adverse events had at least one neurologic manifestation. That works out to be 4,258 people. Fortunately, 99.6 percent were characterized as mild and transient, including symptoms like headaches, weakness, and transient sensory disturbances. From my perspective as someone well versed in a systems approach to biology, I wonder what was going on biochemically and physiologically during symptoms like headaches. Only seventeen serious events were reported: seizures (7), functional neurologic problems (4), Guillain-Barré (3) and transverse myelitis (2).[437]

As a fan of the axiom "first do no harm" and the precautionary principle, I see these reports as a cautionary tale. One in 41,411 people got serious neurologic side effects in the Mexico study, which only included what was reported and published in the first seven months of rollout. Even though that may sound like good odds to people accustomed to gambling, when the intention is to inject seven billion people, multiple doses are recommended, and the cost of a bad bet is neurologic dysfunction instead of gambling debts, the risk bears careful scrutiny.

Lack of Long-Term Follow-Up

It may surprise you to learn that, in many vaccine studies, participants are only asked to monitor their symptoms for the first five to seven days. Such was the case in a South Korea study, where 53 percent of those receiving the first dose of mRNA vaccines (Pfizer and Moderna) and 91 percent of those who got adenovirus vectored vaccines (the J & J product) reported mild adverse reactions, including fever, headache, fatigue, and muscle aches.[438] To put those percentages in perspective, I compared them to other vaccines like influenza, DTaP or Prevnar, which are usually tolerated with fewer side effects. Furthermore, it quickly became apparent that reactions were more common after the second dose.[439]

Seizures

From prior experience with seizures following DPT and MMR vaccines in children, we anticipated that a subset of people exposed to COVID vaccines might have seizures.[440,441,442] A retrospective study of 81,916,351 COVID shot doses revealed only thirty-one new onset seizures, which is rare with a prevalence of 2.73 per million for Moderna, 1.02 million for Pfizer, and 1.01 per million for CanSino.[443] A meta-analysis published in 2023 concluded seizures increased in frequency only 5 percent in persons with epilepsy, which is reassuring.[444] Another meta-analysis of six randomized trials published in JAMA in 2024 showed no significant differences in new onset seizures between 63,521 COVID shot recipients and 54,919 placebo recipients, at least within forty-three days of follow-up.[445] Anecdotal reports from clinicians make me suspect the numbers are worse than reported.

Severe Neurologic Adverse Events: Strokes and Clots

Severe side effects are defined as either life-threatening, requiring hospitalization, or causing severe disability. Conditions like seizures, encephalitis, anaphylaxis, Bell's palsy, and Guillain-Barré make the cut according to the WHO. Early in the vaccine distribution campaign, reports began accumulating in medical journals about arterial strokes, brain hemorrhages, and brain clots. Such post-vaccine vascular events can lead to death or severe disability. It emerged that such events were often associated with severe immune-mediated thrombotic thrombocytopenia (blood clots and bleeding). Low platelet counts occurred five to thirty days after COVID jabs, first reported after the Johnson & Johnson and AstraZeneca adenovirus-vectored vaccines (therefore the nickname of "clot shot" for the Janssen and AstraZeneca vaccines). Antibodies against platelet factor 4 can result in platelet destruction and trigger clotting and bleeding.[446] A *Nature* article in 2021 postulated that leaked genetic material could bind to platelet factor 4 and cause the formation of autoantibodies which destroy platelets.[447,448] It is intriguing to revisit that article for its foresight considering recent concerns about loose DNA fragments in shots scaled up for mass distribution, covered in another chapter.

Changes in Illness Patterns

In the absence of conditions like hereditary clotting disorders or sickle cell disease, it is unusual for young people to have strokes. Clotting was not confined to the viral vectored vaccines. Cases of vaccine-induced clotting and bleeding leading to strokes, some in children, adolescents, and young people (who usually received the Pfizer injection) started appearing in the medical literature. The recommendation emerged that any young person with an arterial stroke after getting a COVID jab should be evaluated for vaccine-induced thrombotic thrombocytopenia and get lab tests for platelet factor 4 antibodies, D dimers, fibrinogen levels, and platelet counts.[449] I think it would have been better not to give a shot that did not prevent transmission to a group at such low risk.

Sometimes strokes can follow abnormal bleeding or clotting in other body parts first. In one case, the patient presented with intractable abdominal pain seven days after an adenoviral vaccine. The work-up revealed necrosis (rotting) of both kidneys, then she died of a massive stroke due to the blocking of her right carotid artery. She had low platelet counts, a high D-dimer, and platelet factor 4 antibodies.[450]

Severe Neurologic Adverse Events: Cerebral Venous Sinus Thrombosis

This is a bad condition. If a patient has a headache that does not respond to pain medicines after COVID vaccination, clinicians should suspect this dreaded complication. Be vigilant! Even though headache is seen in many vaccine recipients, a subset (especially young females some tend to dismiss as hysterical) have this complication. Be skeptical of authorities who reassured the public on the nightly news that uncomfortable side effects show the vaccine is working. Clinicians in Europe reported an uptick in cases of cerebral venous thrombosis, especially after patients got adenoviral-vectored vaccines.[451] 148,792 Danes and 132,472 Norwegians who received vector-based COVID vaccines contributed to the science to determine the excess rates of thrombotic events in the first twenty-eight days after injection. An estimated eleven excess venous thromboembolism events per 100,000 vaccinations and 2.5 excess cerebral venous sinus thrombosis events per 100,000 injections emerged from the analysis.[452] Low-prevalence adverse events, when very severe, can cause significant morbidity and mortality when an intervention is recommended for everyone on the globe.

A group in the Netherlands collected 213 cases with cerebral venous sinus thrombosis post vaccination; eighty-seven cases were after adenoviral shots and twenty-six after mRNA jabs. Thirty-eight percent of the patients in the adenoviral vaccine group died; 20 percent of those in the mRNA jab group died. Forty-six patients died in the first part of 2021 from this complication in this study.[453] Again, the prevalence of this side effect is relatively low, but mortality is high.

Severe Neurologic Adverse Events: Encephalopathy, Encephalitis, and Encephalomyelitis

"Cephalo" means head and "patho" means disease. Acute encephalopathy is a quickly progressing disorder of the brain which can present with delirium, a change in consciousness, or unconsciousness. "Itis" means inflammation, so encephalitis means inflammation of the brain. "Mye" means marrow or spinal cord, so myelitis means inflammation of the spinal cord.

One particularly harrowing diagnosis seen after COVID vaccination is acute disseminated encephalomyelitis (ADEM). The linguistics lesson you just had will lead you to conclude that ADEM is an acute inflammatory disorder of the central nervous system associated with inflammation of the spinal cord and surrounding myelin. Young people can get these severe inflammatory neurologic conditions after COVID vaccines.[454,455,456] One reported case of a twenty-four-year-old female described her limb weakness and change in mental status within two weeks of her COVID jab. Diagnosis after imaging studies: acute disseminated encephalomyelitis.[457] One case series reported patients who presented with cognitive decline, seizures, gait disturbances, or opsoclonus-myoclonus (eye movements like doll eyes and jerky muscle movements) seven to eleven days after adenovirus COVID shots.[458]

Severe Neurologic Adverse Events: Transverse Myelitis

You just learned that myelitis means inflammation of the spinal cord. Add transverse for meaning both sides of the spinal cord and you get transverse myelitis, which comes with various tortures including bowel and bladder dysfunction and paralysis of diverse types depending on the location of the inflammatory lesions. One person who suffered transverse myelitis was an orthopedic surgeon named Joel Wallskog, who can no longer work as a surgeon due to his disability brought on by his COVID injection. He cofounded a group called "React19" which works to provide information and support for people injured by COVID vaccines. The last time I spoke to Joel, React19 was serving over thirty-five thousand people.

Numerous reports have now accumulated about cases of transverse myelitis after various COVID injections. The adenoviral-vectored injections seem implicated more often than the mRNA jabs. Autoimmunity and molecular mimicry are one mechanism of damage. Remember Vojdani and Kharrazian tried to warn everyone? Maybe vaccine ingredients perceived as foreign by the immune system induce inflammation in the spinal cord? Maybe tricking your body into making spike protein, which is pathologic and toxic, is bad for your back?[459,460,461]

Whatever the reason for post-COVID vaccine transverse myelitis, the fact that there were three cases among 11,636 vaccine trial participants should have triggered a red flag. That works out to be one person with transverse myelitis for every 3,876 people jabbed.[462] According to the Worldometer, the world's population is 8,055,125,455 as I write this. If we meet Bill Gates's goal of a vaccine in every arm, that could mean generating 2,078,205 cases of transverse myelitis with all the accompanying disability and human suffering. That assumes the numbers in the trial bear out in real life. Remember the original mRNA trials recruited only healthy non-pregnant people—the Avengers, not couch potatoes or people who already had medical conditions.

Severe Neurologic Adverse Events: Bell's Palsy

Bell's palsy is caused by impairment of the seventh cranial nerve. Humans have twelve sets of cranial nerves. Medical students use the mnemonic "On old Olympus's towering tops, a Finn and German viewed some hops" to remember the twelve cranial nerve names. Many cases of Bell's palsy post COVID injection have been reported. These facial palsies seem more common after mRNA injections.[463,464,465,466] In a case-controlled study in which there were no differences in any clinical parameters between patients who did not receive a COVID vaccine versus those who did, Bell's palsy developed within two weeks of the injection.[467]

What did Pfizer and Moderna know about the risks for Bell's palsy from their initial clinical trials? Four people out of forty-four thousand in the Pfizer trial developed Bell's palsy; all had received the vaccine. In the Moderna trial of 30,400 people, three recipients in the vaccine arm and one person in the placebo arm reported Bell's palsy. The initial trial data evaluated by the FDA showed a risk of Bell's palsy of about 1 in 5,500 (Pfizer) or 1 in 5,000 (Moderna) vaccine recipients.[468,469] One could argue that is a negligible risk that would be worth taking if the vaccine in question did a great job of blocking transmission, preventing hospitalization, and saving lives. But you now know that the vaccine was never proven to stop transmission, despite the hype to the contrary.

Severe Neurologic Adverse Events: Ear Problems

Tinnitus (ringing in the ears), hearing loss, dizziness, and vertigo have all been reported after COVID jabs. Tinnitus can be extremely vexing, driving some patients with unrelenting ringing in their ears to become suicidal.[470,471] To understand the vaccine industry's fascination with mRNA vaccines, Dr. Madhava Setty and I attended the World Congress on Vaccines. Greg Poland, the well-known vaccinologist and editor of the medical journal *Vaccine* who directed the conference, developed severe tinnitus related to his COVID injections, yet remains a staunch promoter of mRNA COVID shots. He told Madhava that his tinnitus had not stopped or improved by the time of that convention, several years after his COVID injection.

Severe Neurologic Adverse Events: Vision Changes

This side effect is particularly important to me because my husband lost part of his vision in one eye soon after his COVID injection. He teaches psychiatry in a residency training program. It seems to me that he loves that job more than any in his lengthy career. The hospital required Mike to get the COVID jab to keep teaching. He struggled with the decision; truly caught between a rock and a hard place. He has a wife strongly embedded in vaccine-safety research who had gone public with her concerns about potential side effects of the vaccines in 2020. He worked for a hospital system that mandated that he comply with the COVID jab or lose his livelihood and the fellowship of the residents and faculty. I wonder how many patients like my husband were evaluated for similar adverse events but never were reported in the published literature. His ophthalmologist suspected a clot in the small vessels of his eye brought on by the COVID vaccine. An early case report of a patient with acute visual impairment along with headache, nausea, and confusion was published in *Inflammation Research* in 2021.[472]

Severe Neurologic Adverse Events: Guillain-Barré Syndrome

Many of you will have heard of Guillain-Barré, a syndrome that sometimes follows viral infection or vaccination and causes an ascending paralysis. Patients sometimes end up on a ventilator because the disease can affect the diaphragm (the muscle that helps you breathe, not the device used for birth control). COVID illness seems to be one of the triggers, and concerns emerged in 2021 that the Johnson & Johnson vaccine might also trigger Guillain-Barré.[473] As more data accumulated, it became clear that all types of COVID vaccines could cause Guillain-Barré syndrome, but adenovector-based vaccines had a stronger association. Post-vaccination ascending paralysis was more common in older adults than children or young adults. Symptoms usually started within two weeks of the injection. Nerve conduction studies in those patients show demyelination, meaning that myelin—the "insulation" around nerves that speeds conduction of messages—was damaged.[474,475,476]

Molecular similarity between structural components of peripheral nerves or myelin and the triggering infection or vaccine component is most likely the root cause of Guillain-Barré. Again, my friend Ari Vojdani tried to warn us that COVID injection-induced spike protein would potentially trigger autoimmune attacks on human tissues because of molecular mimicry.

Severe Neurologic Adverse Events: Small Fiber Neuropathy

I am lucky to know so many amazing doctors. One of my favorites is Suzanne Gazda, a neurologist in San Antonio, Texas. She writes an amazing blog about various neurologic conditions she has seen in association with either long COVID or long vax.[477] Her compassion and commitment have taken her around the globe with Hope for Humans, working where access to care and research is limited. Dr. Gazda has been recognized as a "Texas Super Doc," served as medical director of the Multiple Sclerosis Center of South Texas, been certified in the Bredesen Protocol for reversing neurodegenerative disorders, and is a superb lecturer in integrative and functional neurology. She is widely published in the peer-reviewed literature and an expert in pediatric acute neuropsychiatric syndromes (PANS) associated with infections, a pet interest of mine.

What I know about small fiber neuropathy I have learned from Suzanne. In her blog, she advances the idea that post-vaccine injury and long COVID have replaced syphilis as "the Great Pretender," with diverse and confusing symptoms. What she sees clinically now is vastly different from common neurologic presentations prior to the COVID crisis.[478] She shares my concerns that foreign substances like fragmented RNA, DNA, and spike protein are dispersed throughout the body after the shots.[479,480,481,482,483,484] Dr. Gazda is as worried as I am that autoreactive T cells, driven by lingering viral fragments or spike protein drive neurodegenerative disease. She cites the work of pathologists who found spike protein in the brain, meninges, and skull bone marrow of people after death.[485] She muses,

> No one really knows the long-term implications of lingering mRNA post virus or post vaccine. Since the mRNA vaccines were designed substituting methyl-pseudouridine for all the uridine nucleotides to stabilize RNA against degradation, could lingering modified mRNA post vaccine be even more difficult for the body to break down?[486]

Dr. Gazda cites the Seneff article that helped me understand the consequences of spike protein.[487]

She refers to findings from React19 that over 70 percent of their vaccine-injured have neurological symptoms.[488] She shares research that shows sensory nerves are vulnerable after infection, not to lingering virus but instead to infected neuron-released spike protein, nucleocapsid proteins, and other viral fragments. Very interesting! When small sensory cutaneous nerves are affected, patients experience unusual sensations such as pins-and-needles, tingling, and numbness. Some patients experience burning pains or electric shocks, which may prompt some clinicians to default to the "it's all in your head" hypothesis. After looking for diabetes and B12 deficiency, both known to cause small fiber neuropathies, clinicians would be well served to consider post-COVID infection or injection as potential causes.[489]

Severe Neurologic Adverse Events:
Reactivation of Herpes Zoster

Many readers may have infections with herpes simplex type 1 (which causes fever blisters—also known as "cold sores"—to come and go) or herpes type 2 (which lives in the genital area). After an initial infection, the virus lies dormant in nerve cells, waiting to spring up with stressors or other infections. Herpes zoster is a reactivation of the chicken pox virus, called varicella. Patients present with a classic vesicular rash, that erupts in regions of the skin associated with specific nerve roots. The rash can be extremely painful, which is why I consider it a severe condition, and postherpetic neuralgia happens in one of five sufferers. The pain is often burning, sharp, deep, jabbing, itchy, or some combination of the above. Some patients cannot even bear the light touch of clothing on their skin. In one study of 414 skin reactions to mRNA vaccines, 1.9 percent were diagnosed with herpes zoster.[490] A potential mechanism for the reactivation of herpes zoster is that the COVID vaccines temporarily suppress cell-mediated immunity. The T cells that usually monitor and eliminate circulating cancer cells and various germs like herpes are not able to do their jobs in the first few weeks after the jabs.[491,492,493,494,495]

Genetic Predispositions, Inflammation,
and Muscle Breakdown

When a vaccine is tested on healthy Avenger types and then unleashed on a whole population, people with underlying genetic predispositions or chronic diseases may not fare as well as those in the trial. One example from the medical literature is a case of a patient with a known carnitine palmitoyltransferase-II deficiency, which is medical terminology for an enzyme deficiency that affects energy pathways in the body. Within four hours of the AstraZeneca COVID shot, he developed fever, muscle aches, weakness, and shortness of breath. He started peeing blood. He developed rhabdomyolysis, which is when muscles break down. It is a life-threatening condition that requires rapid diagnosis and intervention. Patients can progress to kidney failure, problems with electrolytes, and the dreaded disseminated intravascular coagulation—widespread blood clotting throughout the body.[496] In the first year after COVID vaccines were released, other reports of muscle inflammation and severe damage emerged.[497,498] Remember, cases that make it to the medical literature are just the tip of the iceberg. Most doctors working in emergency departments during the COVID years were stressed and exhausted and would not have time to fill out reports of vaccine reactions or write up case reports to submit for publication.

Conclusion

Proper neurological functions are valuable assets worth protecting. Motor problems like paralysis and sensory problems like tinnitus or the electric-like zips and zaps that some vaccine-injured people report have huge effects on one's quality of life. People who got COVID injections for altruistic reasons to protect others and now find themselves disabled by life-altering neurologic symptoms are victims who are often left to fend for themselves. As a society and as physicians, we have a moral obligation to help them. Consider contributing to React19 or other organizations trying to help the vaccine injured.

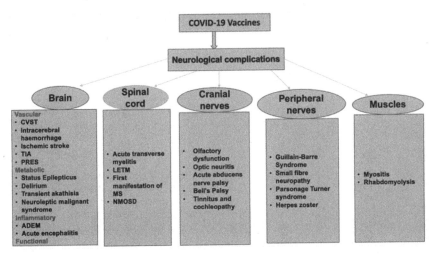

Courtesy of Suzanne Gazda, MD.

FOOD FOR THOUGHT

Do you know someone who developed a new neurologic condition after receiving a COVID injection? Have they had difficulty finding the help they need? If you got a COVID jab, were you told about the possibility of neurologic problems?

A flow diagram depicts the spectrum of severe neurological complications following COVID-19 vaccinations (ADEM, acute disseminated encephalomyelitis; CVST, cerebral venous sinus thrombosis; LETM, longitudinally extensive transverse myelitis; MS, multiple sclerosis; NMOSD, neuromyelitis optica spectrum disorders; PRES, posterior reversible encephalopathy syndrome; TIA, transient ischemic attacks).

CHAPTER 21

What Do We Know About COVID Injection Harms in Pregnancy?

Everything that can be counted does not necessarily count; everything that counts cannot necessarily be counted.

—Albert Einstein

Most published research findings are false.

—John Ioannidis, MD, PhD

Pregnant Women Excluded from Initial Clinical Trials

Pregnant and breastfeeding women were excluded from the original mRNA trials, and participants were cautioned not to get pregnant during the trials.[499,500] Yet, after Emergency Use Authorization was granted, the US government and the American College of Obstetricians and Gynecologists (ACOG) gleefully recommended the injections in pregnancy. Let me restate that: despite having no data about the effects of COVID injections in pregnancy, injections were recommended for pregnant and breastfeeding women! To put this in context, pregnant women are cautioned to avoid taking over-the-counter medications like acetaminophen in pregnancy. Yet obstetricians were recommending a novel, gene-based therapy at a time when babies' organ systems are developing based on exquisite inherent genetic programming. It sounds too horrible to be true, so I invite you to check.

Some have suggested that finances may have played a role in the decision. The US government had pre-ordered 100 million doses, then added another 100 million doses at a cost of $1.95 billion just before Christmas in 2020.[501] Happy holidays to Pfizer and Moderna! Dr. Jim Thorp and his wife Maggie discovered that the American College of Obstetricians and Gynecologists had received $11 million dollars to promote the vaccine.[502]

Pregnancy Trials

In February 2021, a phase II/III trial was undertaken to assess safety and efficacy of the Pfizer/BioNTech injection in pregnancy. Ultimately, just 349 participants were enrolled.[503] As of summer 2023, no data from this trial has been published.[504] If we believe Pfizer representatives, enrollment in the trial decreased once the vaccine was widely recommended. Since so few women were signing up, the study was not large enough to judge whether the vaccine led to enough of an immune response. Since the vaccine had been recommended globally for pregnant women at that point, it was already considered the standard of care. Therefore, continuation of the placebo-controlled trial was not considered justified. It wouldn't be "ethical" to withhold the vaccine from some pregnant women in order to determine whether it was actually safe or effective. The FDA and EMA (European Medical Agency) went along with this scheme, which anyone schooled in logic would label as circular reasoning. You can't make this stuff up.

Manufacturer's Package Inserts: "Insufficient Data in Pregnancy"

Pfizer's FDA-approved Comirnaty vaccine package insert states: "Available data on COMIRNATY administered to pregnant women are insufficient to inform vaccine-associated risks in pregnancy."[505] Moderna's Spikevax package insert says: "Available data on SPIKEVAX administered to pregnant women are insufficient to inform vaccine-associated risks in pregnancy."[506] Yet, officials at the CDC recommended COVID injections for people who are breastfeeding, women of child-bearing age, and those who are pregnant or trying to get pregnant.[507] The Centers for Disease Control, paid for by American tax dollars supplemented by private-public collaborative ties to industry, recommended a new product be used during pregnancy when there was no data to inform such a decision.

Women and Babies: Welcome to the Grand Experiment!

Women have historically gotten the short end of the stick when it comes to medical research.[508] In my opinion, given the absence of data, recommending COVID injections in pregnancy and breastfeeding hit a new low. The initial EUA package inserts for COVID "vaccines" were BLANK.

Dr. Renata Moon, an experienced pediatrician from Washington State, testified under oath during Senator Ron Johnson's "Second Opinion" hearings on COVID. She asked how she could possibly give informed consent to parents in her practice when the package inserts for COVID vaccines were blank. She held up such an insert for inspection. Her participation in that sworn testimony was a factor in her being fired from her job teaching young doctors how to provide care for children. The package insert was the same big size as usual, carefully folded and inserted into the boxes with the vials, but when a doctor opened the document to read about ingredients and side effects, it was BLANK. You can see for yourself in a video Senator Ron Johnson tweeted on December 7, 2022.[509]

Post-Marketing Surveillance for Pregnancy

"Post-marketing surveillance" is the process of discovering adverse events after a new drug product is unleashed on the public. Tragically, serious adverse effects on pregnancy, fertility, and fetal outcomes have emerged. Let's look at the VAERS system, the passive reporting system that is intended to provide early warning of safety "signals" but is plagued by underreporting by a factor of 40 to 99 percent.[510] Menstrual irregularities were reported more than *one thousand times more* per vaccine dose than the historical rates for flu vaccines. What are those abnormal cycles trying to tell us about effects on hormonal regulation?[511,512] The UK equivalent of VAERS is their Yellow Card system. A shocking 23 percent of the 224,960 Yellow Card reports about Pfizer and Moderna COVID injections were about menstrual abnormalities.[513]

At the World Congress for Vaccines in Washington, DC on April 4, 2023, I had the opportunity to confront the medical director of Moderna about these concerning statistics. He tried to reassure me that many of the irregularities amounted to "just one more day" of menses. He voiced no concern about disrupting the elegant system of rhythms that govern women's bodies and their endocrine messages with his profitable drug. Many women might object to a man who never experienced any menses minimizing the significance of extra days of menstrual flow.

Dr. Thorp Is Horrified by Reproductive Effects of COVID Vaccines

An experienced high-risk Ob-Gyn named James Thorp was horrified by what he saw clinically after the COVID injection rollout. That prompted him to analyze VAERS data from January 1998 to June 2022 comparing COVID-19 vaccines to flu shots. The data set revealed significantly more reported menstrual abnormalities, fetal malformations, miscarriages, stillbirths, placental blood clots, premature delivery, preeclampsia, and premature baby death with COVID injections than with flu shots.[514]

While it is impossible to know whether that is a true safety signal from VAERS data alone, it is highly suggestive. Another VAERS analysis, done April 7, 2023, found that, since COVID shots were introduced in December 2020, spontaneous abortions were reported 3.28 times more often than for all other vaccines over the past thirty-two years. Fertility problems were reported 13.38 times more often in the two years after COVID shots began than for all other shots in the previous thirty-two years of VAERS' existence.[515]

Over three thousand case safety reports relating to COVID-19 injections were analyzed from a EudraVigilance, the European Union's adverse event reporting database. Comparing reports of those who received COVID shots versus non-COVID shots, these numbers appeared:

- nearly twelve times the rate of fetal death (0.81% vs. 0.07%)
- almost nine times the rate of hemorrhages during pregnancy (0.62% vs. 0.07%)

- over three times the rate of fetal disorders (2.5% vs. 0.71%)
- nearly three times the number of congenital anomalies (0.11% vs. 0.03%)
- almost four times the rate of premature babies (0.64% vs. 0.17%)
- twice the rate of neonatal deaths (0.06% vs. 0.03%)
- a higher rate of stillbirths (0.22% vs. 0.17%)[516]

It's Just Your Period . . .

Over thirty-nine thousand women who received a COVID-19 "vaccine" from April through June 2021 reported to Lee and colleagues the following concerning information:

- 42% of those with regular menstrual cycles bled more heavily than usual.
- Breakthrough bleeding was reported by 71% of those on long-acting reversible contraceptives who did not normally menstruate.
- Breakthrough bleeding occurred in 66% of postmenopausal women.[517]

Come on, guys. I am appealing to the new billionaires of the COVID crisis like the CEOs of Moderna and Pfizer. In your heart of hearts, you cannot think this is OK. I am calling on women and those who love women to stand up and object to this assault on our reproductive systems and yet-to-be-born babies.

Decidual Cast Shedding

We should learn a lesson from decidual cast shedding, which is the abnormal shedding of the entire uterine lining. Decidual cast shedding is normally so rare there were only forty cases reported in the one hundred years prior to COVID "vaccine" rollout. Between May and December 2021, in a survey of approximately six thousand women who filed VAERS reports, there were 292 cases of decidual cast shedding.[518] Clearly, this method has not identified all the uterine shedding cases.

What If We Compare Vaccinated Versus
Unvaccinated Pregnancies?

The New England Journal of Medicine published a letter describing a CDC-funded retrospective matched-cohort study of pregnant women between sixteen and forty-nine years who either got a COVID "vaccine" or not. This is what they found in the "vaccine" recipients compared to the unvaccinated pregnant women from December 2020 through July 2021:

- 2.85 times (95% CI 1.76–4.61) more likely to experience fever.
- 2.24 times (95% CI 1.71–2.93) more likely to experience malaise or fatigue.
- 1.89 times (95% CI 1.33–2.68) more likely to sustain local reactions.
- 2.16 times (95% CI 1.42–3.28) more likely to experience lymphadenopathy (swollen lymph nodes).

The observation period ended forty-two days after the "vaccine" which means that long-term effects on mother or infant were not evaluated.[519]

A Canadian study of ninety thousand women aged fifteen through forty-nine years included over three thousand pregnant women who got two doses of a mRNA "vaccine." Pregnant vaccinated women were 4.4 times as likely (95% CI 2.4–8.3) to have a significant health event within seven days of the second dose of a Moderna shot compared to unvaccinated controls. Concerning side effects did not appear after one Moderna shot or any Pfizer shot in this study. Common symptoms included malaise, muscle aches, headache, and respiratory tract infection. When the data was further analyzed to account for multiple variables including trimester, age, and prior SARS CoV-2 infection, there was a 2.4 times (95% CI 1.3–4.5) greater risk of a significant health event within seven days of a Moderna jab compared to controls.[520]

Bad News from Israel

A large study (5,618 women) who delivered between December 2020 and July 2021 in Israel found those who were vaccinated with an mRNA vaccine in the second trimester were 1.3 times more likely to deliver prematurely (8.1% vs. 6.2%). The likelihood that those premature births were coincidences was 1 in 1,000.[521] I looked at the number of births during 2021 in the United States and Israel. There were 3.66 million babies born in 2021 in the US[522] and 185,040 in Israel.[523] If the increase in prematurity of 1.9% holds true, there is the potential for approximately 69,540 extra premature babies in the US and an excess of 3,516 premature infants in Israel per year (acknowledging that births rates vary slightly by year). The recommendations by many official medical associations, including the American College of Obstetrics and Gynecology, to recommend this vaccine in pregnancy was shocking to me, since even small increases in premature births mean many families dealing with neonatal intensive care stays and potential developmental impairments in their babies.

Mining Data from Pfizer's Own Documents

Naomi Wolf is a bestselling author with a PhD from Oxford. She spearheaded an initiative by three thousand volunteer doctors and scientists to scour Pfizer documents for data. Dr. Wolf is on record that

> Our volunteers found that a terrifying report went out on April 10th, 2021, showing damage in utero from transplacental or "maternal" exposure to vaccine, Pfizer's words, including death. There are also convulsions and fevers and swelling of babies who are nursing vaccinated moms [sic].[524,525] This report went to CDC, and three days later, Dr. Walensky gave a White House press conference telling the women of America, that especially if they were pregnant, they should take the vaccine; that it was safe and effective, that there was no bad time. "Before you have your baby, during your pregnancy, after having your baby, there is no bad time to take the vaccine."[526]

Dr. Thorp reported what he has seen in his clinic:

> . . . massive disruption of normal menstruation patterns. . . . And that did not occur until after the rollout of the vaccine. I have seen

a marked increase in infertility, male and female. . . . If the [couple] does conceive, then there's a substantial risk of miscarriage. . . . There is an increase in malformations of all organ systems, a substantial increase in fetal death, substantial increase in severe early onset preeclampsia, in preterm premature membrane rupture, an increase in spontaneous preterm labor, an increase in preterm delivery due to vaccine complications. In trying to prevent a death in utero, we are seeing an increase in cardiac anomalies, cardiac malformations, and in the early death of the fetus. I am also seeing an increase in premature delivery and death of the newborns. There is a massive increase in newborns going to the neonatal intensive care unit. There are substantial abnormalities of the placenta.[527]

Dr. Thorp opined that doctors should have expected the vaccines to be pro-inflammatory and thus expected adverse events in pregnancy. Inflammation has been known to be a devastating risk factor in pregnancy. "So of course, [the inflammation caused by the mRNA vaccines] is going to cause damage: miscarriage and malformations and chromosome malformations."[528] Dr. Wolf points out that Report 69 (available from the Daily Clout website) showed Pfizer knew babies would be born with malformations, that placentas showed calcifications and blood clots. I implore you to read that report to see what Pfizer knew and when they knew it.[529]

What About Males?

Pfizer did not test for male reproductive toxicity, not even in animals, according to Pfizer's own documents. Pfizer did not test for adverse effects on their offspring from the semen of men who were given the injection. We know that mRNA ingredients are distributed throughout the body, including into the testes.[530] We know that mRNA vaccines can result in anti-sperm antibodies, which recognize sperm as an invader and try to kill it.[531,532] Some studies, but not all, show mRNA vaccines cause a drop in sperm motility and concentration (which may be transient.)[533]

Conclusion

We have every reason to suspect that the adverse outcomes outlined above represent the tip of the iceberg. The available data supports the immediate cessation of injecting COVID "vaccines" into pregnant women. The pain of losing a longed-for child, the value of a life that never came to be—these are things that cannot be quantified in a scientific study. But to the families affected by stillbirths, miscarriages and infertility, the pain is immeasurable.

FOOD FOR THOUGHT

For those of you who were pregnant or breastfeeding and were advised to get COVID injections, did you receive information that the shot had not been adequately tested for effects in pregnancy or on the development of your fetus? Do you believe you got true informed consent? Do you agree or disagree that we should stop giving COVID "vaccines" to pregnant women?

CHAPTER 22

What Happened to Cancer Trends After Vaccine Rollout?

You have to be willing to give up the life you planned, and instead, greet the life that is waiting for you.

—Joseph Campbell

Cancer didn't bring me to my knees, it brought me to my feet.

—Michael Douglas

Cancer Survivor Turned Oncology Mentor

I met Dr. Nasha Winters when we lectured together at an International Hyperbarics Association conference. She is smart and dynamic and was the center of attention at our evening dinners. When she was a college student, she survived ovarian cancer against all odds and has dedicated her career to helping others with cancer. Here is her story.

She was the first in her family to go to college. She had a volleyball scholarship and worked four jobs to pay for her education. She worked in a library, at a restaurant, and as a certified nursing assistant in a nursing home and detoxification center. Can you imagine her stress level with all those jobs plus a pre-med curriculum? Two weeks before her twentieth birthday, she developed abdominal pain and vomiting and was diagnosed with stage four ovarian cancer causing a complete bowel obstruction. Her doctors told her she would not live to see Christmas; she is in her early fifties now. Due to the size of the tumor and the blockage in her intestines, she could not eat solids for ten weeks. She thinks fasting saved her life. She continued to work in the library and found data on microfiche about the Warburg effect reported early in the twentieth century[534,535,536] and Moresshi's work on fasting and cancer from 1909.[537,538] She became cachectic, with skinny legs and a big belly full of fluid, which doctors had to periodically drain. She made it past Christmas without dying. One of her classmates was romantically interested in her. She did not think it was fair to him to get romantically involved because she would be dead soon. He could not resist being with her. They got married and are still in love.

Findings from a Network of Oncologists

Nasha works with a network of oncologists from around the world who mentor one another through complicated treatment regimens which emphasize nutraceutical and naturopathic strategies. She has partnered with lab companies to develop a whole panel of cancer markers that costs only $450, dramatically less than patients would pay at a typical lab or hospital. Quarterly follow-up labs cost around $250 USD. She worked with imaging departments to develop highly sensitive and highly specific magnetic resonance imaging (MRI) services to avoid the radiation exposure cancer patients typically get from repeat computerized tomography (CT) scans. Her clinical experience is deep and broad. "Now when I look at a patient case," she says, "the gestalt jumps out at me like in those Magic Eye pictures."

COVID Vaccination Impact on Cancer Cases

Nasha reports that the first quarter after COVID vaccines came out her colleagues saw an explosion in large nodes on the same side as the injection.[539,540] By summer of 2021, the oncologists in her network saw an increase of cancer, described as "overwhelming" and "rapid fire." The trend was to find cancer in younger patients and at later stages.

Nasha's group had eighty clinicians when COVID vaccines came out. Now there are over eight hundred clinicians in forty-six countries. Before 2021, 80 percent of the people in the mentoring group got the COVID vaccine; now 100 percent think COVID vaccines cause more harm than good, Nasha says.

Nasha's group reviews lots of labs every month, so they can see objective trends before and after COVID jabs. Their fifty-four-page intake form and lab evaluations yield over ten thousand metric points. This is what the data shows: high IL-6 and IL-8 (pro-inflammatory cytokines), high CRPs (marker of inflammation), abnormal fibrinogen, abnormal IgG4 measures, and high D-dimers (marker of heart damage). Platelet counts, which control clotting, can be high or low. As Dr. Winters puts it, "neutrophil to lymphocyte ratios go bonkers." Usually when a patient with cancer gets an infection it flares and then resolves. After COVID vaccination, patients have a much harder time resolving those infections.

Dr. Winters points out that spike protein impacts complex I of our mitochondria which can lead to the Warburg effect, a shift in cellular metabolism that is associated with cancer.[541,542,543] Nasha explains that there are several studies looking at renal carcinoma, and how spike proteins affect kidney ACE receptors and cancer risk.[544]

Dr. Winters is horrified at the "explosion of prostate cancer in young men." She has also seen patients who were in remission from prostate cancer for fifteen years then developed terminal cancer with metastases soon after a COVID jab. She told me about a case in which a patient's PSA (prostate specific antigen—used as a marker for progression of the cancer) went from 0.8 to 1,080 within one month after a COVID injection! There are identified single nucleotide polymorphisms (SNPs) which are known risk factors for prostate cancer, including TMPRSS2, so some men will be at higher risk than others[545]

Markers used to track cancer are carbohydrate markers related to the epithelium; PIKC3A is a marker elevated in many diverse cancers. Dr. Winters reports data that PIKC3A was seen in 27 percent of network patients before COVID injections rolled out; after COVID injections that number grew to 76 percent.

Dr. Angus Dalgleish was a prominent oncologist from Scotland who did pioneering work on chronic immune activation as a cause of cancer.[546] Dr. Dalgleish was soundly criticized and ostracized after he wrote a letter to the *BMJ*, published in 2022, suggesting that the medical community should pause and think more about the strategy of giving a fourth dose of COVID vaccines. "We believe that the current panic about Omicron is unjustified, that 'vaccinating' children is unnecessary and potentially dangerous and that banning unvaccinated staff from working in health services is not based on science."[547] Dalgleish gave three justifications for his recommendation and supported his position with good science. He cited data that "vaccinating children to protect their parents and grandparents does not work and it cannot be justified on those grounds."[548] Time has proven him right. But during COVID, being right did not get us very far.

Could mRNA Injections Cause Cancer?

Cancer occurs when the body's exquisite immune system does not identify and dispose of abnormally growing cells. You may be surprised to learn that most of us have cancer cells circulating daily. One of the advantages of protecting your immune system is to keep it ready to identify and kill those cancer cells. The process of cancer prevention includes methylation biochemistry to regulate gene expression,[549] T cell identification of abnormal cells,[550] and the immune system deletion of cancerous cells in the early stages.[551] Many of us who perceived the so-called "vaccines" as novel gene therapies worried about the perils of messing with Mother Nature, especially concerning the complicated biologic network that controls gene expression. Proper gene function is crucial for cancer cell identification and destruction.

Two studies raised the horrifying concern that mRNA technology used in Pfizer's COVID "vaccines" can interfere with DNA damage repair, though the first study has since been retracted with the claim (from the paper's first author!) that it used an "improper experiment design."[552] However, quite often retraction has more to do with politics than science. I have read a half dozen retracted papers in which the main flaw was pointing out an inconvenient truth.

Scientists who raised concerns in the summer of 2020 about ingredients in the novel mRNA vaccines entering the nucleus of human cells and causing problems were soundly criticized for raising those concerns. However, an MIT study showed that RNA from SARS-CoV-2 can enter the nuclei of cells and integrate into the genome.[553,554] This raises the concern that the synthetically produced vaccine mRNA, that causes the body to produce toxic spike proteins, could do the same. We were assured by various governments and the pharmaceutical industry that this would not happen.

Let's look at what the Swedish researchers found.

> Adaptive immunity plays a crucial role in fighting against SARS-CoV-2. . . . Clinical studies have shown that patients with severe COVID-19 exhibit delayed and weak adaptive immune responses; however, the mechanism by which SARS-CoV-2 impedes adaptive immunity remains unclear. Here, by using an in vitro cell line, we report that the SARS-CoV-2 spike protein significantly inhibits DNA damage repair, which is needed for . . . adaptive immunity. Mechanistically, we found that the spike protein localizes in the nucleus and inhibits DNA damage repair by impeding key DNA repair protein BRCA1 and 53BP1 recruitment to the damage site.[555]

Let me give some background for those of you who do not have MDs or PhDs in science. Adaptive immunity includes circulating T cells of various types, all marvelously adapted to their specific function. Working as a team, T cells direct B cells to produce antibodies against foreign proteins. When the immune system mistakes proteins in human body parts for foreign proteins, autoimmunity results. The two arms of the adaptive immune system are cellular (T cells and friends) and humoral (antibody formation). When T cells perform poorly, a person is more likely to get bacterial, viral, and fungal infections. When humoral immunity is suppressed, a child is less likely to make effective antibodies against diseases and more likely to have an inadequate antibody response to vaccines. When humoral immunity is upregulated, the child is more likely to get allergies and autoimmunity.

The nucleus is the large organelle that is membrane bound and contains our genetic material in the form of multiple linear DNA molecules, organized into chromosomes. Chromosomes determine our gender, hair color, risk for certain genetic diseases, and a host of other characteristics. An analogy would be to think of the nucleus as the control center of the cell or the library that contains the "book of life." The Swedish researchers found that, looking at cells in a test tube, spike protein got into the nucleus of the cell and made those cells less able to repair DNA damage. Houston, we have a problem (at least until proven otherwise).

I tend to respect Swedish scientists, for many reasons. First, Swedish epidemiologists made the decision to protect the vulnerable but not lock down their society. Schools remained open, and there were no significant differences in COVID cases or deaths when compared with societies that kept their children home from school. Sweden's per capita mortality statistics were just over half that of the U.S.[556] Secondly, in my travels as medical director of the Autism Research Institute, I got to meet several Scandinavian scientists. In my judgment, they seemed smart and dedicated. They were also funny dinner companions. Third, Sweden has the Karolinska Institute, and they give out Nobel prizes in Stockholm. Yes, I realize this is my biased opinion and not an airtight case about the competence of all Scandinavian scientists.

Let's set the stage with some preclinical animal studies of Pfizer mRNA injections. Rats got swollen livers with increased liver enzymes (which reflects at least temporary damage) and vacuoles (holes). The rat studies showed that in rats the mRNA did not stay at the injection site and up to 18 percent of the dose ended up in the liver. The liver is important for elimination of toxins such as those that are known to cause cancers, like the ingredients of tobacco smoke that cause lung cancer. Remember that "tobacco science" was used for decades by cigarette manufacturers to obscure the causal link between smoking and lung cancer. I think we need to be open to the possibility that similar forces are in play to sugarcoat side effects of COVID vaccines.

Swedish scientists from Lund University studied the effect of the Pfizer injection on *human* liver cells called Huh7 in test tubes. They found that the mRNA injection did enter liver cells and that "the mRNA in the shot was reverse transcribed into DNA as fast as six hours after the cells were exposed."[557] The authors thought that endogenous (coming from inside the body) reverse transcriptase LINE-1, which is a big part of our genome, is involved and that protein distribution in the nucleus is elevated by the mRNA injection. In the study authors' words, "Our study shows that [Pfizer's mRNA injection] . . . can be reverse transcribed to DNA . . . this may give rise to the concern if [injection]-derived DNA may be integrated into the host genome and affect the integrity of genomic DNA, which may potentially mediate genotoxic side effects."[558] In my words, "Holy crap! These shots could hurt our genes."

However, careful scientists that they are, they provide this caveat (notice the measured language and call for further studies):

> At this stage, we do not know if DNA reverse transcribed from BNT162b2 is integrated into the cell genome. Further studies are needed to demonstrate the effect of BNT162b2 on genomic integrity, including whole genome sequencing of cells exposed to BNT162b2, as well as tissues from human subjects who received BNT162b2 vaccination.[559]

Clinical Trials of mRNA Injections Did Not Study Gene Toxicity.

Here is an interesting tidbit of information: "In the BNT162b2 toxicity report, no genotoxicity nor carcinogenicity studies have been provided." Did you catch that? No data about toxicity to our genes was provided in the Pfizer toxicity reports to the FDA. Huh? And I mean "huh" in the context of "are you kidding me?" not in the context of Huh7 liver cell lines used in the study. You can read it for yourself in *Current Issues in Molecular Biology.*[560] Here is their conclusion: "Our study is the first in vitro study on the effect of COVID-19 mRNA vaccine BNT162b2 on human liver cell line. We present evidence on fast entry of BNT162b2 into the cells and subsequent intracellular reverse transcription of BNT162b2 mRNA into DNA."[561] One reason I am clinically suspicious that this might be true is that cases of hepatitis after mRNA shots have been described, both in adults and children.[562,563,564]

MIT Study Shows Integration into the Genome

Now let's look at an interesting article from the Whitehead Research Institute at Massachusetts Institute of Technology in collaboration with the National Cancer Institute. Its conclusion is the title of the article,

"Reverse-transcribed SARS-CoV-2 RNA can integrate into the genome of cultured human cells and can be expressed in patient-derived tissues."[565]

In my experience, MIT researchers I have worked with tend to be brilliant and articulate—here's looking at you, researcher Stephanie Seneff and pediatrician Larry Rosen! The MIT alum I married tends to prove the rule. My husband Mike's ability to explain the following concepts in detail off the top of his head still astounds me: the science behind black holes, time warps, and relativity; the psychopharmacology of most psychiatric medications; the complete inner workings of the air conditioner and heating systems in our home; and many more subjects.

MIT scientists "investigated the possibility that SARS-CoV-2 RNAs can be reverse-transcribed and integrated into the DNA of human cells in culture and that transcription of the integrated sequences might account for some of the positive PCR tests seen in patients"[566] who had recovered from COVID 19. They discovered that "DNA copies of SARS-CoV-2 sequences can be integrated into the genome of infected human cells."[26] Then they explained about target site duplications, LINE 1 endonuclease recognition sequences, viral-host chimeric transcripts, subgenomic sequences and the 3' end of the viral genome. Then, of course—they concluded more research is required. The possibility that SARS-C0V-2 sequences can be integrated into the human genome and expressed as chimeric (genetic information from different sources integrated in one organism, like in Greek myth human/horse combinations) RNAs raises this question: Do integrated SARS-CoV-2 sequences express viral antigens in patients that might influence the clinical course of the illness?

Turbo Cancer and the P53 Gene

Pathologist Ryan Cole, MD, noticed a new pattern in the biopsies he examined after COVID vaccines were deployed. He explained to me that he saw a higher volume of cancer than before COVID shots came out, and more cancers were presenting at later stages. There are several potential mechanisms that might be associated with this trend. One is the involvement of the P53 gene, our "guardian at the gate" that keeps cancer cells from growing. This tumor suppressor protein helps regulate normal cell growth and inhibit the division of cells in which the DNA has been damaged.

Clinicians in the trenches, like Ryan, are valuable because they notice changes in patterns. Sadly, during COVID, practicing doctors did not have a seat at the table when decisions about pandemic responses were being made. Experience in any given field of medicine involves developing pattern recognition. For a pathologist, how many cancer biopsies come in per unit time and what stages the cancers are in at presentation are important patterns. For a pediatrician like me, a pattern I notice is how a very sick kid looks compared to a child with a minor illness. Another example is noticing the increase in neurodevelopmental problems that started in the 1990s. For an oncologist, patterns noted include the typical presentation and course of various cancers and the number of new cancer patients they see in each time frame. Anecdotes of increasing numbers and more severe cancers at presentation began circulating in the summer and fall following COVID "vaccine" rollout. Another new pattern oncologists reported was an aggressive recurrence of cancer in patients that had been tumor-free for ten to twenty years.

Such anecdotes provide a foundation for further research. One such research scientist and pathologist is Ute Kruger, MD, DMedSci, a breast cancer specialist. In January 2023, she shared her breast biopsy experiences at a conference in Sweden. Drawing upon her experience evaluating 1,500 breast cancer cases, she had excellent knowledge of the average age of patients, typical tumor size, and degree of malignancy prior to the emergence of COVID-19. By fall of 2021, she had noticed a change in the typical pattern: younger patients, tumors presenting in multiple areas of the body, and more aggressive tumors that were larger. She brought her findings to her colleagues at a conference in December 2021, with little response.[567] Like Dr. Cole, she observed another change: patients who had been cancer free for ten or twenty years had recurrences. In her words, *"Relatively soon after the vaccination against COVID-19, the tumor growth explodes, and there is a pronounced spread of the tumor in the body; and some of the patients die within a few months."*[568] She saw yet another pattern change: "It's not unusual for patients to have several malignancies in different organs at the same time."[569] She calls this anomaly "coincident tumors."

Immune Mechanisms of Increased
Cancer Risk After COVID Jabs

An important paper about repeated immunization for COVID suggests multiple doses induce IgG4 antibodies and tolerance for spike protein. Such immune tolerance could have many negative consequences.[570] Low IgG4 is important for evading cancer. Higher IgG4 is linked with more cancer growth, more metastases, and poorer prognosis. Mechanisms to remove unwanted cells don't work well when IgG4 is elevated. Higher IgG4 tips the scales from tumor suppression to tumor progression, writes Jessica Rose, PhD, whom you met in the chapter about VAERS.

Immunosuppression after COVID vaccines could make one more vulnerable to other infections (like the flu or RSV) and more vulnerable to cancer. Uvershi and colleagues boldly state, "it is conceivable that the excess deaths reported in several highly COVID-19 vaccinated countries may be explained, in part, by this combined immunosuppressive effect."[571] For those of you who want to dive deep into the science, take a look at the references cited here.[572,573]

Shields Up, Scotty!

As early as 2020, published evidence showed that the S2 subunit of the SARS-CoV-2 spike protein would be expected to inhibit p53, our major tumor-surveillance mechanism, and BRCA1/2, which does breast cancer surveillance. P53 is often described as a shield-like mechanism to protect us from cancer,[574] hence the Star Trek reference. The good news is that the S2 segment has not been shown to be present in the body after infections with COVID-19. The bad news is that S2 segments are produced by the billions when instructed to by synthetic mRNA in COVID-19 injections.[575] It would have been nice for the public to know this when they were being mandated to take the shots.

World Council of Health Insights

On October 9, 2023, the World Council of Health gathered experts from around the world to discuss the implications of DNA plasmid contamination of COVID vaccines and included a discussion of the implications for cancer risk.[576] They identified the following concerns: increased mutagenesis with dsDNA contamination; chronic insults to the immune system; inhibiting guardians of the genome like p53 and BRCA.

Death records in Massachusetts show a clear increase in cancer mortality. Janci Lindsay, PhD, who is a toxicologist, warned that one should expect oncogenic potential from COVID injections. She pointed out that prior gene therapies were not brought to market for widespread use because latent cancer appeared two to four years later. She mused that she could think of nine potential ways mRNA COVID vaccines could induce cancer:

1. Lipid nanoparticles take mRNA and DNA to all cells and transfect stem cells.
2. LNPs cause cancer cells to spread by inducing endothelial leakiness.
3. SV40 "super promoter" enhances gene expression of oncogenes.
4. SV40 enhancer region, a nuclear targeting sequence, is designed to take DNA to nucleus for insertional gene therapy.
5. Spike protein can inhibit p53, an important tumor suppressor.
6. Plasmid DNA can go to nucleus even without SV40.
7. Insertional mutagenesis causes frame-shift mutations leading to aberrant proteins that lead to cancer.
8. mRNA is reverse transcribed to DNA and integrated into our genome, which can cause cancers.
9. T cells that hold back cancer clones from expanding and metastasizing are destroyed.[577]

Professor Alexandra Henrion Caude, from the French equivalent of NIH, reported how no genotoxicity or carcinogenicity studies were done by manufacturers of COVID injections.[578] She correctly pointed out that it should not be the responsibility of independent scientists to demonstrate toxicity after widespread use in the public.[579] Jessica Rose, a smart and sassy PhD, has been using her bioinformatics training to analyze VAERS data during the COVID crisis. VAERS reports after COVID injections show odd cancer patterns, like breast cancers in males, and acute lymphocytic leukemia in old people (ALL is usually a childhood blood cancer).

SV40 Revisited

You may remember hearing about simian virus 40 (SV40) contamination in polio vaccines. SV40 particles were subsequently found in tumors. Scientists later determined SV40 could cause several types of cancer. We now know that SV40 promoter contamination is present in COVID injections.[580] SV40 oncogene promoters inhibit DNA repair. SV40 enhancer is a strong initiator of somatic hypermutation which could contribute to carcinogenesis. It is crucial that humans be able to repair DNA, or cancer can result.

What If You Got a Jab and Got Cancer?

In individual cases, correlation alone cannot prove causation, so you may not be able to decipher which environmental and genetic factors were involved in your cancer. In most cases, cancer is a treatable condition. Innovations in cancer research are overturning some older models of cancer as due to genetic mutations only. New thoughts include envisioning cancer as a metabolic disease, which opens avenues for therapies targeting the metabolism of cancer cells. Dr. Winters has written a book about this, *The Metabolic Approach to Cancer*,[581] and mentors clinicians who want to learn that approach. Dr. Paul Marik, of the FLCCC, has written a monograph about cancer care.[582] Data about the role of repurposed drugs and the value of dietary changes like intermittent fasting or low-carb, high-fat diets is intriguing and encouraging.

What Is a Good Price to Pay for Short-Term Protection?

It is crucial to understand that immunity from COVID "vaccines" is short lived, lasting only around four months in many people. You may have figured that out already, with the push for more and more boosters. A study in Sweden showed "progressively waning vaccine effectiveness against SARS-CoV-2 infection of any severity across all subgroups."[583] Surely it is time to assess whether the short-term immunity induced by COVID jabs (which morphs into negative immunity later) is worth giving up intact immune function and effective cancer surveillance. The immune system is so spectacular when left to its own devices. Personally, I think we should minimize interfering with our own exquisite immune mechanisms.

FOOD FOR THOUGHT

Have you seen the cancer patterns described above in your friends or family? Did you see fact checkers proclaim there was no connection between cancer and COVID injections? Did you check to see if those claims were backed by any data? Did you know that diet and lifestyle can be an important part of healing cancer?

CHAPTER 23

What Does Excess Mortality Mean?

If only one man dies . . . that is a tragedy. If millions die, that's only statistics.

—Joseph Stalin

Facts are stubborn things, but statistics are pliable.

—Mark Twain

Deaths in Young Adults

I hope you have visited the Vietnam memorial in Washington, DC. A black slab of granite is engraved with the names of 58,318 American soldiers who died in Vietnam over a decade. In year two of the pandemic, the virulent original Alpha and very contagious Delta strains had passed through and milder strains were emerging. Once we were into widespread distribution of COVID vaccinations, about sixty-one thousand extra millennials died.[584] Sixty-one thousand above and beyond what was expected! It was a scale of death like Vietnam—and happened in one year instead of ten! Have you heard about this?

More People Died After COVID Than During the Peak

Every Thursday for the last several years, I have been Zooming with a group of scientists, clinicians, lawyers, and activists from around the world to discuss all things COVID and global health related. One of my colleagues on those calls lives in Israel. Shortly after the deal between Pfizer and the Israeli prime minister, vaccine distribution began and he reported hearing more ambulance sirens than ever before, even more than during the worst days of the pandemic. He told stories of young people dying quickly and unexpectedly. His anecdotes proved to be a harbinger of what has now been proven: COVID mRNA vaccines are associated with risks of death which should have been anticipated.

Report from the Israeli People Committee

The Israeli People Committee (IPC) is a civilian group of Israeli health experts. In April of 2021, they reported side effects from the Pfizer vaccine showing damage in almost every bodily system.[585] They interpret the results as catastrophic. See what you think. Notice that this information was reported in May of 2021, only *four months* after the big rollout in the United States began in December 2020.

- *According to Central Bureau of Statistics data during January– February 2021, at the peak of the Israeli mass vaccination campaign, there was a 22% increase in overall mortality in Israel compared with the previous year.*
- *January–February 2021 have been the deadliest months in the last decade.*
- *Amongst the 20–29 age group the increase in overall mortality has been most dramatic—an increase of 32% in overall mortality in comparison with previous year.*
- *Statistical analysis of information from the Central Bureau of Statistics, combined with information from the Ministry of Health, leads to the conclusion that the mortality rate amongst the vaccinated is estimated at about 1: 5000 (1: 13,000 at ages 20–49, 1: 6,000 at ages 50–69, 1: 1,600 at ages 70+).*
- *There is a high correlation between the number of people vaccinated per day and the number of deaths per day, in the range of up to 10 days, in all age groups.*
- *There appears to be an age effect on timing of death: Ages 20–49—a range of 9 days from the date of vaccination to mortality, ages*

*50– 69—5 days from the date of vaccination to mortality, ages 70
and up—3 days from the date of vaccination to mortality.*

- *The risk of mortality after the second vaccine is higher than the risk of
  mortality after the first vaccine.*
- *These reports indicate damage to almost every system in the human body.*
- *26% of all cardiac events occurred in young people up to the age of
  40, with the most common diagnosis in these cases being myositis or
  pericarditis.*
- *A high rate of massive vaginal bleeding, neurological damage, and
  damage to the skeletal and skin systems has been observed.*
- *A significant number of reports of side effects are related, directly
  or indirectly, to Hypercoagulability, Myocardial infarction,
  stroke, miscarriages, impaired blood flow to the limbs, pulmonary
  embolism.*[586]

Power to the people of Israel! This spectacular work by the Israel People
Committee should have informed vaccine decisions around the world.
Surely their findings should have prevented the Biden administration
from implementing COVID vaccine mandates. I honor these committee
members as fellow Cassandras, seeing what the future could hold, documenting what did happen, and trying their best to warn officials. But their
warnings were not heeded. Many governments continued to push the vaccines. Travel restrictions were in place for those who did not succumb to
the jabs. Many people only agreed to the shots under coercion when told
they would lose their jobs.

Excess Deaths: The Actuaries Sound the Alarm

Actuaries who work for insurance companies analyze data about age of
death and help inform price points for insurance buyers of various ages.
Typically, insurance costs are low when risk of death is low. The price is
higher as we age toward our inevitable meeting with the Grim Reaper. On
August 17, 2022, The Society of Actuaries Research Institute published
a report on Group Life COVID mortality. Their data showed a striking
increase in mortality in the third quarter, during the same period vaccine mandates were ordered by the Biden administration.[587] Could it be a

coincidence? Maybe. But their data was replicated by others, and no one
has brought forth a reasonable alternative hypothesis to explain how so
many previously healthy young working people ended up dead.

Hedge Fund Guru Sounds the Alarm, but Did You Hear It?

In 2021 and 2022, there was a huge spike in deaths among young people.
This is documented in Ed Dowd's remarkable book, *Cause Unknown*.[588]
I have read the book twice, and it is a heartbreaker. In many cases, the
deaths occurred in the first weeks after a COVID injection in a previously healthy person or athlete. Ed is a hedge fund financial guru with
a talent for spotting trends ahead of his competitors. He used the CDC
All-Cause Mortality Data—a different database and a different population—to find the same trend in excess deaths in working-age adults
that the insurance actuaries had noticed.[589,590,591] In 2021, he saw news
reports of previously healthy young people dying in their sleep or while
playing sports. He is a numbers guy, so he looked at the statistics, comparing baseline numbers of deaths in 2017, 2018, and 2019 to 2020,
then again to 2021 and 2022. Deaths went up in 2020, the first year
of the pandemic, but not as much as you might think and were primarily in older people. Young people at peak of productivity, however,
started dying in terrifying numbers, often in their sleep or during ath-
letic performances, in 2021 and 2022.[592] Something new and terrible
was happening. . . .

Cause Unknown?

Dowd's book, *"Cause Unknown": The Epidemic of Sudden Deaths in 2021
and 2022*, is a must-read. Dowd's talent for detailed analysis and dogged
pursuit of truth lays out horrifying data that mainstream media has not
broadcast. His book is a tribute to those whose lives were cut short and a
call to action to prevent this disaster from happening in the future. His
publisher includes QR code links to a list of a thousand media reports and
scientific papers, published by the end of 2022, about COVID vaccine
injuries. Dowd's treatise also lists QR codes to hundreds of news articles

about deaths in young people from around the world. *Cause Unknown* features stunning graphic charts illustrating the trends he describes.

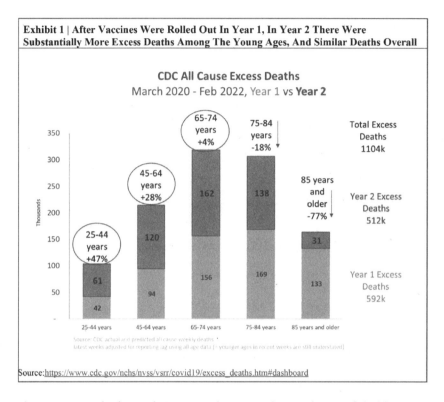

Exhibit 1 | After Vaccines Were Rolled Out In Year 1, In Year 2 There Were Substantially More Excess Deaths Among The Young Ages, And Similar Deaths Overall

Chart courtesy of Ed Dowd, Cause Unknown: The Epidemic of Sudden Death in 2021 and 2022, *page 71.*

40 Percent Rise in Deaths in Working-Aged People in 2021 & 2022

In the third and fourth quarters of 2021, OneAmerica Financials' CEO, Scott Davison, reported a shocking 40 percent increase in deaths in the eighteen- to sixty-four-year age group. To put that percentage in context, a disaster-related 10 percent increase in deaths would be expected about once every two hundred years. A 40 percent increase is unheard of and horrific.[593,594,595] Why did these previously healthy people die in such unprecedented numbers? What was new and different?

These Excess Deaths Cannot Be Blamed on COVID Itself

COVID has low mortality in previously healthy working-age folks, as we discussed in earlier chapters. By the third quarter of 2021, COVID had evolved to a less serious, although more contagious, virus. Also, by that time, governments and employers were mandating experimental COVID injections. Are COVID "vaccines" the elephant in the room?

I think that there has been an attempt to discount and normalize the deaths of these young people, which is an atrocity. Do you recall starting to see news reports of "sudden adult death syndrome" (SAD), analogous to sudden infant death syndrome (SIDS)? Do you recall the unprecedented efforts to use social media and entertainment outlets to advertise the novel injections as safe and effective? Did you need proof of vaccination to enter concerts or restaurants, attend college or keep your job? Lots of young people did.

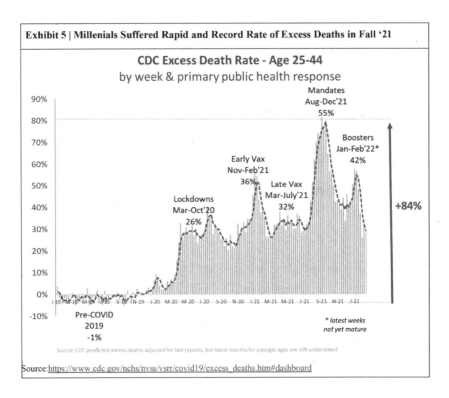

Chart courtesy of Ed Dowd, Cause Unknown: The Epidemic of Sudden Death in 2021 and 2022, *page 70.*

Dramatic Shift in Ages of Death

Another brilliant numbers guy joined Ed Dowd to testify under oath in a congressional hearing that I attended. Josh Stirling documented a dramatic shift in age of death from the elderly (which is the norm) to college and working-age adults from 2020 to 2021. He put the data into a chart that tracked deaths by week and public health response. The data shows marked spikes in death rates in working-aged people with lockdowns, early vaccines, vaccine mandates, and boosters. You may be shocked by what the chart above says compared with what you were told at the time.

Too Little, Too Late

In November 2023, Dr. Robert Califf, FDA Commissioner, took to Twitter to bemoan the decline in US life expectancy: "We are facing extraordinary headwinds in our public health with a major decline in life expectancy. The major decline in the US is not just a trend. I'd describe it as catastrophic."[596] Reading his post, which covered topics like chronic illness, access to health care, smoking, and poor nutrition, I was disappointed but not surprised that he did not include novel COVID injections developed at Warp Speed and marketed aggressively in his list of suspects. Health statistics documenting 158,000 excess deaths in Americans in 2023 compared to the same time period at baseline year 2019 are readily available.[597] Where are the urgent investigations to figure out what is happening?

Culling the Herd

To use a grotesque analogy that is offensive when applied to human losses from the disease of COVID, one could postulate that deaths should be relatively lower for a few years after the pandemic, because "the herd" was "culled" when COVID killed so many who were already elderly or infirm with multiple comorbidities. I am indebted to my friend and colleague Pierre Kory and investigative journalist Mary Beth Pfeiffer for reporting that insurers are facing a crisis, with mega payouts for deaths in those who would usually be paying their premiums for decades before collecting.[598] Actuarial reports continue to confirm what Ed Dowd first noted in 2021—young, working-aged people are dying in unprecedented proportions.[599]

Is the CDC tracking and investigating this horrifying phenomenon? It appears not. In September 2023, searching the excess death datasets yielded this message: "this web page is archived for historical purposes and is no longer being updated."[600] What? I have working-age kids and friends who got COVID shots, so I have dogs in this fight. Inquiring minds need to know what is taking down our best and brightest in the prime of their lives.

In parallel with the uptick in working-age deaths, there's been a decline in deaths in the elderly, which were down six percent below pre-pandemic normal in the second quarter of 2023. Mortality was 19 percent higher in those twenty-five to thirty-four years old and 26 percent higher in those thirty-five to forty-four years old. This is not the natural order of life.[601] Analysis of claims confirms that COVID deaths dropped 84 percent from early 2021 to early 2023.[602]

What Could Be Causing These Deaths?

Spokespeople are tempted to blame some deaths in the young on drug overdoses and other deaths of despair, including suicide, and there may be some element of truth to those hypotheses. But overdoses happened in other age groups compared to working-aged people.[603] The actuarial analysis of government data shows that the young are dying from liver and kidney diseases, diabetes, and cardiovascular events.[604] What is likely to be causing such diseases? You have just read about some possible mechanisms of harm from COVID vaccines in those arenas of organ function.

Our colleagues in the UK are noticing similar trends, but they seem to be looking into the phenomenon. The BBC is now starting to be critical of governmental countermeasures, noting "With each passing week of the COVID inquiry, it is clear there were deep flaws in the way decisions were made and information provided during the pandemic."[605] Billions of people have received COVID vaccines under emergency use authorization, including pregnant women and babies.[606] I am frightened by the actuary predictions that excess deaths will continue in working-aged people with life insurance through the year 2030.[607]

Money Talks

In 2020, the first year of the COVID crisis, death claims increased 15.4 percent, the biggest jump since the 1918 flu epidemic. The insurance industry paid out $90 billion to their insured.[608] Insurance claims hit $100 billion in 2021. Now insurers are trying to track post-COVID and post-vaccine risk factors, to see if interventions can mitigate the risk of death, which translates to the risk of big pay-outs.[609]

Excess Deaths Associated with COVID "Vaccines" in Seventeen Countries

This section is likely to be shocking and depressing for the typical reader. Let me take you through an extensive research effort published in a 180-page paper with more than 175 scientific references. The conclusion is striking. The researchers examined data from seventeen countries on four continents in the Southern Hemisphere. The average vaccination rate was 1.91 injections per person of all ages.[610] The researchers discovered that all seventeen countries showed an increase in all-cause deaths that was consistently associated with the rollout of vaccines.[611] Nine of the seventeen countries did not have any detectable increase in deaths during the first year of the declared pandemic, including times when the original strain was circulating.[16]

Read that again. Cliff notes version: "pandemic" = no increase in overall deaths. Vaccine rollout = increased all-cause mortality. The authors had previously published articles about the "anomalous" increase in all-cause mortality in Australia, Canada, India, Israel, and the US which correlated with COVID vaccination campaigns.[612,613,614,615,616]

For those of you who have traumatic memories of refrigerated trucks in places like New York and overflowing ICU wards in Italy, stay with me while we work through this research. During the pandemic, it seems that more people may have died from COVID, but there were fewer deaths from other causes like influenza. Some reclassification of mortality cause for 2020 may be associated with the financial incentives to list COVID as a cause of death on official death certificates. In addition, the authors of the paper under discussion argue that increases in mortality in hot spots like New York, Los Angeles, and Italy were related to ineffective protocols hastily rolled out during the crisis. Let's evaluate the evidence.

Epidemiology 101 from an Expert

First, a quick lesson on epidemiology, which is a medical specialty that studies the incidence and distribution of health and disease. The foremost epidemiology authority in the world in terms of publications and citations is John Ioannidis, whose work I mentioned elsewhere. According to a 2015 lecture Ioannidis gave on revisiting the nine Bradford-Hill criteria of causation, there are only three that are truly important: "experiment remains important and consistency (replication) is also very essential. Temporality also makes sense, but it is often difficult to document."[617]

Rancourt and colleagues make a compelling argument that they satisfy these "robust criteria for proving causality" in their analysis of data from seventeen countries (Argentina, Australia, Bolivia, Brazil, Chile, Colombia, Ecuador, Malaysia, New Zealand, Paraguay, Peru, Philippines, Singapore, South Africa, Suriname, Thailand, and Uruguay) in the following ways:

- Experiment: The same phenomenon is independently observed in distinct jurisdictions, for distinct age groups, and at different times, which constitutes ample verification in independent real-world large-scale experiments.
- Temporality: The many stepwise increases and anomalous peaks in All-Cause Mortality are synchronous with vaccine rollouts; including in jurisdictions in which excess mortality did not occur until vaccination was implemented after approximately one year into the declared pandemic.
- Consistency: The phenomenon is qualitatively the same and of comparable magnitude each time it is observed."[618]

Canadian Scientists Find Compelling Data

By my interpretation, they have the graphs and supporting data to prove their conclusions. The authors get into the nitty-gritty details in each country. For example, detailed mortality and vaccination data for Chile and Peru allow resolution by age and by dose number. It is unlikely that the observed peaks in all-cause mortality in January–February 2022 (and additionally in: July–August 2021, Chile; July–August 2022, Peru), in each of both countries and in each elderly age group, could be due to any cause other than the temporally associated rapid COVID-19-vaccine-booster-dose rollouts. Likewise, it is unlikely that the transitions to regimes of high all-cause mortality, coincident with the rollout and sustained administration of COVID-19 vaccines, in all seventeen Southern Hemisphere and equatorial-latitude countries, could be due to any cause other than the vaccine.[619]

In other words, after carefully examining age of "vaccine" receipt and number of doses, considering other possibilities to explain the excess deaths, plus comparing the timing of "vaccine" rollout and booster dose administration, they came to the inescapable conclusion that the shots killed people. A lot of people.

In Peru and Chile, the vaccine-dose fatality rate increased exponentially with age. With more doses, the likelihood of death increased. In fact, the risk of death doubled every four years of age, reaching about 5 percent in the over-ninety age group. This evidence means that the shot was responsible for the deaths of one in every twenty people ninety years old or above in those countries.[620] Now, one might argue that, for the extreme elderly, one in twenty odds are not so bad if the shot helped them. Sadly, Rancourt and colleagues could not find any evidence that COVID "vaccines" prevented deaths. In India and for those about eighty years old, the vaccine-dose fatality rate was 1 percent, meaning one in a hundred people died because of the injection.[23]

A New Phenomenon in Nursing Homes

This epidemiology rings true to me as I correlate it with clinical experiences in my hometown. In one nursing home close to where I live, nurses and medical aides noticed a new phenomenon. In contrast to watching the elderly die slowly over a period of days to weeks following a predictable pattern of increasing weakness, loss of desire to eat, then multisystem organ failure, they were confronted by sudden deaths of people who had seemed vigorous hours before. They reported elderly people who seemed fine during community lunch, but were found dead by dinnertime. Such death experiences are not typical.

Were COVID Shots an Iatrogenic Killing Event?

Now for the scientists' shocking conclusion that COVID injections were "a mass iatrogenic event that killed (0.213 ± 0.006) percent of the world population (one death per 470 living persons, in less than three years), and did not measurably prevent any deaths."[621] Let their statement sink in. They contend that something humans manufactured and distributed killed one in 470 fellow humans, without affording any discernible benefit. Furthermore, they contend that the risk of death by COVID injections is "globally pervasive and much larger than reported in clinical trials, adverse effect monitoring, and cause-of-death statistics from death certificates, by three orders of magnitude (1,000-fold greater)."[622] For those of us who have voiced concerns from the beginning that data from the Vaccine Adverse Reporting System probably only captured one in one hundred actual adverse events (or one in forty, when we were being careful not to overstate) this is a shock. Could it really be that we are dealing with undercounting by a factor of a thousand?

As I think about how one could possibly undercut their arguments (which, if true, reveal a horrific truth), it is important to note that epidemiologic data carefully collected by geographic region, gender, and age is not subject to reporting bias the way other types of studies can be. In fact, the authors quote over seventy-five independent scientific papers to support their methodology. Yet Dr. Mandy Cohen, the current CDC director, seems unaware of this damning data as she continues to promote the "safe and effective" mantra.

Conclusion

Excess deaths in young people in the years after COVID shots were rolled out are real and deserve scrutiny. I highly recommend you get Ed Dowd's book, *Cause Unknown*, about excess deaths in working-aged people because it presents the faces and human stories behind the statistics. It is a sad read, but we owe it to those who made the ultimate sacrifice, not recognizing the risks they were taking by getting the COVID injections.

FOOD FOR THOUGHT

Do you know young people who have died suddenly without explanation in the past few years? Did you read news articles about elementary school children dying of "heart attacks?" Did you read news articles about colleges and universities that had increased deaths of seniors who never got to walk the stage during graduation? Did you know that there was excess mortality after the shots were released before you read this chapter?

CHAPTER 24

Are There More Disabled People Since COVID Shots?

The USA is really mean to sick people!

—Steven Magee

Just because a man lacks the use of his eyes doesn't mean he lacks vision.

—Stevie Wonder

Spike in Disabilities

The best current evidence suggests there was a sharp increase in disabilities which started in spring of 2021 after the rollout of COVID injections began in December of 2020. Phinance Technologies commissioned work performed by the Humanity Projects Team, which found a strong correlation "between the mRNA inoculations and levels of disability within the US labor force."[623] Now, as we all know "correlation does not equal causation." Even if we didn't know it already, we've been hearing that song from government agencies and medical societies for over twenty years as more and more parents of children with autism reported their observations that symptoms of autism began shortly after childhood vaccinations. Notice the careful language the Humanity Projects team used. This gives me confidence in their work and conclusions:

Although a range of factors may be at play, the timing and sudden nature of the increase in disabilities suggest that rollout of vaccination programmes could have caused a significant impact. Other factors (those related to a return of more normal economic and healthcare activity) would be more likely to cause a gradual change in disability rates, beginning earlier in the post-lockdown recovery phases of mid-late 2020.[624]

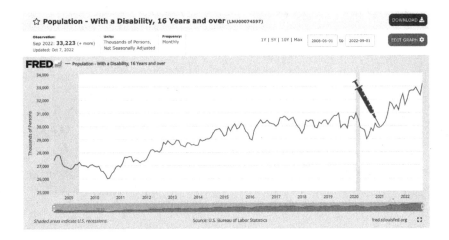

The chart above shows a sharp rise in disabilities well above the five-year average. (Chart courtesy of Ed Dowd, Cause Unknown: The Epidemic of Sudden Deaths in 2021 and 2022, *page 117.)*

Impact of Disability on Jobs and Economy

Friends of mine in the military have worried that disabling effects on our soldiers from COVID injections might rise to the level of being a national security issue.[625] Frequent flyers have worried more about their own safety since there have been more reports of pilots having heart attacks and strokes while in the air after the vaccines were mandated.[626] Restaurant owners have reported problems finding waitstaff, and it is unclear how much of that new problem might be related to people being unable versus unwilling to work.[627]

Time Trend Analysis Shows a Peak in Disability After COVID Jabs

A chart in *Cause Unknown* shows the baseline from 2008 to 2020, then an uptick in disabilities in those sixteen years and older starting in May 2021. The uptick coincided with increased exposure of the population to novel gene therapy products called vaccines. In fact, the disability level is higher than at any time in the past fifteen years. The data was published by the Federal Reserve Bank, but it comes from the US Bureau of Labor Statistics.[628]

Dowd notes that 3.5 million more Americans have become disabled since the COVID injection rollout than the previous five-year average.[629] In fact, healthier employed people suffered a greater increase in disability than the total US population—a whopping 22.6 percent increase, more than three times the increase of 6.6 percent for the whole population.[630] People like Ed are trained to look at trends and deal with the facts. The numbers, which represent lives lost or diminished, compelled Ed to conclude that vaccine makers and government agencies are in cover-up mode.[6316] It is absolutely appropriate to ask whether the novel injections and employer-based mandates contributed to the horrific story the numbers tell.

Careful Analysis of mRNA Data Predicted Serious Side Effects

Peter Doshi, a senior editor at the *BMJ* and associate professor at the University of Maryland, led a group of independent researchers who analyzed Pfizer and Moderna data on serious adverse events of special interest, which were identified separately by the Brighton Collaboration. This is what they found:

> Combined, the mRNA vaccines were associated with an excess risk of serious adverse events of special interest of 12.5 per 10,000 vaccinated (95 % CI 2.1 to 22.9); risk ratio 1.43 (95 % CI 1.07 to 1.92). The Pfizer trial exhibited a 36 % higher risk of serious adverse events in the vaccine group; risk difference 18.0 per 10,000 vaccinated (95 % CI 1.2 to 34.9); risk ratio 1.36 (95 % CI 1.02 to 1.83). The Moderna trial exhibited a 6 % higher risk of serious adverse

events in the vaccine group: risk difference 7.1 per 10,000 (95 % CI −23.2 to 37.4); risk ratio 1.06 (95 % CI 0.84 to 1.33). Combined, there was a 16 % higher risk of serious adverse events in mRNA vaccine recipients: risk difference 13.2 (95 % CI −3.2 to 29.6); risk ratio 1.16 (95 % CI 0.97 to 1.39).[632]

Changing the Trajectory of a Young Girl's Life

Meeting Maddie de Garay and her mom was one of the most disheartening and heartbreaking experiences during my "COVID time." Maddie's mom testified at the January 2022 Senate hearing held by Ron Johnson. You met Maddie in the chapter about COVID "vaccine" complications. Thank you, Senator Johnson, for taking the reports of adverse COVID injection reactions seriously and giving a voice to the vaccine injured.

Maddie was a healthy twelve-year-old girl who volunteered for the COVID vaccine trial being done in conjunction with Cincinnati Children's hospital. Maddie, I applaud your altruistic motives, and I am so sorry about what happened to you. As the science has progressed since those dark days when you were in and out of the hospital with no clear answers or effective treatments, we now have a better understanding of what may have happened to you. I do not have many good ideas about effective treatments for you yet, but a lot of amazing doctors and scientists are working furiously to figure it out. Sadly, you are one of thousands.

Within twenty-four hours of her shot, Maddie ended up in the emergency room with abrupt new symptoms including neurologic symptoms. She was in and out of the hospital five times in the next few months. Let's look at Maddie's medical problems in detail. She developed a paralysis or inflammation of her muscles (hard to know since I was not there) which landed her in a wheelchair. She vomited multiple times per day after eating to the extent that she had a feeding tube when I met her. I wonder if she had toxin-induced gastritis or gastroparesis or mitochondrial damage which affected the motility of her GI tract (speculation on my part, but deserving of investigation). At one point, she was referred to an eating disorders clinic, with the assumption that her problems were "functional," which is medical code for no organic cause, must be psychiatric. She fainted ten to twenty times a day beginning shortly after the vaccine. I think she had autoimmune POTS syndrome (positional ortho-static tachycardia syndrome). In this condition, the body makes antibod-ies against various tissues and cells that are associated with automatic functions of the autonomic nervous system, including heart regulation and stabilization of blood pressure when you stand up. POTS has now been reported numerous times with COVID injections. Maddie's mom has reached out to the hospital that conducted the trial, the CDC, and the FDA and has not been able to get help for her daughter.

Changing the Trajectory of a Surgeon's Career

Joel Wallskog is an orthopedic surgeon from Wisconsin. Since 2002 his practice has focused on hip, shoulder, and knee replacements. He was exposed to numerous patients and coworkers who tested positive for COVID-19. Although he remained without symptoms, his blood showed a positive antibody response, indicating he had been infected. He fol-lowed the CDC recommendations at the time to get a COVID injection anyway, which he did three months after his antibody test. Shortly after his first Moderna vaccine on December 20, 2020, he developed numb-ness, weakness, and problems keeping his balance. He was diagnosed with transverse myelitis, an inflammation of the spinal cord, which you learned about in chapter 20. Imaging showed he had lost the protective myelin at the level of his eighth and ninth thoracic vertebrae. Despite high-dose steroids, intravenous immunoglobulin, and physical therapy, his symp-toms are unchanged. He will no longer be able to perform the surgery he loves and trained so long to perform. Prior to his Moderna injection, Dr. Wallskog was an active father of four children who loved hiking, biking, wake surfing, and water skiing. Now he can only walk short distances. He cofounded React19 to help the vaccine-injured find medical help.

Preschoolers with Learning Challenges
Lose a Talented Teacher

Joel's cofounder of React19 is a previously healthy preschool teacher and mother of two young children. Brianne Dressen has always been driven to strive for excellence; she graduated college at the age of twenty. After specialized experience in child development programs, she founded her own preschool to help children who struggle in a typical classroom environment. She participated in the clinical trial for the AstraZeneca COVID shots and got one injection in November which changed the trajectory of her life. Before her COVID injection, she climbed mountains for hours and was a "supermom" to her children. After the shot, Brianne developed blurry vision, brain fog, memory loss, and extreme sensitivity to touch and sound. Her extremities were weak, she had motor problems, and she lost control of her bladder. She developed symptoms of POTS; her heart raced and her blood pressure dropped when she stood up. She described to me sensations of internal vibrations and electric shocks. Since she is a beautiful woman who puts on a brave face, one cannot always notice her discomfort. Despite all the neurologic symptoms, she was diagnosed with anxiety.

Brianne spearheaded an effort to educate officials at the CDC, FDA, and NIH to research COVID injection side effects in hopes of developing treatment strategies for the vaccine-injured. I heard her speak eloquently at two congressional hearings convened by Senator Ron Johnson, who has demonstrated a commitment to those injured by the COVID shot rollout. Her pleas for help did not yield adequate responses from government agencies. Therefore, despite her significant disabilities, she cofounded React19, the first 501c(3) to focus on helping those injured by the "vaccines." She is amazing.

Peer-Reviewed Articles Document Disabilities

The list of side effects and significant adverse events from COVID injections is daunting. Pfizer compiled a list of adverse events that were anticipated. There are more than thirty-five thousand articles in the medical literature documenting the cases of unfortunate "vaccine" recipients who developed life-changing symptoms associated with COVID injections. Here is a sampling of wide-ranging adverse events—again—drawn from over thirty-five thousand publications:

- Abicic, A., et al. (2022). "New-Onset Ocular Myasthenia Gravis after Booster Dose of COVID-19 Vaccine." *Cureus* 14(7): e27213.

- Asaduzzaman, M., et al. (2022). "COVID-19 mRNA Vaccine-Associated Encephalopathy, Myocarditis, and Thrombocytopenia with Excellent Response to Methylprednisolone: A Case Report." *J Neuroimmunol* 368: 577883.
- Bidari, A., et al. (2023). "Immune Thrombocytopenic Purpura Secondary to COVID-19 Vaccination: Systematic Review." *Eur J Haematol* 110(4): 335-353
- Brito, S., et al. (2021). "A Case of Autoimmune Hemolytic Anemia Following COVID-19 Messenger Ribonucleic Acid Vaccination." *Cureus* 13(5): e15035
- Gao, J. J., et al. (2021). "Acute Transverse Myelitis Following COVID-19 Vaccination." *Vaccines* (Basel) 9(9)
- Gonzalez-Enriquez, J. O. (2022). "[Bell's Palsy Secondary to COVID-19 Vaccine Pfizer: Case Report]." *Rev Med Inst Mex Seguro Soc* 60(2): 224-228
- Hermel, M., et al. (2022). "COVID-19 Vaccination Might Induce Postural Orthostatic Tachycardia Syndrome: A Case Report." *Vaccines* (Basel) 10(7)
- Erdem, N. S., et al. (2021). "Acute Transverse Myelitis after Inactivated COVID-19 Vaccine." *Ideggyogy Sz* 74(7-08): 273-276
- Filfilan, N. N., et al. (2023). "Effects of Different Types of COVID-19 Vaccines on Menstrual Cycles of Females of Reproductive Age Group (15–49): A Multinational Cross-Sectional Study." *Cureus* 15(5): e39640
- Kolahchi, Z., M. Khanmirzaei and A. Mowla (2022). "Acute Ischemic Stroke and Vaccine-Induced Immune Thrombotic Thrombocytopenia Post COVID-19 Vaccination; A Systematic Review." *J Neurol Sci* 439: 120327
- Salunkhe, M., et al. (2023). "Spectrum of Various CNS Inflammatory Demyelination Diseases following COVID-19 Vaccinations." *Acta Neurol Belg*
- Sirisuk, W., et al. (2023). "Incidence and Clinical Characteristics of Adverse Neurological Events and Stroke-like Syndrome Associated with Immune Stress-Related Response after

COVID-19 Vaccination in 2021 from Thailand." *Clin Neurol Neurosurg* 231: 107804

- Bolletta, E., et al. (2021). "Uveitis and Other Ocular Complications Following COVID-19 Vaccination." J Clin Med 10(24)
- Fraiman, J., et al. (2022). "Serious Adverse Events of Special Interest following mRNA COVID-19 Vaccination in Randomized Trials in Adults." *Vaccine* 40(40): 5798-5805
- Gulumsek, E., et al. (2023). "Minimal Change Nephrotic Syndrome with Acute Kidney Injury after the Administration of Pfizer-BioNTech COVID-19 Vaccine." *Case Rep Infect Dis* 2023: 5122228
- Maltezou, H. C., et al. (2023). "Anaphylaxis Rates following mRNA COVID-19 Vaccination in Children and Adolescents: Analysis of Data Reported to EudraVigilance." *Vaccine* 41(14): 2382-2386.
- Nakagawa, A., et al. (2023). "Acute Pulmonary Hypertension Due to Microthrombus Formation following COVID-19 Vaccination: A Case Report." *Eur Heart J Case Rep* 7(8): ytad353.

Human Stories Behind the Numbers

Heart-wrenching stories of "vaccine" injuries and deaths are recorded on the React19 website. It is a tough read, but it's important to know the range of damage the vaccines can do. The pharmaceutical industry is excited and mobilized to bring out new mRNA vaccines within one hundred days of the next viral outbreak if they can streamline the regulatory process. To prepare to make decisions for yourself and your children, you need to learn from the experiences of those who did not fare well during this "vaccine" rollout. When I asked Moderna executives about vaccine adverse events, they replied that their post-marketing data was not showing significant problems. They claimed they had not heard anything about the insurance industry reporting increases in deaths or disability.

The Children's Health Defense Bus:
Collecting Injury Histories

Polly Tommey, whose son Billy regressed after the MMR vaccine, traveled the roads of America for nearly a year. She and her crew interviewed people who were injured from COVID vaccines or had relatives who died in the hospital under standard treatment protocols. You can watch the videos and the movie *Vaxxed III* on the Children's Health Defense website.[633]

FOOD FOR THOUGHT

Do you think the Moderna reps were being honest when they told me they had not heard of the insurance statistics? Do you know anyone who developed a new chronic illness or disability that seems connected to their COVID injections?

CHAPTER 25

Where Have All the Athletes Gone?

It's not what you look at that matters, it's what you see.

—Henry David Thoreau

If you don't read the newspaper, you're uninformed.
If you read the newspaper, you're misinformed.

—Mark Twain

Cardiac Arrests in Athletes While Competing

The popularity of NFL football led to widespread awareness of twenty-four-year-old Damar Hamlin's cardiac arrest and dramatic resuscitation on live television on January 2, 2023, during a Buffalo Bills game. He had been tackled by a Cincinnati Bengals player and his official diagnosis was "commotio cordis," which can occur after a hard hit to the breastbone. This condition is relatively common in sports that use projectiles, like baseball or hockey, but is not typically seen in football.[634] Fortunately, Damar recovered. The drama raised speculation about whether his arrest could be a sign of a COVID vaccine adverse event. I do not know his vaccine status, nor did I review his medical records, so I should not speculate about his health. But I do want those of you who watched Hamlin's collapse or viewed other athletes collapsing on the court or field in the past few years to ask yourself these questions: Is this something that has always occurred, and we did not notice? Or could something new and horrible be happening?

Sports Video Montages: Concerning or Misleading?

There are several video montages of athletes collapsing suddenly on the soccer field or basketball court in 2021 and 2022. The independent web-based show *The Highwire* has raised concerns about athletes dying suddenly. They are careful to cite the limitations in their reporting when they do not know the vaccination status of the athletes they show collapsing, and their due diligence about removing videos from their montages when they find it is unlikely the player had received a COVID vaccine reassures me that we should pay attention to the questions they raise.[635,636]

The pushback against merely asking the question about possible associations with COVID shots was strong, with dozens of articles claiming that "deceptive videos" were being used unfairly to imply an association with COVID shots. As late as January 23, 2023, "experts also told Reuters Fact Check that there is still no evidence of an increase in deaths or serious cardiac events among athletes, nor evidence that known effects of the vaccines have led to the type of cardiac events seen in these players."[637] But that statement is hard to reconcile with the peer-reviewed scientific papers published in 2021 and 2022 indicating an increase in cardiac damage from COVID vaccines.

When I watched those videos, the patterns of collapse struck me as very odd. Most of the athletes did not clutch their chests or try to protect themselves when they fell. Instead, the typical incident involved an athlete on the playing field suddenly face planting on the ground or falling backward. As a doctor, I was very curious about what had happened so suddenly that their basic protective responses did not have time to kick in. Yes, I know that some athletes faint on the field because they are dehydrated or have vaso-vagal syncope. Yes, I know some athletes die because they have cardiac arrhythmias that were undetected previously. These cases looked different to me.

Looking at the Evidence

One cardiologist who was very intrigued by this phenomenon is the widely published Dr. Peter McCullough, whom you met in the myocarditis chapter. Shortly after Dr. McCullough started saying that a variable new to 2021 and 2022 was that most athletes had received a novel genetic product in the name of vaccination, he was soundly attacked by "fact checkers" who proclaimed there had been no increase in sudden cardiac deaths in athletes.[638,639]

You may recall my own grudge against fact checkers, who declared my fifteen-year clinical research project fake news. Since I know what I found and stand by my results, I am predisposed to be skeptical of what non-medical, non-scientist fact checkers say. When I looked at Dr. McCullough's experience as a cardiologist to see what he might know about cardiac events in athletes, I found he'd published numerous papers on the subject prior to COVID.[640,641,642,643,644]

Dr. McCullough used his hypothesis that the widespread deployment of COVID vaccines (which had been shown to be associated with myocarditis and pericarditis by the time of Damar Hamlin's cardiac arrest) to argue for "rational harm-benefit assessments by age group" when making decisions about COVID-19 vaccines. Absolutely reasonable request in my opinion. So why such a backlash? His broad and deep understanding of cardiology, especially in the context of young athletes, led him to worry that adverse events that were not reported in the original trials might be killing young people. Dr. McCullough suggests that if we have reason to suspect COVID vaccines pose greater risks than benefits for some groups, like young athletes, we should honor the precautionary principle:

An independent secondary analysis of serious adverse events reported in phase 3 clinical trials of Pfizer and Moderna, found that the mRNA vaccines combined were associated with an excess risk of serious adverse events of 1 per 800 vaccinated individuals . . . Nevertheless, indiscriminate COVID-19 vaccination has been expanded to include age groups and naturally immune with minimal chance of suffering major complications due to COVID-19. In these groups COVID-19 vaccination is not clinically indicated nor medically necessary.[645]

He references the paper on adverse events of "special interest" led by Peter Doshi that I mentioned earlier.[646]

Dr. McCullough and I share the same concern that "one size fits all" strategies have led to giving COVID-19 "vaccines" to age groups for whom the virus poses little risk and people with natural immunity who have minimal risk of suffering badly with the illness. When I went to medical school, it was considered unethical to give a medical intervention to someone who was unlikely to benefit, especially if there were known harms to the medication or procedure. And there are always unknown harms to consider in every medical intervention.

Risks, Benefits, and Harms Vary by Age and Health Status

One large-scale risk-benefit analysis concluded that between 31,207 and 42,836 young adults aged eighteen to twenty-nine years would need to receive a third mRNA vaccine dose to prevent one COVID-19 hospitalization over a course of six months."[647] Let those numbers sink in. Using the lower range numbers, more than thirty thousand young adults would need to take a vaccine that has a *one in eight hundred* risk of a serious side effect to prevent *one* hospitalization. That means nearly forty *healthy* people could experience a serious or long-lasting side effect in order to prevent *one* hospitalization. That's an objectively terrible risk-to-benefit ratio.

Kevin Bardosh and colleagues estimate (using a much more conservative rate for serious adverse events gleaned from Pfizer's small clinical trial) that at least 18.5 serious adverse events could occur for every COVID-19 hospitalization prevented.[648] Do you think that is an acceptable tradeoff? According to the V-safe app, seven percent of COVID vaccine recipients visited an emergency department or doctor's office soon after their COVID shots. Applying that rate to the Bardosh study, to prevent one hospitalization from COVID as many as 2,590 people would end up seeking medical help. Consider the large number of people who must get shots to prevent hospitalizations or deaths and the now-documented serious side effects.

Historical Trends and Recent Sudden Deaths

Swiss researchers published a paper in 2006 researching the thirty-eight-year timespan from 1966 to 2004 asserting that globally 1,101 athletes under age thirty-five died due to heart conditions. That works out to a baseline of twenty-nine per year. The authors report that 50 percent had cardiomyopathies or congenital (from birth) anatomical changes in their hearts. About 10 percent had early onset atherosclerotic heart disease, the typical heart condition that leads to myocardial infarctions (heart attacks) as we get older.[649] We know that number is much higher for 2021 and 2022. "A small team of investigators, news editors, journalists, and truth seekers," began collecting reports of sudden deaths in athletes. They determined that *more than 1,650 professional and amateur athletes collapsed due to cardiac events in 2021 and 2022; 1,148 died.*[650,651,652] *Over one thousand died!*

Are There New Heart Problems in Athletes?

One study that horrified me was published in 2022. Thirty percent (OK, 29.24 percent) of teens between thirteen and eighteen who had received two doses of the Pfizer injection reported heart symptoms like tachycardia (fast heart rate) or palpitations (sensation the heart is fluttering or beating irregularly). In this research a horrific 2.33 percent got myopericarditis. (heart inflammation).[653] That is roughly one in fifty teens!

Myocarditis is often caused by a viral infection, so one could argue that during 2021 and 2022 there was an uptick in virally induced myocarditis (inflammation of the heart) or pericarditis (inflammation of the heart sac). If that were the case, we would not expect the unvaccinated to be spared. Yet according to this study, no statistically significant increase in myocarditis or pericarditis was seen in unvaccinated people after SARS-CoV-2.[654] Or perhaps the increase in heart inflammation was due to COVID-19 itself? This is a reasonable hypothesis and one that is quoted as a concern by those advocating COVID injections for children and teens. But data from twenty-seven countries in Europe show increasing deaths in children and teens *after* vaccines, not during the initial COVID waves.[655]

Fact checkers insist there is no increase in deaths for athletes in 2021 and 2022. It's easy to find those posts online. When I read them, I note they often quote or paraphrase unidentified experts. When experts are identified and I follow the source cited, I find it is often prose written for a website and does not include references. I am just as worried as Dr. John Ioannidis that much of what is published in the medical literature does not stand up to full scientific scrutiny, but these posts don't even cite bad science. I am very worried that corporate capture of web-based media makes much of what so-called fact checkers say unreliable. I like to see the raw data and look at the charts for myself.

Team Vaccination Rates Could Be a Clue

To be clear, assuming all sudden cardiac deaths in athletes are due to COVID "vaccines" would be irresponsible and counterproductive. There is a baseline of cardiac deaths in athletes, which is one reason children and teens must get sports physicals to be cleared to play. Physicians do a history and exam to detect risk factors worrisome for sudden cardiac death. We may order EKGs (heart tracings) or cardiac echocardiograms (ultrasound images of the heart) to be safe. We usually do not know the COVID injection status of the athletes we watch perform, but we have been told some of the vaccination percentages in various sports. NBC sports even posted some percentages with the editorial comment "may the most vaccinated team win."[656]

NBC reported in September 2021 that the Commanders in DC have a 90 percent COVID vaccination rate. Overall, the NFL's vaccination rate in fall 2021 was reported to be 93 percent (much higher than the general population).[657] Major league baseball and hockey have rates around 85 percent. Major league soccer rates are reportedly 95 percent. The WNBA has the highest reported rate at 99 percent.[658] I'm a huge basketball fan, and I've been admiring their point guard Steph Curry's dedication and athleticism since he was at Davidson College in North Carolina. The Golden State Warriors, like some other teams, had not announced vaccination rates by fall of 2021, but said the unvaccinated would need special authorization to enter Chase Center.[659]

Cause Unknown and Lack of Scrutiny

Ed Dowd starts *Cause Unknown* with a chapter about sudden deaths in athletes. He quotes prominent sports figures saying that in their long careers, they did not see sudden cardiac arrests at anything approaching the scale of the last few years. He suggests the reader do their own Google search, using any combination of search items related to sudden deaths in athletes and compare what you find pre- and post-COVID injections. Compared to baseline over decades, the past few years are worse by orders of magnitude.

Conclusion

My research on COVID over the last four years has convinced me there are plausible physiological mechanisms behind the increase in deaths in fit young people who received COVID injections and died during a sports event, but those of us calling for an unbiased investigation into this senseless loss of life are infuriatingly gaslighted. It makes me so sad to think of all those young people, who spent so much time and effort getting into excellent physical condition, falling victim to a human intervention that was misrepresented from the beginning. I am angry that the vaccine profiteers profess ignorance of the serious side effects from the products they made.

FOOD FOR THOUGHT

Do you think more athletes and young people have died in their sleep or on the field than you remember from previous years? Do you have the gumption to look into this phenomenon even if it makes you angry and sad?

PART IV
LESSONS FROM THE PANDEMIC

CHAPTER 26

What Is the Cell Danger Response and Why Do Clinicians Need to Know About It?

Rarely do we find men who willingly engage in hard, solid thinking. There is an almost universal quest for easy answers and half-baked solutions. Nothing pains some people more than having to think.

—Martin Luther King Jr.

If you learn to use adversity right, it will buy you a ticket to a place you could not have gone any other way.

—Tony Bennett, UVA basketball coach

Danger on a Cellular Level

Many times, when reading about people in the hospital who were getting extremely sick and dying from COVID, I thought "all those treatments like remdesivir and ventilators are locking patients into cell danger responses and preventing true healing." Remdesivir has the unlucky moniker of "run, death is near" due to its truly horrific side-effect profile, including a high prevalence of kidney failure.[660,661]

In studies for Ebola, which has a 50 percent mortality rate in most series, remdesivir was dropped from the trials since it led to more deaths than Ebola itself [662]—and that is saying something horrific!

Early in the spring of 2020, ICU doctors I respect noticed that ventilators often made patients worse; they adopted prone positioning, which seemed less harmful and more therapeutic.[663] Could it be that simple basic interventions make a big difference? It made me think of Dr. Bob Naviaux's lectures on his complicated research excavating the innermost workings of cells in earthworms that were stressed. *Earthworms under Duress*—sounds like an interesting book title or sci-fi movie, I think.

Merging Academic Expertise with Clinical Experience

Many years ago, I was shocked when Bob wanted to visit my practice to see how I take care of patients. Bob Naviaux is a world-renowned expert in mitochondrial genetics. I am a general pediatrician trying to figure out how to help patients recover in the midst of a tsunami of chronic illness, neurodevelopmental disorders, and now long COVID and COVID vaccine injuries. What could such an eminent academic scientist learn from me? As it turns out, Bob is an extraordinarily curious medical scientist who respects and learns from a clinician's perspective. In my view, this makes his work even more valuable and translatable to medical practice.

Bob and I met at one of many Autism Research Institute Think Tanks I have attended since 2004. ARI Think Tanks feature about thirty people from around the globe meeting once or twice a year (except during the years of SARS-CoV-2—another unfortunate consequence of lockdowns). My colleagues and I look forward to the ARI Think Tank as one of the most intellectually interesting, collaborative, and rewarding professional events of the year. I cannot overstate how exciting it is to meet with clinicians, researchers, and parents of sick children to generate collective wisdom about the causes and potential treatments for the epidemic of neurodevelopmental disorders that has swept over us in the past few decades. This synergy between research scientists and clinicians leads to scientific breakthroughs that would not happen in settings where bench scientists only talk to fellow lab scientists and clinicians only talk with other practitioners. ARI's research agenda was often shaped by parents who had become experts in the medical problems of their children with autism and shared their extraordinary insights.

Chronic Cell Danger Response and Chronic Illnesses

Dr. Naviaux's excellent research about the cell danger response connects environmental factors with the function of our mitochondria and the rising tide of chronic illness. It changes the entire paradigm of medical science and opens opportunities for a new approach to chronic disease that focuses on causes and developing effective strategies for healing.[664] It speaks to me as a student of integrative and functional medicine in a way that traditional allopathic medicine does not.

The Need for New Paradigms in Medicine

Bob Naviaux writes that the *First Book of Medicine* contains medical knowledge from the past five thousand years. During this timeframe, doctors learned how to treat acute illnesses caused by injury, infections, and poisonings. He proposes we need the *Second Book of Medicine* to focus on chronic diseases that last longer than six months. He surmises that an entirely new class of treatments will be needed to help doctors adjust the setpoint of their patients' cell danger responses so they are no longer hypersensitive to environmental triggers that perpetuate their chronic illnesses.[665,666]

Hang in there as we explore the rich treasures of Cell Danger Response biology. The cell danger response (CDR) is a universal response to environmental threats or injuries. As Dr. Naviaux explained to me, after the cell danger response is triggered, healing is incomplete until choreographed stages of mitochondrial response return the cell to a state of readiness. This cellular response can change development, fertility, resilience, and susceptibility to chronic illness. OMG! Do you recognize the incredible implications of this statement? Dr. Naviaux brings his experience as an evolutionary biologist, pediatric geneticist, and molecular biologist to gain incredible insights into the fundamental nature of biologic functions across species. Yes, he is one of the most brilliant people I know; I hope he gets a Nobel prize for his work.

The Nitty Gritty Science

One of the things I enjoy most is learning about complex concepts and teaching them in ways that are easy to understand and, on a good day, enjoyable. The mechanisms of the cell danger response are vitally important for all people on earth. I am thrilled to take on the challenge of translating this exquisite biologic phenomenon into a language everyone can understand. With luck, it can even be fun.

Mitochondria are the powerhouses of the cell. Long ago, a bacterium was incorporated into human cells and evolved into mitochondria. The seminal work establishing this science was done by the brilliant evolutionary biologist Lynn Margulis,[667,668,669] who was the mother of Jennifer Margulis, Paul Thomas's coauthor of *The Vaccine-Friendly Plan* and a friend of mine. A primary function of mitochondria is to make cellular energy in the form of ATP (adenosine triphosphate). Mitochondria are fundamentally responsible for cellular biochemistry.

Mitochondria Are Persnickety

For organelles that are so crucial to maintaining cellular function and therefore life, mitochondria are quite persnickety. They operate on the Goldilocks principle—not too cold, not too hot. Everything must be just right. For example, if the cell is too salty or not salty enough, the cell danger response can be triggered. If the cell is a little bit too acidic or is not acidic enough, the CDR can be activated. Lockdown! Guard the perimeter! Kill the enemy! Mitochondria are extremely sensitive to a wide variety of environmental toxins. Viruses, bacteria, fungi, and parasites can all damage mitochondrial function.[670,671] Long COVID is associated with mitochondrial dysregulation and oxidative stress.[672] Heavy metals such as aluminum, mercury, and lead can be disastrous for mitochondrial function.[673,674,675] Many vaccines have contained or currently contain aluminum and/or mercury, but the act of vaccination itself has effects on the mitochondria.[676]

As Dr. Naviaux explains, "Chemical, physical, and microbial changes that surround all multicellular life on Earth are translated into changes in mitochondrial structure and function. These changes in mitochondria are used to signal safety or danger in the cell. . . ."[677] Such changes can alter the expression of our genes. For example, if you are a newborn baby, you want to turn on genes for growth and development; if you are a woman who has breast cancer, you want to turn those genes off.

Environmental changes include toxins, pollution, vaccinations, infections, and heavy metals—all of which can trigger the cell danger response. The environmental chemicals that are new to nature have overwhelmed the ability of our mitochondria to adapt. More and more people, with a wide variety of chronic illnesses, are living their lives with mitochondria trapped in cell defense mode.

Take Away What Harms; Give What Heals

One of my most valued mentors, Sidney Baker MD, encapsulates a basic approach to complex chronic illness as follows: "Take away what harms, and give what heals." Cleaning up our environment so that the individual's toxic load diminishes, especially for children and the elderly, is a crucial part of that valuable strategy. Otherwise, we will never break out of the vicious cycle of the CDR stuck in defense mode and unable to carry out critical jobs. The staff and supporters at Children's Health Defense are working hard to roll back some of the damage that has been inflicted on our planet since the industrial age began. When COVID vaccines are added to the childhood vaccination schedule for babies down to six months of age, the adults in the room need to make sure we are not giving "what harms" to this generation of children. Hence, my passion for writing this book.

Giving what heals includes adopting lifestyle changes such as better nutrition, intermittent fasting, exercise, and gratitude practices that can activate the cell healing response. The Institute for Functional Medicine has been teaching strategies for the *Second Book of Medicine* for several decades. Cleveland Clinic clinical case management trials comparing a functional medicine approach to standard of care (which is higher at the Cleveland clinic than some other medical facilities) found that the functional medicine approach was superior for complex chronic illness management.[678] The Cleveland Clinic also did an important study during COVID that showed no benefit of a second shot for those with natural immunity.[679]

Originally, the term cell danger response included all aspects of an organism's response to stress, including epigenetics, immunity, metabolism, inflammation, and changes in the microbiome. When an environmental stressor activates the cell danger response, multicellular organisms shift their priority away from growth and development to a defensive posture that optimizes survival.

Sleep When You're Sick

Sickness behavior is one example of the cell danger response playing defense. Most people have personal experience with so-called "sickness behaviors." When we are fighting a virus, our instinct is to sleep more, disengage from our families, and withdraw from creative or energy-requiring activities. Evolutionary biologists view sickness behavior as a way to protect your tribe. This is when it helps your neighbors if you "lock down." Lockdowns do not protect the tribe when they remove healthy people from interacting with one another, as happened during COVID. In fact, isolating *healthy* people suppresses the immune system and causes mental health problems.

At a cellular level, the CDR cannot be deactivated until the cell receives an "all is well—crisis over" signal. If the cell does not get the message that it is OK to go back to work making ATP and promoting growth and development, the cell gets stuck in a repeating loop that blocks further healing. I fear this is happening to thousands of people who have long COVID or bad vaccine reactions, who may not be able to find the help they need to recover. I am furious that they are so often ignored and gaslighted.

Medieval Forts

Dr. Naviaux uses the analogy of mitochondria functioning like a fort in medieval times. When the peasants revolted and charged the castle, the royals inside pulled up the bridge over the moat and staffed the castle walls with warriors bearing bows and arrows. Non-urgent activities like feeding, farming, and f— procreating—were put on hold so the castle could be defended against invasion. Similarly, when mitochondria sense an invading organism or pollutant, they stiffen their cell's membranes, squirt out bleach to kill the invading organisms, and send messages to the neighboring cells warning of the attack.

Learning Complex Science Through Basketball

I like to use college basketball analogies to explain complex medical concepts. Like many middle-aged women in Virginia (I use that term even though technically I am "elderly"), I have a crush on Tony Bennett, the coach of the National Championship–winning University of Virginia basketball team. Bennett's coaching strategy relied on a "suffocating defense." It is hard to score against UVA; many years the average point tally by opposing teams is in the fifties. However, Bennett knows his team can't win a basketball game without scoring points, so they must be skilled at offense also. Similarly, the cell, and therefore the creature that contains it, cannot "win" (in this case, heal) if it stays locked in a defensive posture by an ongoing cell defense response. To score points—to generate energy, maintain normal metabolism, and turn the right genes on and off—the cell *needs* to play offense. Just like Kyle Guy shooting three free throws back-to-back in the last three seconds against Auburn to win a championship game, the mitochondria are capable of performing seeming miracles when they are not locked in defense mode 24/7.

Mitochondrial Stressors in the Future

Human mitochondria have had to work harder and harder since the Industrial Revolution began in the 1700s. Since World War II, the rise of large-scale industries and the change from homestead farming to industrial farming has presented ever-increasing challenges to mitochondrial function. Children are analogous to canaries in coal mines. Children's health is likely to be the first impacted by environmental changes and accumulating chemical exposures. Receiving new-to-nature, synthetic, and inadequately tested genetic therapies that have no long-term safety studies and concerning evidence of harm could potentially be disastrous for this generation of children.

Currently, ever-increasing exposures to electromagnetic radiation, ubiquitous chemicals that are new to nature, and industrial pollution combined with decreasing exposure to the healing powers of nature contribute to chronic illnesses that allopathic medicine is ill-equipped to manage. Add the unprecedented stressors associated with novel mRNA technology, and our cells are very likely to feel overwhelmed, attacked, and surprised. One cell walks into a bar and says to another, "Hey, I've never seen this before. What the hell should we do?" The other cell says, "We do what my daddy and grandpa always told me—protect yourself." In response to what humankind did at Warp Speed with suboptimal analysis, our cells have millennia of evolutionary adaptation shaping their response based on prior experience. In response to humans with hubris messing around with the very essence of our genetics, a cell's got to do what a cell's got to do. In response to inadequate management by regulatory agencies, cells go into defense mode and stop doing the good things we depend on them to do. Unless we restore a nurturing environment, many people with chronic illnesses *will not heal.* Chronic environmental stressors will keep humans and animals on our planet in a state of perpetual cell defense. This concept goes way beyond COVID and its aftermath.

A Call for Action

The mapping of the human genome opened the possibility of using precision medicine and the individual constellation of a person's genomics to identify areas of weakness that can be compensated for and strengths that can be capitalized on in the journey towards healing. That is one reason I resist a "one size fits all" solution like giving the same vaccine to seven billion people without taking their individual risk factors into account. In the model of personalized care, functional and integrative medicine that utilizes lifestyle modifications such as nutrition, exercise, stress management, emotional resilience, and time in nature is much more effective than traditional medicine, which seems locked into a model of "a pill for every ill," targeting symptoms rather than underlying causes.

Long COVID, the Cell Danger Response, and How to Heal

So, to recap a complicated concept, how does Bob Naviaux's innovative research illuminate insights about COVID and COVID vaccines? My hypothesis is that those who get very sick with COVID, develop long COVID, or have bad reactions to COVID injections may be locked in cell defense mode. These folks may need a combination of healing therapies to get back to playing offense, which for the cell means making proteins and generating energy to carry out healthy growth and development. These people deserve our compassion and support. Medical schools should move away from teaching pharma-sponsored "pill for every ill" principles and teach young clinicians how to promote healing on a cellular level.

FOOD FOR THOUGHT

Have you heard about functional medicine before? Does the phrase "integrative medicine" make you think "not real medicine," or do you like the concept? Acknowledging that many health-care workers tried hard to save lives, have you been happy with the way mainstream medical institutions performed during the COVID crisis?

CHAPTER 27

What Did We Learn from FOIA? Oy!

If people let the government decide what foods they eat and what medicines they take, their bodies will soon be in as sorry a state as are the souls who live under tyranny.

—Thomas Jefferson

The truth will set you free, but first it will piss you off!

—Gloria Steinem

Freedom of Information Act

According to the Oxford dictionary, *oy* is derived from the British informal exclamation *oi*, which means to attract someone's attention, especially in a rough or angry way. It's also a Yiddish exclamation (often "Oy vey!") used to convey that someone is upset, shocked, disappointed, or worried. That describes my reaction to the various information brought to light after conscientious individuals used Freedom of Information Act requests (FOIAs).

Full Transparency Versus Waiting Seventy-Five Years for the Reveal

Dr. Peter Doshi, an editor at the *BMJ* and another person I have admired throughout COVID, argued that we need full transparency about the information from the Pfizer mRNA genetic modification products. Since Pfizer's COVID vaccines were developed "at the speed of science" and seemed to be a frontrunner when "shots in every arm" became a stated goal of public health authorities, it seemed prudent to know the results of the clinical trials in detail. Science by press release is becoming more common, but sticklers for detail like to see the nitty gritty data.

When Pfizer and the FDA received FOIA requests for clinical trial information, their first response was to request the information be hidden for seventy-five years.[680] In other words, shine no light on the data until most of the people asking for it are long dead. After delays and legal wranglings, a judge ruled that the data had to be released at a rate of fifty-five thousand pages a month. In another request for data on the trial which enrolled twelve- to fifteen-year-olds, in which the now paralyzed Maddie de Garay took part, the FDA and Pfizer requested 23.5 years to release the data. Leading off with "democracy dies behind closed doors," a judge ordered that the files be released at a rate of 180,000 pages a month.[681] The information being revealed is shocking! A good place to learn more is on the DailyClout website.[682]

What Did FOIA Reveal About V-Safe?

V-safe is the smartphone app that you learned about in part two that was designed and released to track health impact after COVID injections. Users were asked to report their status in real time, initially daily and then spaced out as time marched on. This app is one of the reasons the public was convinced that COVID jabs were monitored by the most intense safety monitoring efforts in history. However, collecting data alone does not guarantee safety; one must analyze the data and react to warning signals.

Aaron Siri, a lawyer with the firm Siri and Glimstad, works on behalf of the Informed Consent Action Network (ICAN). ICAN exists to protect the population's right to true informed consent consistent with the principles embodied in the Nuremburg Code. As you learned in part two of this book, on behalf of ICAN, Siri's firm requested all de-identified data sent to V-safe since January 1, 2020, shortly after the vaccine rollout began. No health agencies are allowed to release any data that includes identifying information, so it was appropriate to ask for the de-identified data. Siri (the lawyer, not the app on your phone) knew the CDC had already de-identified the data because their protocol revealed they had already provided Oracle access to the de-identified data. Oracle is the largest database management company in the world. Larry Ellison, the CEO, earns 700 million USD per month; his net worth was 108 billion dollars as of January 2024.[683]

Siri's first response from the CDC was "a search of [CDC's] records failed to reveal any documents pertaining to your request."[684] One must wonder: was that response a result of incompetence on the part of the person answering FOIA requests, or was it part of a strategy to delay reporting the data? Mr. Siri appealed the response and pointed out that since Oracle had the data, responsive documents did indeed exist. But, despite many phone calls to the Department of Justice (which supplies lawyers for the CDC) and many time-consuming court filings, the DOJ maintained the CDC's position that there was no de-identified data—nothing to see here, keep moving along!

According to Mr. Siri, these CDC claims were categorically false. He says the CDC could have supplied the requested data within minutes by "simply downloading the five files it eventually provided."[685] You can read Mr. Siri's eight-part series on the V-safe data on his Substack column. After lots of expensive legal maneuvering and paperwork, the CDC and DOJ finally capitulated. The court ruled under Federal Rule of Civil Procedure that the CDC must produce the requested documents on or before September 30, 2022, which is the date the data was released, over 440 days after the original, valid request. It took multiple legal demands and the filing of two federal lawsuits to get the data we are entitled to. Even then, only the check-the-box V-safe data were released; the free-form, fill-in-the-blank data was not turned over until 2024.

High Rates of Vaccine Recipients Seeking Medical Care for Side Effects

Nine of the ten million initial V-safe users signed up between December 2020 and April 2021. They were the "early adopters" and probably did not expect significant adverse health effects from the shots. If anything, these shot recipients, which included vaccine enthusiasts and health-care workers, might be inclined to under-report their symptoms. V-safe data revealed why the CDC might have been reluctant to release it: 7.7 percent of the approximately 10 million users reported having to receive medical care after receipt of a COVID injection.[686]

Let's analyze the V-safe process in more detail. After a person registers, they are asked to complete a health check on the day of the injection. They are asked to answer questions, mostly from pre-populated choices. There are a limited number of free text fields. Every day for the first week, users are prompted to submit a "check-in." Then every week for six weeks, users are prompted to submit more check-ins. Then, at six months and one year, the prompts repeat. V-safe collects limited, pre-selected information systematically. The following symptoms are included in the check boxes during the first week: chills, headache, joint pain, muscle or body aches, fatigue, nausea, vomiting, diarrhea, abdominal pain, and rash. V-safe users also can pick from three health impact choices: unable to perform normal daily activities, missed work or school, and needed medical care. If "needed medical care" is selected, the user is asked whether hospitalization, emergency room, urgent care, or telehealth was used.

As I examined the symptoms asked for from V-safe users, I wondered why certain symptoms like chest pain, shortness of breath, neurologic changes, seizures, or bleeding problems were not options. In the summer and fall before the rollout, the medical literature already included those serious potential side effects.

- In a CDC presentation dated October 30, 2020, titled "*CDC post-authorization/post-licensure safety monitoring of COVID-19 vaccines*,"[687] a preliminary "list of VSD pre-specified outcomes for RCA [rapid cycle analysis]" and "list of VAERS AEs [adverse events] of special interest" both included *acute myocardial infarction, anaphylaxis, convulsions/seizures, encephalitis, Guillain-Barré syndrome, immune thrombocytopenia, MIS-C, myocarditis/pericarditis, and transverse myelitis,* among others.[688]
- A preliminary report in *The New England Journal of Medicine* highlighted thirty-five adverse events that were related to the

mRNA vaccination, including eye disorders, gastrointestinal disorders, musculoskeletal and connective tissue disorders, *and* nervous system disorders. It was published in July 2020, before the vaccines were rolled out.[689]

- An October 2020 *JAMA* article said that likely adverse events included *"allergic, inflammatory, and immune-mediated reactions, such as anaphylaxis, Guillain-Barré syndrome, transverse myelitis, myocarditis/pericarditis, vaccine-associated enhanced respiratory disease, and multisystem inflammatory syndrome in children."*[690]

Aaron Siri reports in his Substack series, "the CDC's own protocol for V-safe, at least as early as January 28, 2021, . . . identified "Adverse Events of Special Interest," which were serious conditions that should be tracked after COVID vaccination. The list included myocarditis, pericarditis, acute myocardial infarction, stroke, GBS, transverse myelitis, and other serious illnesses. "Yet v-safe was launched *without* including any check-the-box fields for these conditions and v-safe was *never* subsequently updated to include any check-the-box fields for these conditions."[691] If the CDC had included these conditions in the check-box options, they would have been easy to track and report. There would be a clear denominator with which to calculate the rate of occurrence of significant side effects, just as it is easy to determine what percentage required medical care. Out of 10,108,273 users, 782,913 sought medical care at least once after the jabs. That works out to 7.7 percent using sixth-grade math.

If you are so inclined, you can search through the V-safe data through the use of ICAN's data mining dashboard.[692] Anywhere from 1 to 3 percent of V-safe users (depending on factors like age, brand, and dose) needed medical care **in the first week alone.** While 782,913 people sought medical care, there were 2,108,022 instances of medical care. Therefore, each medical-care seeker had an average of two to three visits, implying their condition was not trivial or transient. Moreover, 70 percent of those seeking care went to either urgent care centers, an emergency room, or the hospital. Only 30 percent chose to call a doctor, visit an outpatient office, or make a telehealth appointment. This suggests their symptoms were perceived by the sufferer as serious. Twenty-five percent of V-safe users reported being unable to perform normal activities, including missing school or work. As a pediatrician who has seen thousands of children after immunizations, I can attest that this is a very high rate. I have to wonder: Was the system designed to assess the true safety of these injections or merely to show the usual minor, transient side effects?

The ten million V-safe users reported more than 71 million symptoms. Of these, 4.1 million were reported as severe, and 23.3 million symptoms were characterized as moderate, limiting normal daily activities. An average of over seven symptoms were reported per V-safe participant out of a total of only ten symptoms. In my experience, this is a much higher tally than typically occurs with earlier vaccines. There were 13,963 infants and toddlers under age three in the V-safe database. About thirty-three thousand symptoms were reported in those babies and children. Remember, symptoms of distress like prolonged or high-pitched crying, sleep disturbances, febrile seizures or poor feeding were not solicited.

FOIA Sheds Light on Real Risks of COVID in Healthy People

Paul Thacker is a true investigative reporter and former investigator for the U.S. Senate. I have learned a lot from his Substack, *The Disinformation Chronicle*.[693] Paul writes about true information that has been labeled as misinformation per government guidelines. He posted a newsletter by Rav Arora and Jay Bhattacharya, the Stanford MD, PhD professor you met in an earlier chapter. In response to a freedom of information request asking how many people in Israel died of COVID without underlying health conditions, the answer was zero. No deceased people between the ages of eighteen and forty-nine without underlying conditions.[694,695] Oy!

Remember that the prime minister of Israel made a deal with Pfizer to immunize the whole population of Israel, including young healthy people. Remember that, even as this book goes to press, the CDC is recommending everyone get COVID injections and emphasizing that healthy people can die from COVID. The childhood immunization schedule now includes "one or more doses of updated" COVID vaccine starting at six months of age.[696] Vaccine manufacturers would profit from a dose of "updated vaccine" as an annual recommendation just like flu vaccines.

Lack of Efficacy Data for COVID Boosters

The *New York Post* published a story indicating that the CDC was withholding COVID data from the public over "fears of misinterpretation."[697] Dr. Pierre Kory in testimony to the Maryland legislature said that when the CDC was asked under a FOIA request to provide "all data concerning or reflecting the efficacy of COVID-19 'booster shots' for people 12–49 years of age," their response was that a "search of our records failed to reveal any documents pertaining to your request."[698] Isn't that shocking and chilling? "While this might be written off as errors in judgment and legal compliance on the part of a FOIA office," Kory continued, "the consistency with which the CDC has failed to provide data, and that it could or would not identify studies in support of booster shots for those 12–49 years of age, means that policy makers are deprived of supportive data under which any mandate might be reasonable."[699]

Pregnant Women, Babies, and ACOG

The American College of Obstetricians and Gynecologists provides guidelines for the care of pregnant women in the United States. Its recommendations are utilized in other countries. Dr. James Thorp, whom you met in the chapter about reproductive harms, used the expert legal skills of his wife Maggie to submit an airtight Freedom of Information Act to the CDC, ACOG, and the Department of Health and Human Services. Thorp alleges that a Deputy Assistant Secretary for Public Affairs/Human Services at HHS was complicit in a $13-billion advertising campaign (some would say propaganda) to use social media influencers and faith leaders to promote COVID vaccines, even for pregnant women.[700] Remember, pregnant women were excluded from the COVID vaccine clinical trials so there was no data to support safety and efficacy during gestation.

By February 28, 2021, Pfizer recorded 1,223 deaths within the first ten weeks. The American College of Obstetrics and Gynecology received over $11 million to promote the COVID injection in pregnancy.[701] The Thorps collected 1,400 pages of correspondence between ACOG and HHS, more than 50 percent of which is redacted, as a result of their FOIA request.

ACOG is not allowed to deviate from the messaging promoted by HHS, under huge financial penalties.[702]

The Pfizer Papers and *Pfizer Documents Analysis Reports*[703] provide huge amounts of data that were not shared with the public. Government agencies promised transparency at the start of "vaccine" development but did not deliver on that vow. You will be shocked at what Pfizer knew and when they knew it!

FOOD FOR THOUGHT

Have you already heard the information revealed in this chapter? How much data should be deemed proprietary and withheld from public view for products people are asked to inject into their bodies? Were you surprised at the long and convoluted legal maneuvers that were necessary to compel Pfizer to share data from their trials? What lessons did we learn from these FOIA documents that could guide the way for policies for the next "crisis vaccine"?

CHAPTER 28

What Do You Mean –
Negative Efficacy?

The important thing is to not stop questioning.

—Albert Einstein

When morality comes up against profit, it is seldom that profit loses.

—Shirley Chisholm

Wishing and Hoping

Deborah Birx, MD (the infectious disease doctor who wore scarves during White House COVID briefings) and Rachel Walensky, MD (former CDC director) both alluded to their "hopes" that the novel COVID injections would work. "Wishin' and hopin' and dreamin'" might work well if you are Dusty Springfield singing Hal David's lyrics in a Burt Bacharach song, but it is not a good strategy for mandating novel injections into billions of people. What does the real-world data show now that these products have been injected into most of the world's humans?

Negative Efficacy and Original Antigenic Sin

Horrifically, real-world data suggests that mRNA vaccine benefit is quite short lived and not very good. Furthermore, with increasing doses a person actually becomes more likely to get COVID![704] You read that right. Researchers looked at participants for twenty-six weeks after COVID injections. Risk of COVID infection increased 17 percent with each injection![705]

Now we have numerous studies that show all-cause mortality is higher in the vaccinated than unvaccinated. Let's examine what that means. Let's assume for a moment that COVID injections protect you from dying from COVID disease. If those injections cause your death by heart attack or stroke, you are just as dead. I suspect you may find that your experience being dead is the same whether caused by a disease or a vaccine side effect.

One reason vaccines may lead to negative efficacy after a while is sometimes called "original antigenic sin."[706,707,708] In a letter to the *New England Journal of Medicine*, Dr. Paul Offit, a developer of the rotavirus vaccine, a frequent member of the Advisory Committee on Immunization Practices (ACIP) and a chair of the Vaccines and Related Biological Products Advisory Committee, stated his concerns about COVID vaccines causing immune suppression by that mechanism. You may remember that Dr. Offit stated that the only reason he voted "no" when the ACIP considered boosters for children is that "hell, no!" was not an option. Dr. Offit is famously pro-vaccine and has been a thorn in the side of parents who thought their child's autism had been triggered by certain vaccines or too many vaccines at once.

Lower Efficacy Standards for Children for COVID Shots

Pfizer shots in five to eleven-year-olds exhibited 31% efficacy initially, according to the CDC. Usually, a vaccine is required to demonstrate at least 50 % efficacy, but Peter Marks, MD PhD, the Director of the Center for Biologics Evaluation and Research (CBER) at the FDA, told reporters prior to the ACIP meeting that even if the vaccines did not meet the standard of 50% efficacy, "we will probably approve it anyway."[709] Why, Peter? Our children deserve the best, most efficacious products. Why cut corners when children have their whole lives ahead and the most to lose from side effects that turn out to be long lasting? A metanalysis published in *The Lancet* found two doses of COVID shots were only 41.6% effective against Omicron.[710]

The Need to Track Real-World Effectiveness in Subgroups

Hungerford published an article in the *BMJ* (*British Medical Journal*) in August 2021, arguing for tracking real-world effectiveness during COVID vaccine distribution. "Vaccine performance is highly context dependent, influenced by population risk of infection and disease, so assessment is required among multiple different subgroups. . . . The need to evaluate different covid-19 vaccines against multiple endpoints and variants in a range of subgroups means that effectiveness studies will be a staple of public health and academic workload for the foreseeable future."[711] He shares my perspective that decisions that may be right for the elderly or those with coexisting medical conditions are not appropriate for healthy children and adolescents.

Let's look at a study from Brazil in adults aged seventy or older. Effectiveness after two doses of CoronaVac was 42% against symptomatic COVID-19 of any severity. Two doses in seventy to seventy-four-year-olds were 80% effective against hospital admissions, but effectiveness declined with increasing age.[712] Not the 95% efficacy we were told to expect, but 42% against COVID symptoms and 80% against hospitalization seems like good odds for those around seventy years old. A study in Canada of two mRNA vaccines (BNT162b2 and mRNA-1273) showed one-dose efficacy of 41–63% and two-dose effectiveness of greater than 80%. The end point was symptomatic COVID-19 infection with either alpha, beta, or gamma strains.[713] Again, not 95% as promised, but still good for some age groups. Hungerford points out that since severe COVID-19 is rarer in children than adults and elderly, "it is imperative that rigorous long-term safety and effectiveness studies are conducted in children to quantify risk-benefit balance and inform vaccine policy and choices."[714]

Not as Effective as Claimed

Here is the bottom line: COVID injections were not nearly as effective as we were told. A test negative case control study in England examined adults aged seventy and above between December 2020 and February 2021. Cases of PCR-confirmed COVID-19 were linked to the National Immunization Management System. Participants aged eighty years and older vaccinated with BNT162b2 before 4 January 2021 had higher odds of testing positive for COVID-19 in the first nine days after vaccination (odds ratio up to 1.48)[715] than those vaccinated later. In those age 70–80 years, who received BNT162b2, vaccine effectiveness reached 61% (51% to 69%) from twenty-eight to thirty-four days after vaccination, then plateaued.[716] This seems consistent with what we know now about immunosenescence in the elderly and the decline in T-cell function in the first two weeks after COVID injections discussed earlier. They report vaccinated individuals aged seventy and older had a similar underlying risk of COVID-19 as the unvaccinated. One dose of BNT162b2 reduced risk of death by 37%–62% and hospitalization by 37%–62%.[717]

A study in Sweden was a register-based cohort study that looked at adults above eighteen years of age with COVID-19. Of nearly three hundred thousand vaccinated and unvaccinated individuals, the rate of COVID infection was 0.4% in the vaccinated and 1.4% in unvaccinated. Effectiveness was 21% after one dose and 59% after two doses.[718] Less than expected. Looking at rates of serious side effects from the vaccines in various age groups versus the risks of serious illness in that age group is important when interpreting the efficacy reports. Let's break down the science.

Natural Immunity

A paper published in October 2021 in *Morbidity and Mortality Weekly Report* (*MMWR*) suggested that "unvaccinated adults with prior infection, or natural immunity, were more likely to be hospitalized with COVID-19 than fully vaccinated people with no prior infection."[719] Catherine Bozio and colleagues wrote "All eligible persons should be vaccinated against COVID-19 as soon as possible, including unvaccinated persons previously infected with SARS-CoV-2."[720]

When I read the *MMWR* report, I noticed that there were seven limitations to the study listed, which introduces significant doubt about the conclusions. Zachary Stieber, an intrepid reporter for the *Epoch Times*, reported what was not disclosed in the *MMWR* report: "The raw data for the study 'is owned by external partner organizations and was maintained by a contractor.'"[721] The CDC's Freedom of Information Act (FOIA) office told a requester recently, "CDC subject matter experts did not receive copies of the raw data prior to the contract termination."[722] This means that a report that was repeatedly cited as a reason people who had already had COVID should get an experimental injection not only had seven limitations on interpretation of the data, but the raw data was *never actually analyzed by CDC employees.*

Vaccine Failure

Vaccines can fail if they do not generate a certain level of immunity in the recipients, if the induced immunity wanes quickly, or if the side effects outweigh the benefits. Let's see if COVID injections are currently meeting the mark for success or if they are failing as time goes on. We are indebted to Zachary Stieber for shedding light on the issue of breakthrough COVID infections in the vaccinated, based on data from the CDC.[723] Examining the data released in July 2023, he noted that more than five million COVID cases occurred in 2021 in people who were vaccinated.[724] At that time, government officials were still claiming that if you get the vaccine you won't get COVID or pass it to your neighbors.

In July 2021, President Biden said, "If you're vaccinated, you won't get COVID-19."[725] Dr. Fauci said in May 2021 that vaccinated people become "dead ends" for the virus.[726] He went on to say, "So even though there are breakthrough infections with vaccinated people, almost always the people are asymptomatic, and the level of virus is so low it makes it extremely unlikely—not impossible but very, very low likelihood—that they're going to transmit it."[727] CDC Director Rochelle Walensky stated that data showed vaccinated people do not carry the virus in March 2021.[728]

Breakthrough infections were noticed within the early days of COVID shot rollouts, but the scale of vaccine failure was higher than originally reported. By April 30, 2021, the CDC had reported 10,262 cases of COVID in the vaccinated. After that, the CDC transitioned to monitoring only those breakthrough infections that occurred in those who were hospitalized or died, leaving out numbers of people who handled their COVID infection at home, which is the vast majority.[729] *Epoch Times* reports, "Cases began spiking in July 2021, when more than 219,000 were recorded among the fully vaccinated. Across all vaccinated people, cumulative case counts eclipsed 500,000 by August 2021 and one million by the following month. The month with the most cases was December 2021. In that month, 2.7 million cases were identified among the fully vaccinated, and another 266,736 were recorded among the partially vaccinated."[730]

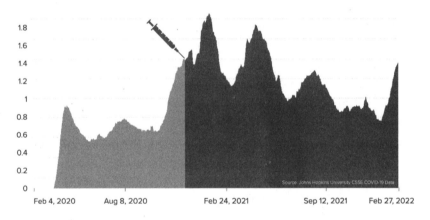

Daily New Confirmed COVID-19 Deaths per Million People
7-day rolling average. Due to varying protocols and challenges in the attribution of the cause of death, the number of confirmed deaths may not accurately represent the true number of deaths caused by COVID-19.

Chart courtesy of Ed Dowd, Cause Unknown: The Epidemic of Sudden Deaths in 2021 and 2022, *page 81.*

In the appendix of Dowd's book, he documents how deaths went up dramatically after the shots were rolled out in the following countries: South Korea, Thailand, Malasia, Uganda, Nepal, Portugal, Mongolia, Zambia, Paraguay, Bahrain, Uruguay, Tunisia, Sri Lanka, Afghanistan, Taiwan, Israel, and Vietnam.

Public Messaging Had to Change

Eventually, officials abandoned their prior reassurances about prevention of illness and transmission and began talking about how if you got COVID after being vaccinated, you would not get as sick. Here is my editorial opinion about the ethical use of messaging in matters of public health: The message should conform to the facts as best we know at the time. If you want to manipulate me through advertising that buying your clothes and cosmetics will make me look better, buying a shiny new car will make me feel more important, or getting your miracle supplement will help me lose weight, well, that is the American way of driving consumerism. But, in my opinion, distorting the facts to manipulate people to make health decisions is not ethical.

As the Virus Mutated, Shots Were Less Protective

The Cleveland Clinic analyzed all employee's data, vaccinated or not, and incidence of COVID-19 infection for twenty-six weeks after the bivalent COVID injections shots became available. Protection from the bivalent vaccine was analyzed as a time-dependent covariate using Cox proportional hazards regression. They looked at the effect on protection as the virus mutated to less virulent forms analyzing multiple variables. Among 51,017 employees, the estimated vaccine effectiveness was 29 percent during BA.4/5, 20 percent during BQ-, and 4 percent during XBB dominant phases. "The risk of COVID-19 also increased with time since the most recent prior COVID-19 episode and *with the number of vaccine doses previously received.*"[731] (emphasis added)

More Shots, More COVID Infections

The Cleveland Clinic looked at 39,766 employees to determine whether boosting previously infected or vaccinated individuals with injections designed for earlier variants protected against Omicron. They found boosting protected against Omicron if it had been six months or more since prior infection or vaccination. However, for the previously infected, two doses were associated with a higher risk of COVID infection compared to one dose. They concluded, "There is no advantage to administering more than 1 dose of vaccine to previously infected persons."[732]

A study in Iceland found similar results. "During an Omicron wave in Iceland, individuals who had previously received two or more doses were found to have a higher odds of reinfection than those who had received fewer than two doses of vaccine, in an unadjusted analysis.[733] A large study found, in an adjusted analysis, that those who had an Omicron variant infection after previously receiving three doses of vaccine had a higher risk of reinfection than those who had an Omicron variant infection after previously receiving two doses of vaccine.[734] Another study found, in multivariable analysis, that receipt of two or three doses of an mRNA vaccine following prior COVID-19 was associated with a higher risk of reinfection than receipt of a single dose.[735] Immune imprinting from prior exposure to different antigens in a prior vaccine[736,737] and class switch towards non-inflammatory spike-specific IgG4 antibodies after repeated SARS-CoV-2 mRNA vaccination[738] have been suggested as possible mechanisms by which vaccination after prior vaccination may provide less protection than expected. We still have a lot to learn about protection from COVID-19 vaccination; in addition to a vaccine's base effectiveness, it is important to examine whether multiple vaccine doses given over time is having a less beneficial effect than is generally assumed."[739] *COVID viruses change faster than we can make new vaccines to fit them.*

Since 2022, vaccinated individuals are more likely to get one of the COVID variants that continue to evolve. Did you know that vaccines being given in the first half of 2023 were constructed using the Wuhan strain of virus that has been extinct since 2020? COVID shots that were given in the first half of 2023 included the Wuhan strain and BA variants that were common in the summer of 2022. Booster doses recommended for fall of 2023 were for a strain of the virus that comprised less than 4 percent of the circulating virus in early September. Furthermore, from the data released by Pfizer to justify the shots, an experiment with ten mice in each of two groups was reported. The mice were injected twice, twenty-one days apart. There were no human studies reported. Then the booster was authorized for humans.[740]

Evolving Mutations Tend to Be Less Deadly

SARS-CoV-2 is a highly mutating virus. Dozens of variants have emerged since the start of the pandemic. Fortunately, in nature, subsequent variants of viruses tend to be less pathogenic—which means they are usually not as serious as previous strains. The original COVID jabs directed the body to make antibodies against the original spike protein. Since then, major mutations in the spike protein have occurred over time. Adaptive immunity—the type induced by vaccines utilizes antibodies to "remember" infections—is highly specific. Changes in the spike conformations can introduce an "Achilles heel" dynamic and lead to a "Whack-a-Mole" situation where we are chasing the latest strains, which may well outsmart or elude us. Nature and germs are smarter than humankind. Those of you who are as old as I am may remember the "it's not nice to fool Mother Nature" commercials. We could find ourselves in a game of playing catch-up, trying to outrun evolution of new variants with ever-changing vaccines.

The Perils of Being Primed

Omicron and its subvariants dominated beginning in January 2022. Omicron is phylogenetically different from Wuhan and Delta, meaning it is on a different branch of the virus family tree. Researchers from Stanford, one of the premier medical research institutions in the United States, found that "prior vaccination with Wuhan-Hu-1-like antigens followed by infection with Alpha or Delta variants gives rise to plasma antibody responses with apparent Wuhan-Hu-1-specific imprinting manifesting as relatively decreased responses to the variant virus epitopes compared with unvaccinated patients infected with those variant viruses."[741] In other words, your immune system does a better job of protecting you from new variants if you have not been "primed" to respond to an extinct version.

In a peer-reviewed paper, researchers found that, at the country level (and US county level), there seems to be no real relationship between the percentage of the population fully vaccinated and new COVID-19 cases.[742] In fact, mass vaccination policies may paradoxically lead to more cases. The evolutionary biologists who warned that vaccinating during a pandemic was likely to prolong the pandemic may have been on to something! In fact, the trend line in a *European Journal of Epidemiology* study suggests that countries with a higher percentage of people who received COVID injections have higher COVID cases per 1 million people.[743]

Conclusion

In any human endeavor in which people are subjected to a drug, vaccine, or intervention, it is crucial to examine data as it emerges over time. In my practice of pediatrics, I try to remember to tell my patients the following: "If what I recommend causes side effects or makes him worse, let me know. That can be a crucial clue that I am analyzing his case incorrectly, and we need to change paths. It will not hurt my feelings or offend me." Things get more complicated when big money and loss of face on an institutional level are involved. We are much less likely to hear, "Whoops, we got it wrong."

To put this in context for parents trying to make COVID vaccine decisions for their children, at this point in time, with the current evolution of variants, getting a COVID vaccine for your child seems likely to *increase* his or her risk of getting COVID. But remember, your child has probably already had COVID and has natural immunity anyway.

FOOD FOR THOUGHT

Did you lose faith in vaccines or boosters at some point? When? Have you heard of negative efficacy before? Do you think vaccines are our best weapon against COVID infections?

CHAPTER 29

What Can We Learn from Other Countries?

Too often we enjoy the comfort of opinion without the discomfort of thought.

—John F. Kennedy

In questions of science, the authority of a thousand is not worth the humble reasoning of a single individual.

—Galileo Galilei

The Swedish Choice

When the Swedes decided to avoid lockdowns (while the rest of the world was shutting down), they understood that transmission and infection depend on many social determinants. For example, the chance that an elderly person who lives alone, gardens as a hobby, and has groceries delivered would infect many people is quite low. The chance for transmission would be higher for a nurse working without adequate personal protective equipment in a hospital COVID unit who lives with eight other people in a 1,000-square-foot home with poor ventilation.

Countries tended to use four distinct pandemic response strategies: nudge, mandate, decree, and boost. Different responses to the same threat evolve from the cultural practices in each country. By March 2020, national governments had employed isolation, quarantine, social distancing, and various community containment strategies to combat viral transmission.[744] Oxford University created a stringency index to track and compare the scale and scope of government policy interventions across countries.[745]

The degree to which power and authority are centralized versus decentralized in a country influences strategy. Centralized states emphasize the authority of the national government, "making it easier to adopt and implement policies in a top-down fashion"[746] In decentralized governments, actual authority is given to local authorities to make policy decisions.[747] Decentralized countries provide recommendations but avoid mandates.

Sweden employed a "nudge" strategy early on. Their chief epidemiologist, Nils Anders Tegnell, served from 2013 until March 2022, but received harsh criticism from around the world for opposing travel restrictions, lockdowns, and mandates. However, when we look at his decisions in the context of Swedish culture, I think his thinking was sound. Swedish culture views the individual as an independent and autonomous entity with unique internal attributes; thus, Sweden encouraged individual flexibility and choice about risk-taking.[748] Sweden's COVID-19 response was designed to change behaviors without prohibiting options or imposing upon individuals' freedom of choice. Swedish gyms, schools, restaurants, and shops all remained open throughout the spread of the pandemic. There were no regulations restricting citizens' mobility. Instructions concerning individually targeted self-protection techniques were delivered daily from the Swedish Public Health Agency, state epidemiologists, and the Prime Minister.[749] The key here is the "sense of individuals' self-responsibility and high level of trust in Swedish society."[750]

Sweden's response to COVID-19 has received both praise and criticism, with American commentators heaping on more of the latter. Sweden's goals were to minimize morbidity and mortality in Swedish citizens, minimize negative health effects of containment measures, and safeguard societal services. Tegnell explained his strategies in the article below, starting with governmental structure. "Sweden has three levels of domestic government: national, regional, and local. The county councils are responsible for tasks such as health care, and they levy income taxes to cover their costs. The twenty-one regions of Sweden are autonomous when it comes to management of health care, including disease prevention and control. Each region has its own County Medical Officer who is responsible for communicable disease control in the region." [751]

Epidemiology of the Pandemic in Sweden

During the winter of 2019–2020 approximately one million Swedes traveled abroad.[752] Genome sequencing of SARS-CoV-2 early on showed three distinct genetic groups of viruses circulating in Sweden. Initially, most transmission chains from infected travelers were broken by the classic public health measures of testing for COVID, tracing contacts and asking people to self-quarantine. There was a peak of COVID illness in March and April, then a significant decline in gravely ill patients and new cases. In October, more cases emerged in young adults. Local governments could respond fluidly to introduce temporary restrictions when indicated by high numbers of cases or severity.

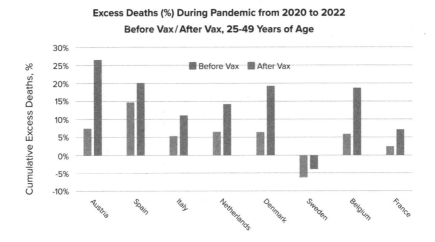

Excess Deaths (%) During Pandemic from 2020 to 2022
Before Vax / After Vax, 25-49 Years of Age

Chart courtesy of Ed Dowd, Cause Unknown, The Epidemic of Sudden Deaths in 2021 and 2022, *page 105.*

Without locking down society, Sweden suffered losses from COVID but the mean age of death was eighty-four years. Ninety percent of deaths in

March and April were in those more than seventy years old. Furthermore, during the spring peak, 50 percent of deaths in the elderly resided in nursing homes.[753] Compared to the US death rate of 341 per 100,000 cases, Sweden had 235, according to the Johns Hopkins Coronavirus Resource Center.[754] Seems like we should not have been throwing stones from our glass houses at the Swedes for their choices. Our per capita death rates were worse. We paid for our lockdowns with the economic ruin of small businesses, mental health crises in our population, and educational backsliding in our children.

How Did the Open School Policy Affect COVID in Children?

Jonas F. Ludvigsson, an MD, PhD scientist from the prestigious Karolinska Institute, wrote the following: "Despite Sweden's having kept schools and preschools open, we found a low incidence of severe Covid-19 among schoolchildren and children of preschool age during the SARS-CoV-2 pandemic. Among the 1.95 million children who were 1 to 16 years of age, 15 children had Covid-19, MIS-C, or both conditions and were admitted to an ICU, which is equal to 1 child in 130,000."[755]

Let's look at the data more closely. COVID cases peaked in spring 2020. Face masks were not encouraged or mandated. Social distancing was encouraged but not enforced. All children who were admitted to the intensive care unit with laboratory or clinically confirmed COVID from March 1st to June 30th, 2020 were followed. The number of deaths among nearly two million children from any cause in the four months prior to COVID (65) was not significantly different from the deaths in the first four months of COVID (69). From March through June 2020, fifteen children were admitted to the ICU with COVID. None died. Fewer than ten preschool teachers and twenty schoolteachers received intensive care for COVID. The preschool teachers were at slightly higher risk for COVID ICU hospitalization than people in other non-health care occupations (relative risk of only 1.1 with 1.0 being equal risk). Schoolteachers were at about half the risk of other non-health care occupations (relative risk 0.43).[756]

I admire the Swedish epidemiologists who followed solid epidemiological principles to protect the vulnerable and keep schools open. At the time, they received lots of cruel criticism. Follow-up data supports their decisions. Good for them.

Australia

I love the Aussies! It has been a pleasure to mentor clinicians in Australia on more than a dozen trips down under. I found Australian clinicians to be kind, intelligent, and effective during their care of children. One of the tragedies of how COVID was managed in Australia is that several of these fine doctors chose to surrender their medical licenses rather than face the prospect of being deregistered for treating their patients in ways that worked but were not part of the government's directives.

One of my colleagues, whose name I will refrain from disclosing, was a spectacular doctor who had developed extra expertise in treating mold toxicity. Sadly, many Australians were exposed to mold after the torrential floods that hit Queensland, Victoria, and New South Wales in 2010–2011. She took care of lots of children with neurodevelopmental disorders, and I learned as much from her as she may have learned from me, the official mentor.

In Melbourne where she lived, the countermeasures were draconian. This doctor told me she could only travel five kilometers from her home and had to prove to any authorities that stopped her that she was buying groceries, exercising, or going to the doctor. Physicians were threatened with deregistration if they suggested treatments like vitamin C or D or natural anti-inflammatories or antivirals. Peaceful protesters on the streets of Melbourne were met by police.[757] After coping with the horrific situation for months, this talented young physician in great demand in her community went "off the grid," moving to the Outback where she could tend to her own family and be off the radar of government officials who were sanctioning doctors. This truncated her career, is a waste of her considerable healing talents, and is a terrible loss to the families in her community.

Another terrific clinician practiced on the Gold Coast, famous for its spectacular surfing beaches. She provided clinical care to families and took

exceptional care of children with neurodevelopmental or neuropsychiatric
problems. She published research on the value of B6 and Zinc supplements
for aggressive teenagers. She left a void of medical care in her community
when she surrendered her license rather than risk being deregistered. Make
no mistake—she was not providing advice that was wrong or ineffective.
But her advice did not jive with the official government declarations.

One of my heroes in the aftermath of COVID is Australian Senator
Gerard Rennick. During hearings on October 26, 2023, a representative
of the Office of the Gene Technology Regulator was asked: "Why weren't
the mRNA vaccines tested for genotoxicity?" and "Why didn't the Office
of the Gene Technology Regulator look at it in terms of a gene technology?" It is worth watching the Australian bureaucrat squirm on video as
she tries to evade the question.[758]

A look into files released under a FOIA request[759] reveals the following:

Genotoxicity

No genotoxicity studies were conducted for the vaccine. This is in line with relevant guidelines for
vaccines. There were also no genotoxicity studies with the novel excipients. The sponsor stated that
the novel lipid excipients are not expected to be genotoxic based on *in silico* analysis (Derek Nexus
6.1.0, Derek Knowledgebase 2020 version 1.0 and Sarah Nexus 3.1.0, Sarah Model 2020.1 Version
1.8) of the novel lipids and their primary metabolites (reports not provided).

To translate: because it was called a vaccine, genotoxicity studies were not
required because traditionally vaccines do not have to undergo genotoxicity experiments. To put her statements in perspective, the reality is that,
for purposes of drug safety, this "vaccine" was gene therapy, known to
carry genotoxicity and carcinogenic risks. mRNA vaccines contain four
thousand nucleotides, far more than other gene therapy products that
were required to have genotoxicity and carcinogenicity studies. You can
put lipstick on a pig, but it is still a pig.

Furthermore, Dr. Rajumati Bhula, representing OGTR, testified
under oath that COVID "vaccine" products did not transfect cells of the
good people of Australia. Transfection is the process of deliberately intro-
ducing naked or purified nucleic acids into eukaryotic cells (cells with a
nucleus and organelles—like humans have). In fact, clinical documents
available to the Australian regulators mention transfection repeatedly.[760]
Oh, what a tangled web. . . .

Value of Nutrition and Immune Support

Ian Brighthope brought a message of enlightenment and hope to Australia when he testified about the value of nutritional and immune support to allow patients to recover from COVID infections. He was trained in medicine and surgery and practiced in five places in Australia. He cofounded the Australasian College of Nutritional and Environmental Medicine, known as ACNEM, and served as president for twenty-six years. The organization utilizes scientific data about the value of nutrients, minerals, and vitamins for a wide variety of human conditions. Dr. Brighthope argued appropriately that the use of treatment modalities well known to Asian and Australian practitioners could have been of immense value for treating COVID. He correctly pointed out that masking, social distancing, and lockdowns were based on flawed science and caused damage while being ineffective.[761] He spoke on behalf of other Australian clinicians who were afraid to speak out. I agree with his perspective that early treatment and nutritional immunology could have saved lives lost to COVID.

Canada: Private Citizens Examine
Their Government's Actions

Private citizens banded together to fund, investigate, and draft a five-thousand-page report examining Canada's COVID pandemic response. The conclusions were quite critical of the government's actions. More than three hundred Canadian citizens and expert witnesses testified on the record. More than sixty government authorities and regulators were subpoenaed but none complied to give testimony.

The report documents that politicians, public officials, and the media repeatedly characterized COVID injections as "safe and effective" while stating that restrictions would not end until everyone did their part. Canadians who were hesitant to get this vaccine, even if they had gotten all prior recommended vaccines, were demeaned as "anti-vaxxers." The citizens allege that the government and media worked to spread fear in the population to manipulate Canadians to take the experimental gene therapy injections. A social dynamic was set up whereas those who took the injections felt morally superior to their neighbors who did not. Those who resisted the shots were shamed and blamed for ongoing restrictions to freedom.[762]

The Freedom Convoy

You may remember media coverage of the Canadian truckers protest. Truckers whose job required travel between the United States and Canada wanted to eliminate mandates for COVID "vaccines." Hundreds drove to Ottawa, often met by Canadian citizens waving red maple leaf flags along the way in a show of support. The report states that the Freedom Convoy participants protested peacefully. The Canadian government used its power to freeze some of the protestors' bank accounts. The report alleges that the government either colluded with the media to portray the truckers as racist anarchists or the media accepted the government's narrative that the protestors were dangerous at face value. The authors of the report call for further investigation into the matter. Nonetheless, well-honed propaganda techniques were used to discredit those who dissented from government recommendations and mandates.[763]

Vaccine Mandates

Canadians who were fearful of the side effects of COVID shots faced mandates and often capitulated because they could not afford to lose their jobs. Those who remained steadfast without jabs often lost their unemployment insurance, the report states. Vaccine mandates for federal public servants and people working in federally regulated institutions were imposed in October of 2021. Medical or religious exemptions were hard to obtain, Canadian citizens discovered. Educational institutions denied unvaccinated students access to classes. The report points out that an announcement that unvaccinated people would not be allowed next to vaccinated people on airplanes was made at a time when transmission of COVID from the vaccinated had already been scientifically demonstrated.[764] Vaccine passes blocked the unvaccinated from going to gyms, restaurants, sports arenas, and movies. Unvaccinated parents were blocked from visiting their children's schools.

The *Epoch Times* has done a remarkable job of reporting on the underbelly of the COVID crisis. They reported that Ken Drysdale, chairman of the National Citizens Inquiry, said the testimonies from Canadian citizens and experts "provide irrefutable evidence that an unprecedented assault has been waged against the citizens of Canada."[765] Authors of the report call for an immediate cessation of COVID genetic vaccines and a full judicial investigation into the process by which the products were approved and mandated. Let's see what happens.

Bermuda

Eugene Dean and his colleagues in Bermuda are some of my favorite people. They have a superb appreciation for nature and its rhythms. Living on a small sunny island with beautiful beaches fosters a sense of well-being and promotes healthy lifestyles. When COVID came, the Bermuda government deferred to Dr. Fauci and the CDC for guidelines. The shutdowns were draconian. Bermuda relies on tourism and the financial services industry to maintain its economy. Eugene told me that when tourism dried up, many businesses closed.

Eugene and his colleagues noticed that only ten people in Bermuda allegedly died from or with COVID during the crisis prior to the rollout of the Pfizer jabs. After people took the injections, "the number of deaths increased exponentially," he says. In response to mandates and pressure to get "vaccines," several community groups acted in 2021 to provide alternative perspectives, express their concerns, and raise awareness of the true nature of COVID and the injections. You can meet the remarkable Eugene Dean and hear about the work of his group on an August 14, 2023, show on the Children's Health Defense website.[766]

One of the most compelling stories from Bermuda concerned a couple with proof of natural immunity from COVID (a positive antibody test showing they had the disease and recovered) who wanted to go home instead of quarantining in a Bermuda hotel (which would have cost them ten to twelve thousand dollars). The Bermuda government refused to recognize natural immunity. In fact, Eugene says, they "consistently implemented policies which imply that natural immunity does not exist against

COVID." Dr. Christina Parks, Dr. Peter McCullough, and Dr. Paul Alexander all provided expert testimony on the couple's behalf, detailing the science behind the couple's proven natural immunity and why they were not a threat to the community. Two years after the fact, when I went to Bermuda to lecture, the criminal charges were still pending. I got to witness an incredible performance by Dr. Parks under cross examination by a government lawyer who was baffled by her "unmovable stance" that natural immunity, encompassing innate and adaptive immunity with exquisite T-cell responses, was superior to the more transient, limited immune response from COVID injections. Good for you, Christine, for poise under pressure in a courtroom setting.

If you watch the CHD TV link referenced above and below, you can see the Bermuda swat team surrounding the couple's house and see Eugene and his colleagues asking why they were there. Sophia and Michael, whom I met, are wonderful, smart people with firm courage of their convictions. They put themselves through years of scrutiny and risked criminal sentences to stand for truth and justice. I admire and support them.

US Health-Care Systems Perform Poorly So Do Not Follow Us

Other countries followed the guidance of the CDC and Dr. Fauci during the COVID crisis. They may have had better outcomes developing their own local guidelines, because the United States has a history of poor health outcomes. When compared to other developed countries, we usually rank in the mid 30s for many markers of health. In a 2023 ranking of health systems worldwide the United States ranked around 70th.[767]

Here are seven unsettling facts about the United States compared to other developed countries:

- Highest infant mortality
- Highest maternal mortality
- Lowest life expectancy at birth
- Highest death rates for treatable conditions
- Among the highest suicide rates

- Highest obesity rate
- Highest percentage of people with multiple chronic conditions[768]

Our per capita COVID mortality data was the **worst** in the world, so we did not earn the trust many other countries gave us when they followed our guidelines.

Comparing Africa and USA for COVID success

Country	Deaths per million
Tunisia	2,198
South Africa	1,582
Botswana	1,084
Egypt	219
Zambia	214
Somalia	86
Congo	69
Ethiopia	63
Chad	11.6
Burundi	3.3

USA: deaths per million = 2,554

FLCCC
ALLIANCE

FRONT LINE COVID-19 CRITICAL CARE ALLIANCE · FLCCC.NET
PREVENTION & TREATMENT PROTOCOLS FOR COVID-19

Chart courtesy of the Front Line COVID Critical Care Alliance.

When health-care systems were compared in eleven high income countries for access to care, equity, efficiency and outcomes, the US ranked dead last. Australia, Norway, and the Netherlands ranked at the top.[769]

You may have heard that the United States spends more of its gross domestic product on health care than any other country, which is true.[770] The US burns a lot of money in administrative overhead costs. Top performing countries provide for universal health-care coverage, removing the access to care and financial obstacles that are present in the US. Countries with good health outcomes invest in primary care services and social support services, especially for children and parents. Countries which reduce administrative hassles have healthier citizens.[771]

Unless America learns from our mistakes made during COVID and makes major changes in our alphabet agencies, other countries should not follow our lead during the next proclaimed infectious disease crisis.

Peterson-KFF
Health System Tracker

Introduction | Long-Term Health Outcomes | Treatment Outcomes | Patient Safety | Patient Experiences | Discussion

Price Transparency   Affordability   Prescription Drugs

SEARCH

Country	Under 50 years, Total	Under 50 years, per 100,000	Overall, Total	Overall, per 100,000
Austria	14K	244	17K	217
Belgium	17K	217	22K	224
Canada	203K	786	278K	836
France	119K	279	167K	306
Germany	317K	604	789K	1,129
Japan	16K	21	31K	32
Netherlands	40K	344	56K	371
Sweden	17K	239	44K	498
Switzerland	15K	263	18K	243
U.K.	334K	723	885K	1,530
U.S.	5M	2,034	9M	3,139

Note: Australia is excluded from years of life lost from the COVID-19 analyses due to negative excess mortality.

Source: Peterson-KFF Health System Tracker.

Many Countries Showed an Increase in COVID Cases After "Vaccines"

In multiple countries, cases of COVID illness were lower before and increased dramatically after COVID "vaccines" were distributed. There is a clear trend of increasing cases temporally associated with vaccine introduction.[772]

FOOD FOR THOUGHT

How do you feel about the United States having a position of leadership during COVID? Did you hear criticism of the Swiss for their epidemiological choices not to lock down? Did you see videos of Australian police and protesters? Did you know about the Canadian truckers? Had you heard that other countries had less per capita deaths from COVID than the United States? Did you know that many countries showed increasing COVID cases after shots were introduced?

CHAPTER 30

What Is the Role of Propaganda in Decision-Making?

The most potent weapon in the hands of the oppressor is the mind of the oppressed.

—South African Steve Biko

Each time a man stands up for an ideal, or acts to improve the lot of others, or strikes out against injustice, he sends forth a tiny ripple of hope, and crossing each other from a million different centers of energy and daring those ripples build a current which can sweep down the mightiest walls of oppression and resistance.

—Robert F. Kennedy

Defiance in Childhood Saved Her Life

Vera Sharav was born in Romania in 1937. In 1941, her family was deported to a concentration camp in Ukraine, where her father died of typhus. She was rescued in 1944 just as the Final Solution, namely the liquidation of the Jews of Europe, was being frenetically implemented. At the age of six, she defied authority and refused to get on a boat with other orphan children. She insisted on joining a family she had befriended who were assigned to one of the other boats. What happened to the boat she refused to board? A submarine torpedoed it, killing all the orphaned children.

That seminal event cemented her distrust of authority and reinforced her inclination to trust her instincts. She came to New York City to live with her mother and grew up there. Another tragedy struck when her son died of an adverse drug reaction to clozapine. She became an activist exposing unsafe and unethical aspects of biomedical research and founded the Alliance for Human Research Protection. She is a strong advocate for informed consent and ethical treatment of children in research trials. I think she is fabulous.

How Propaganda Works

Vera taught me a lot about how propaganda is used to shape human behavior. She had a front row seat observing how propaganda was used to instill fear in a population by demonizing ethnic groups (Jews and Romani) and the disabled ("useless eaters"). Early in the COVID crisis, Vera recognized many of the same techniques were being used to shape compliance with COVID countermeasures like lockdowns, masking, and social distancing.

Vera taught me about Edward Bernays, the father of modern propaganda, who considered propaganda like an "invisible arm of the government."[773] He studied how powerful psychological techniques could be used to manufacture consent among populations for wars or compliance with government agendas. An important part of propaganda was to divert attention away from hidden financial incentives. Vera explains how it is important for propagandists to control the flow of information and suppress other ideas.[774] She draws an analogy between how Jews were demonized and how governmental officials criticized the unvaccinated during the COVID vaccine campaign. For this she was soundly criticized. I think her deep understanding of twentieth-century history fortifies her insights.

Propaganda Used to Make War Acceptable to the Population

As World War I was fomenting, big corporations like J. P. Morgan, Carnegie, Ford, and Dupont used propaganda to lobby President

Wilson to relent from his stance of opposing US involvement in World War I. Vera cites J. P. Morgan as a financial beneficiary of the war effort, with 341 directorships around the globe leading to the equivalent of $775 billion in today's dollars. Morgan was the sole munitions financier to Great Britain and France, leading to a one percent commission on three billion dollars.[775]

The Committee for Public Information

In America, the newly formed Committee for Public Information was tasked with "firing up Americans to join us" in the war effort. Bernays, who incidentally was Sigmund Freud's nephew, was instrumental in creating the slogan "Make the World Safe for Democracy." As part of the campaign, the iconic poster "Uncle Sam Wants You" was distributed; seventy-five thousand "four-minute men" were recruited to give four-minute speeches that exaggerated potential threats to manipulate people's emotions and promote support for the war.

Propaganda Morphs into Public Relations and Advertising

After the war, propaganda was rebranded as public relations. The focus turned away from war to consumerism. As one example of how opinions about products were manipulated, cigarettes were marketed to women as "torches of freedom" to exploit the growing movement for women's right to vote and hold wider positions in the workplace. Tried and true propaganda techniques were redirected to promote a culture of unlimited acquisition and planned obsolescence. Think of your iPhone or smart TV—bigger and better is just around the corner! The American dream, initially focused on social equality issues, was repurposed as conspicuous consumption. With pressure to consume came rising debts. In the 1950s, Victor Lebow declared, "Our enormously productive economy demands that we make consumption our way of life."[776] Currently, pharmaceuticals and sick care facilities make up a huge chunk of the economy. Do you think Lebow's vision came true? The BBC did a piece on how the world embraced consumerism.[777]

Rethinking the COVID Crisis
Through a Lens of Propaganda

Thinking about the COVID crisis in the context of propaganda, consider what messages rose to the top of collective human consciousness and what voices were suppressed. Recall that John Ioannidis, previously the most cited epidemiologist in the world, warned in March of 2020 that emerging data was utterly unreliable. Denis Rancourt, a Canadian scientist, warned prior to the "vaccine" rollout, that there was no increase in overall mortality in 2020 anywhere in the world.[778] Yet the fear persisted and the myth of COVID vaccines as the only savior persevered. Remember Dr. Fauci telling us "the calvary is coming"?[779]

How Messaging About COVID Vaccines Was Targeted

When I attended two conferences by the World Vaccine Congress and the annual American Academy of Pediatrics Convention in 2023, one of the hot topics was how to combat vaccine hesitancy. Many millions of dollars were spent on researching which type of message resonated with which segment of the population. Messages were massaged to target specific ethnic groups. Social media influencers were paid to post messages to their followers encouraging vaccine uptake. Here is an excerpt from a clinical trial looking at how to tailor the message to get COVID shots by appealing to various aspects of human motivation.[780]

(1) Baseline Informational Control	To end the COVID-19 outbreak, it is important for people to get vaccinated against COVID-19 whenever a vaccine becomes available. Getting the COVID-19 vaccine means you are much less likely to get COVID-19 or spread it to others. Vaccines are safe and widely used to prevent diseases and vaccines are estimated to save millions of lives every year.
(2) Self-Interest	Stopping COVID-19 is important because it reduces the risk that you could get sick and die. COVID-19 kills people of all ages, and even for those who are young and healthy, there is a risk of death or long-term disability. Remember, getting vaccinated against COVID-19 is the single best way to protect yourself from getting sick.

(3) Community Interest	Stopping COVID-19 is important because it reduces the risk that members of your family and community could get sick and die. COVID-19 kills people of all ages, and even for those who are young and healthy, there is a risk of death or long-term disability. Remember, every person who gets vaccinated reduces the risk that people you care about get sick. While you can't do it alone, we can all protect everyone by working together and getting vaccinated.
(4) Community Interest + Guilt	(3) + Imagine how guilty you will feel if you choose not to get vaccinated and spread COVID-19 to someone you care about.
(5) Community Interest + Embarrassment	(3) + Imagine how embarrassed and ashamed you will be if you choose not to get vaccinated and spread COVID-19 to someone you care about.
(6) Community Interest + Anger	(3) + Imagine how angry you will be if you choose not to get vaccinated and spread COVID-19 to someone you care about.
(7) Not Bravery	Soldiers, fire-fighters, EMTs, and doctors are putting their lives on the line to serve others during the COVID-19 outbreak. That's bravery. But people who refuse to get vaccinated against COVID-19 when there is a vaccine available because they don't think they will get sick or aren't worried about it aren't brave, they are reckless. By not getting vaccinated, you risk the health of your family, friends, and community. There is nothing attractive and independent-minded about ignoring public health guidance to get the COVID-19 vaccine. Not getting the vaccine when it becomes available means you risk the health of others. To show strength get the vaccine so you don't get sick and take resources from other people who need them more, or risk spreading the disease to those who are at risk, some of whom can't get a vaccine. Getting a vaccine may be inconvenient, but it works.

(8) Trust in Science	Getting vaccinated against COVID-19 is the most effective means of protecting your community. The only way we can beat COVID-19 is by following scientific approaches, such as vaccination. Prominent scientists believe that once available, vaccines will be the most effective tool to stop the spread of COVID-19. The people who reject getting vaccinated are typically ignorant or confused about the science. Not getting vaccinated will show people that you are probably the sort of person who doesn't understand how infection spreads and who ignores or are confused about science.
(9) Personal Freedom	COVID-19 is limiting many people's ability to live their lives as they see fit. People have had to cancel weddings, not attend funerals, and halt other activities that are important in their daily lives. On top of this, government policies to prevent the spread of COVID-19 limit our freedom of association and movement. Remember, each person who gets vaccinated reduces the chance that we lose our freedoms or government lockdowns return. While you can't do it alone, we can all keep our freedom by getting vaccinated.
(10) Economic Freedom	COVID-19 is limiting many people's ability to continue to work and provide for their families. People have lost their jobs, had their hours cut, and lost out on job opportunities because companies aren't hiring. On top of this, government policies to prevent the spread of COVID-19 have stopped businesses from opening up. Remember, each person who gets vaccinated reduces the chance that we lose our freedoms or government lockdowns return. While you can't do it alone, we can all keep our ability to work and earn a living by getting vaccinated.
(11) Community Economic Benefit	Stopping COVID-19 is important because it is wreaking havoc on our economy. Thousands of people have lost their jobs and are unable to pay their bills. Many others have been laid off by their employers and do not know when they will be called to return to work. Remember, every person who gets vaccinated reduces the risk that someone else gets sick. While you can't do it alone, we can all end this outbreak and strengthen the national economy by working together and getting vaccinated.[8]

Marketing, Social Media, and Peer Pressure

Many billions of dollars and countless hours of work by government officials, pharmaceutical executives, and marketing gurus coalesced to bring you the messages you got about taking COVID vaccines. Bobby Kennedy Jr. understood the power of these messages early on in 2020 and tried to educate the populace through a series of podcasts, for which he was soundly rebuked as a conspiracy theorist. Let's dive into the strategies used to convince global populations to take novel genetic biologic products and see why they worked.

Nudge Theory

Behavioral economics is a field that looks at human decision-making and cognitive biases. My favorite book on the subject is by Dr. Dan Ariely, *Predictably Irrational*.[781] *Misbehaving*[782] by Richard Thaler, a professor at University of Chicago, picks up where his book *Nudge* left off.[783] The bottom line goes something like this: Humans make irrational decisions all the time. Our enemies can use our subconscious biases to manipulate us into making decisions we would not otherwise make. Marketing appeals to our emotions to nudge us toward decisions that meet the goals of the company or organization.

No One Is Immune to Propaganda

Falling for propaganda or being nudged does not make you stupid. It makes you human. In fact, believing propaganda is a phenomenon that affects all humans, despite educational levels. People born in poverty and robbed of opportunities for higher education may have an instinctive resistance to being told what to do even when situations are confusing and rapidly changing. People with the opportunities to get advanced degrees may still be nudged to go along with expert recommendations or believe majority opinions and denigrate minority opinions.

The Milgram Experiment

Stanley Milgram was a social psychologist at Yale who conducted a series of experiments in the 1960s on the theme of obedience to authority. Perhaps Milgram's intense interest in the subject was molded by his Jewish heritage and the occurrence of the Holocaust just a few decades before. He recruited nearly eight hundred people (only about forty were women) to participate in an experiment for which they were paid $4.50. As often happens in psychology experiments, the participants were not told the true objective of the research—to measure compliance with authority figures. They were told the experiment would study the effects of punishment on learning and memory. Those giving electric shocks were called "teachers" and those receiving shocks were "learners."[784]

The conductor of the experiment played the role of an authority figure. He instructed the participants to administer a series of electric shocks to people who were in another room whenever they answered a question wrong. The gauges were labeled with emotion-inducing descriptions like "slight shock," "intense shock," and "danger: severe shock," and ranged from 15 to 450 volts. The participants were given a 45-volt shock so they would understand what the shocks they were giving would feel like. The "learners" were all actors who responded to the sham shocks with varying cries, moans, and screams related to the sham shock strength. One of Milgram's hypotheses was that Americans, coming from ancestors who had the gumption and individuality to stand against their king and go across an ocean on a dangerous journey for more freedom, would be less susceptible to obeying authority than the German people had been. He guessed that 1–3 percent of "teachers" would administer the maximum shock for wrong answers. In fact, in the initial experiment, 65 percent gave what they thought *could have been a lethal shock.*[785] Sixty-five percent!

Milgram's experiments would not be able to be conducted today and have been criticized for lack of informed consent and failure to debrief the teachers after the trauma they experienced. However, these experiments yielded valuable insights about how most people's behavior is influenced by authority figures.

Willful Blindness

A wife ignores the hotel receipts in her husband's jacket pocket many times before catching him with his mistress. An employee rationalizes the expenses her boss turns in for reimbursement until he is indicted for embezzlement. A father overlooks large pupils and signs of estrangement in his son until the teenager overdoses on cocaine. We all have blind spots.

Margaret Heffernan has written a fascinating book about the phenomenon of willful blindness, called *Willful Blindness: Why We Ignore the Obvious at Our Peril.*[786] She collated studies from neuroscientists and psychologists and interviewed criminals and whistleblowers to determine how we protect ourselves from seeing evidence of impending domestic tragedies, corporate malfeasance, and crimes that hurt innocent people. Willful blindness serves a purpose: it makes us feel more secure and helps us avoid conflict.

You may have seen Margaret Heffernan's TED Talk "Dare to Disagree" (June 2012)[787] or her TED Talk on the dangers of willful blindness (March 2013). She argues that by encouraging debate, challenging our own biases, and refusing to conform, we can make better decisions. She indicts the type of groupthink that was used during COVID to implement economically and psychologically dangerous lockdowns and social isolation policies. She analyzes why intelligent and powerful people often set aside evidence right in front of their eyes when making decisions with big impacts. How can smart scientists and doctors turn a blind eye to fraudulent studies and fatal errors in research papers? They are human and subject to willful blindness. Her book is full of cautionary tales. We should take her insights seriously if we want to prepare for the next medical crisis that threatens to impinge on informed consent and bodily autonomy.

FOOD FOR THOUGHT

Do you feel manipulated to buy things? Do you worry about how much Siri and your GPS software seem to know about you? Have you ever been pressured by your peers into doing something that hurt someone else? Have you ever experienced a personal or professional setback and thought to yourself, "Why didn't I see that coming?" Did you feel manipulated or pressured to get a COVID injection?

CHAPTER 31

Does Censorship Affect
Health Information?

*Once a government is committed to the principle of silencing the voice
of opposition, it has only one way to go, and that is down the path of
increasingly repressive measures, until it becomes a source of terror to all
its citizens and creates a country where everyone lives in fear.*

—Harry S. Truman

*Censorship, in my opinion, is a stupid and shallow way of approaching
the solution to any problem.*

—Dwight D. Eisenhower

Research in My Pediatric Practice

My personal censorship trauma occurred in May 2020. After I failed to
convince the CDC, AAP, and NIH to compare health outcomes of children based on their vaccination status, I decided to do it myself. I started
an ongoing clinical research project in my general pediatrics practice in
2005. The project had two aims: to see if some cases of autism could be
prevented and to see if there was a difference in chronic health conditions
comparing under vaccinated and fully vaccinated children.

I started advising parents to work on strategies to protect their infant's
good gut flora, maintain good vitamin D levels, and avoid neurotoxic
processed food, since gut dysbiosis, low vitamin D levels, and bad diets

were common in children who were diagnosed with autism. As I learned more about what environmental factors seemed implicated in the autism epidemic from the Autism Research Institute, I advised parents in my practice to avoid those triggers. Full disclosure, I did not have any tracking mechanisms to determine to what extent parents followed my advice. However, within seven years I was able to show a lower-than-typical prevalence of autism for patients born into my practice. One in 297 children at Advocates for Children developed autism, better than the one in fifty children who developed autism at that time in the US. My project, "Can Awareness of Medical Pathophysiology in Autism Lead to Primary Care Autism Prevention Strategies?" was published in the *North American Journal of Medicine and Science* in 2013.[788] It seemed like I had found some important things that clinicians could do to prevent at least some cases of autism. I sent a copy of my outcomes to the AAP and was disappointed when they did not reply or ask more about my strategies.

More Vaccines Correlate with More Chronic Illness

By 2020, we had studied various chronic diseases in relationship to vaccination status. In that study, children who got even one vaccine were counted in the vaccinated group. We found that children with the lowest number of vaccines tended to have fewer chronic illnesses. That data revealed more ear infections, asthma, and developmental disorders across percentiles as children received increasing numbers of vaccines. Our data showed that the unvaccinated had significantly fewer chronic illnesses.[789] Wow! My clinical impression about potential downsides of vaccines had been confirmed by actual data and statistics and published in a peer-reviewed publication. For those who would argue that I just proved correlation and not causation, I agree. But, when combined with the clinical histories of the patients, the known mechanisms of vaccine injury, and the dose-dependent relationship we demonstrated, we fulfilled lots of the Bradford Hill criteria for causation. The research raises important concerns that should be taken seriously.

I wrote about my years of experience trying to help children with autism for Children's Health Defense here.[790] I had spent fifteen years instituting various strategies trying to stem the tide of childhood chronic illness and neurodevelopmental disorders. Actually, we are in a tsunami, a tidal wave that will dwarf all other considerations. It took Facebook fact-checkers thirty-six hours to deem my work "false." I was devastated. I obsessed about all the parents who were being told that there was nothing they could do to decrease their child's chance of autism.

Federal Government Colluded with Social Media to Censor Medical Information

As I write this, a federal judge has ordered the Biden administration to end government directed censorship on social media.[791] Many of my colleagues and I have been on the receiving end of these initiatives. Science moves forward as clinicians and scientists publish our findings. Over the course of time, we should get closer and closer to the true facts, as replication studies either confirm or disprove our findings. One study that found something different should not negate my fifteen years of work. I stand by my findings.

Federal Judge Ruling

I quote attorney Aaron Siri, who wrote on Substack, "Thank you, Judge Terry A. Doughty, for having the courage to breathe significant renewed life back into the promise of free speech, and hearty congratulations to the Attorneys General of Missouri and Louisiana who led the charge in this case!"[792] From the ruling: "If the allegations made by Plaintiffs are true, the present case arguably involves the most massive attack against free speech in United States' history."[793] The plaintiffs allege "The CDC became the 'determiner of truth' for social-media platforms, deciding whether COVID-19 statements made on social media were true or false. . . . If the CDC said a statement on social media was false, it was suppressed, in spite of alternative views."[794]

From Judge Doughty: "The Biden Administration is 'hereby enjoined and restrained from taking the following actions as to social-media companies: (1) meeting with social-media companies for the purpose of urging, encouraging, pressuring, or inducing in any manner the removal, deletion, suppression, or reduction of content containing protected free speech posted on social-media platforms; (2) specifically flagging content or posts on social-media platforms and/or forwarding such to social-media companies urging, encouraging, pressuring, or inducing in any manner for removal, deletion, suppression, or reduction of content containing protected free speech; (3) urging, encouraging, pressuring, or inducing in any manner social-media companies to change their guidelines for removing, deleting, suppressing, or reducing content containing protected free speech . . ."[795] Unfortunately, on June 25, 2024, the Supreme Court ruled that the government could indeed censor the information its citizens receive.

I realize this is a complex issue. One of my son's good friends from childhood works for Google, trying to make decisions about what information should not have a world-wide audience. The internet has clearly been used for all sorts of nefarious purposes. I argue that the medical information about COVID from many of my colleagues has aged well and been proven correct. Yet, these clinicians and scientists were censored, maligned, and denigrated for pointing out inconvenient truths.

Censorship of Front-Line Doctors and Good Scientists

It turned out that being denigrated put me in very good company. Lots of clinicians I respected were being shut down by powerful forces whose agendas were not served by our clinical observations. Two non-profit organizations I supported—the Front Line COVID Critical Care Alliance and Children's Health Defense—were attacked for spreading disinformation. Some of my friends and colleagues were included in the notorious "Dirty Dozen of Misinformation" by the Center for Countering Digital Hate (which sounds like a good thing on the surface, but turned out to be quite a misleading moniker).[796]

Documenting the full extent of censorship of sincere and hard-working front-line doctors is beyond the scope of this book. I believe that America's response to COVID suffered because practicing clinicians were not included in the decision-making process. In fact, good doctors' clinical treatment strategies were actively suppressed, and they were personally attacked despite their successful outcomes. Decisions that affected the whole world were made by bureaucrats who were not treating patients as a regular part of their workdays.

What Is Labeled "Misinformation" Can Be True

Evidence has emerged that the government instructed Facebook and Google to label physician recommendations that ran counter to government advice as misinformation or disinformation. Many censored recommendations have now been proven to have been true all along.[797] The government-censored medical doctors, scientists, and investigative journalists, including Jay Bhattacharya, MD; Paul Marik, MD; Pierre Kory MD; Peter McCullough, MD; Alex Berenson, and Naomi Wolf. Wikipedia entries were rewritten to characterize people who raised concerns about side effects of the new products as "anti-vaccine." Aggressive legislation interfered with doctor/patient communications and informed-consent decisions in California, where doctors were threatened with loss of their medical licenses if they said anything that deviated from CDC guidelines when discussing COVID with their patients.[798,799,800] To be clear, doctors who practiced in California could lose their licenses if they stated their concerns about COVID shot side effects, since that was counter to the "safe and effective" government messaging. In September 2023, the California gag legislation was quietly made null and void, allowing doctors to resume true informed-consent discussions about COVID with their patients.[801]

Calling Epidemiologists from Harvard, Stanford, and Oxford "Fringe"

Earlier in this book you learned about the Great Barrington Declaration, which I signed along with more than sixty-three thousand other scientists and medical professionals. The authors wrote about their concerns for collateral effects from widespread lockdowns and argued for focused protection for the vulnerable. It was very disturbing for me to read the email correspondence between Dr. Francis Collins, Director of NIH, and Dr. Tony Fauci, Director of NIAID, characterizing the eminently credible authors from Harvard, Stanford, and Oxford as "fringe epidemiologists" and planning how to take them down.[802] As this book goes to press, I am thrilled that Dr. Bhattacharya from Stanford has been nominated to be the new Director of NIH.

Censoring Reports of Adverse Effects from COVID Injections

It took numerous filings of Freedom of Information Act requests to uncover the extent to which adverse side effects from COVID novel technology products were censored. Important safety signals for myocarditis, blood clots, menstrual abnormalities, and Bell's palsy were detected soon after the vaccines rolled out, but not shared with the pubic.[803]

Let's look at how the information about myocarditis as a side effect of COVID injections was managed by government agencies. On April 26, 2021, Rochelle Walensky, CDC director, told the public in press conferences that the CDC had not detected any safety signals for myocarditis in their databases.[804] Let's look more closely at the timeline of what the CDC knew about myocarditis with the shots and when they knew it. My thanks again to Zachary Stieber of the *Epoch Times* for putting together this timeline. Evidence emerged that the CDC withheld statistically significant myocarditis signals for three months, during which time millions of people received the vaccines.[805]

In fact, on September 22, 2020 (before the injections were distributed), the CDC identified myocarditis as a condition of special interest; on October 30, 2020, so did the FDA. There was a flurry of communication between Israel and the CDC in February through April of 2021. The Israelis reported a teenager hospitalized for myocarditis after the Pfizer injection on February 1. By February 18, a myocarditis signal was triggered in the VAERS system using the Proportional Reporting Ratio. On February 19, the Fisher's Exact Test triggered another safety signal for myocarditis in VAERS. By February 28, Israeli officials had privately alerted the CDC to large numbers of myocarditis reports in young people after Pfizer jabs. March 20, 2021, the first postvaccination myocarditis case report was published; a second case report followed on March 31,

2021, the same day a previously health twenty-two-year-old woman died of postvaccination myocarditis. By April 2021, 158 cases of myocarditis or pericarditis were reported to VAERS. In early 2021, after COVID vaccines were mandated for the military, myocarditis cases spiked in soldiers. On April 12, 2021, US military officials briefed the CDC on postvaccination myocarditis cases.[806] Yet, Dr. Walensky told the press and therefore the public that there were "no signals of myocarditis" on April 26, 2021![807]

Dr. Peter McCullough has written extensively about myocarditis after the novel COVID shots. He got censored while Dr. Walensky got CNN interviews. McCullough noted a publication documenting YouTube's censorship of people attempting to report their adverse reactions from the COVID injections.[808] Yee Man Margaret Ng and colleagues analyzed YouTube's COVID-19 content and concluded they were successful in wiping out content reporting side effects and replacing it with pro-vaccine comments.[809] Given that these experimental injections generated more VAERS reports than all other vaccine side effect reports combined for the past thirty years and that 7.7 percent of vaccine recipients in the first 10 million people to use the V-safe app had to seek medical care after their COVID shot, does it really serve the public interest to shut down public discussion of these very real medical complications?

Fauci's Proclamations Went Outside His Expertise

I was uncomfortable with the folk hero status that Dr. Fauci achieved during the early days of COVID. His proclamations did not make sense to me. As a pediatrician vested in prevention, the idea of advising people to "stay home until your lips turn blue then go to the Emergency Room" seemed absurd. As time went on, it became clear that Fauci misled the public about important issues like the value of natural immunity, the efficacy of repurposed medications, and COVID injections' inability to block transmission of the virus.[810,811,812,813]

In my opinion, by presenting a vaccine as the only way out of the pandemic and pushing for lockdowns and school closures, he ignored other reasonable options. By advocating censorship of those who advocated traditional public health measures instead of lockdowns and isolation,

Fauci ignored the known literature about social determinants of health and well-being. Speaking on behalf of American children, COVID countermeasures advocated by Fauci and exported to other countries adversely affected kids' development, academic progress, and mental health.[814,815,816] A Brown University ongoing longitudinal study of child neurodevelopment examined general childhood cognitive scores in 2020 and 2021 versus the preceding decade, 2011–2019. They discovered that

> children born during the pandemic have significantly reduced verbal, motor, and overall cognitive performance compared to children born pre-pandemic. Moreover, we find that males and children in lower socioeconomic families have been most affected. Results highlight that even in the absence of direct SARS-CoV-2 infection and COVID-19 illness, the environmental changes associated COVID-19 pandemic is significantly and negatively affecting infant and child development.[817]

FOOD FOR THOUGHT

How should social media monitor content? Should the government be able to instruct Big Tech to take down medical information from doctors and other health-care providers who do not work for the government? Do you think the CDC has preserved its reputation as a source of truthful health information? Were you surprised that what the government called "misinformation" could actually be true? Do you think it is ethical for the CDC to withhold information about a potentially fatal heart complication in the name of fighting vaccine hesitancy?

CHAPTER 32

Is There Evidence to Suspect Fraud in mRNA Trials?

Did you ever stop to think, and forget to start again?
—Winnie the Pooh

Always tell the truth. That way you don't have to remember what you said.
—Mark Twain

Corporate Fines for Fraud Built into Pharma Budgets

In 2012, a doctor and professor at the Nordic Cochrane Center in Copenhagen investigated crimes committed and fines paid by pharmaceutical companies between 2007 and 2012. He found crimes that included "marketing drugs for off-label uses, misrepresenting research results, hiding data on harms, and Medicaid and Medicare fraud. Doctors were often complicit in the crimes, as kickbacks were common. The crimes were repetitive."[818] After an extensive review of multiple drug companies, he concluded that

the crimes persist because crime pays. Harder sanctions are therefore needed, including prison sentences for CEOs and other senior executives. Doctors and their organisations should consider carefully whether they find it ethically acceptable to receive money that may have partly been earned by crimes that are harmful to patients.[819]

Pfizer Fraud

In 2009, Pfizer agreed to pay $2.3 billion to settle a case of fraud.[820,821] At the time, it was the largest health-care fraud settlement in Department of Justice history. Pfizer had promoted four drugs for uses not approved by drug regulators: Lyrica (pregabalin, used for epilepsy), Geodon (ziprasidone, an antipsychotic), Bextra (valdecoxib, an anti-arthritis drug), and Zyvox (linezolid, an antibiotic). Part of the settlement included a clause that Pfizer would enter a Corporate Integrity Agreement with the Department of Health and Human Services and be on good behavior for five years.[822]

Other Pharmaceutical Companies Pay Fines

Pfizer is not the only company found guilty. In 2011, GlaxoSmithKline outpaced the 2009 Pfizer fine when they agreed to pay $3 billion for fraud.[823,824]

Here is a small sample of other pharmaceutical fines for fraud:

- In 2009, Eli Lilly agreed to pay more than $1.4 billion for a wide range of marketing strategies to convince doctors to use Zyprexa, an antipsychotic drug with significant side effects, off-label. At the time of the fine, Lilly had made $40 billion from Zyprexa.[825]
- In 2009, Sanofi-Aventis paid more than $95 million.[826]
- In 2010, AstraZeneca paid $520 million to settle fraud case.[827]
- Novartis agreed to pay $423 million in 2010.[828]
- In 2012, Johnson & Johnson was fined more than $1.1 billion. A jury found that the company and its subsidiary Janssen had downplayed and hidden risks associated with the antipsychotic drug Risperdal (risperidone). In addition, the judge discovered nearly 240,000 violations for Medicaid fraud in one state alone.[829]

Is the Pharmaceutical Industry Dangerous to Health?

Fabien Deruelle, an independent researcher in France, thinks the pharmaceutical industry is dangerous to health and cites proof from the COVID debacle. He writes that COVID highlighted a problem that has been brewing since the days of tobacco science. His study concluded that knowledge is manipulated by powerful corporations for profit. He is concerned that Big Pharma works to suppress individual choices in order to facilitate global control of public health. He reports his results as follows:

> Since the beginning of COVID-19, we can list the following methods of information manipulation which have been used: falsified clinical trials and inaccessible data; fake or conflict-of-interest studies; concealment of vaccines' short-term side effects and total lack of knowledge of the long-term effects of COVID-19 vaccination; doubtful composition of vaccines; inadequate testing methods; governments and international organizations under conflicts of interest; bribed physicians; the denigration of renowned scientists; the banning of all alternative effective treatments; unscientific and liberticidal social methods; government use of behavior modification and social engineering techniques to impose confinements, masks, and vaccine acceptance; scientific censorship by the media.[830]

I share Deruelle's concerns that by presenting only one side of the science while suppressing alternative hypotheses, choices, and treatment strategies, the public was disinformed during the COVID crisis. He cites grave concerns that vaccination laws were imposed based on "industry-controlled medical science," which "led to the adoption of social measures for the supposed protection of the public but which became serious threats to the health and freedoms of the population."[831]

Brook Jackson Alleges Fraud in Pfizer COVID Trial

Brook Jackson impressed me as someone who is intelligent, ethical, and willing to suffer personal indignities to do the right thing. Brook had been working as a regional director for a Texas-based company hired to conduct research trials for Pfizer's mRNA COVID product. On September 25, 2020, Ms. Jackson filed a complaint to the FDA about irregularities she had witnessed at three trial sites. She reported "falsified data, unblinded patients, and inadequately trained vaccinators who were slow to follow up on adverse events."[832] She expected the FDA to intervene, but the FDA did not inspect the trial sites as Brook had hoped. A *BMJ* investigation of regulatory documents revealed that only nine of the one hundred fifty-three Pfizer trial sites were inspected before the vaccine was licensed. Only ten of ninety-nine Moderna trial sites were inspected.[833] When the FDA does not notify the public about failed inspections or protocol violations, the medical community cannot learn about scientific misconduct affecting clinical decision making. Would most doctors recommend a product they knew had been approved despite inadequate safety trials? The *BMJ* quotes David Gortler, a pharmacologist who worked as a senior adviser to the FDA commissioner in 2019–2021, as saying "The lack of full transparency and data sharing does not allow physicians and other medical scientists to confirm the data independently and make comprehensive risk-benefit assessments."[834]

The *BMJ* investigation revealed a long history of inadequate oversight of clinical trials by the FDA. A 2007 report by HHS's Office of the Inspector General discovered that "the FDA audited less than one percent of clinical trial sites between 2000 and 2005."[835] In fact, at that time the FDA did not have a database of all clinical trials in the nation.[836] In fact, even as I write this, the FDA does not have a complete database of all the clinical trials in the United States, citing lack of resources. The FDA reported it had created a task force to improve the conduct of clinical trials, but when the *BMJ* asked to interview a member of the task force, the FDA declined.[837]

A journalism professor at NYU conducted a mind-blowing analysis of clinical trials over a fifteen-year period. In fifty-seven published clinical trials in which the FDA found evidence of problems,

> 39% had falsification or submission of false information, 25% had problems with adverse events reporting, 74% had protocol violations, 61% had inadequate or inaccurate recordkeeping, and 53% failed to protect the safety of patients or had problems with oversight or informed consent. Furthermore, only 4% of the trials that were found to have significant violations were mentioned in the study's journal publications.[838]

An investigation in 2020 concluded that the FDA rarely leveled sanctions or demanded proof that problems were fixed.[839] Peter Doshi was first author on a paper documenting that the FDA does not usually alert medical journals or the public about research misconduct in clinical trials.[840]

Argentina Lawyer in Pfizer Trial Alleges Misconduct

Thanks to James Lyons-Weiler for bringing this case to my attention. Jack's dogged pursuit of intellectual truths is inspiring. Jack tells the story of Augusto Roux (subject #123129982, study C45921001), a lawyer who volunteered for Pfizer's stage 3 trial because he wanted to protect his mother who had emphysema. You can find more details in Dr. David Healy's blog.[841] Dr. Jack writes about numerous topics at his "Popular Rationalism" Substack. Dr. Healy and Dr. Lyons-Weiler believe there is massive evidence of fraud in the Pfizer trial. This is important because there is loophole in the PREP Act's protection of vaccine manufacturers from liability—evidence of fraud. See if you think the following qualifies.

After his second dose of Pfizer vaccine on September 9, 2020, Augusto developed a high fever and felt very sick. He fainted on September 11, an ominous omen. He went to the hospital on September 12 and got a CAT scan of his chest. It showed an abnormal collection of fluid in the sac around the heart, which is called pericarditis. At discharge on September 14, his doctor wrote that he had suffered an adverse reaction to the vaccine. Augusto reports that hospital staff said they had seen a huge influx of people from the clinical trial seeking medical care at their hospital (not the one associated with the clinical trial). One nurse estimated around three hundred patients had been seen for illness soon after their injections, but this has not been verified.

Argentina was a crucial component of Pfizer's clinical trials. In a few weeks they enrolled about three thousand volunteers. That is about seven percent of the total of approximately forty-four thousand subjects. Since he was a lawyer, Augusto sued to get his medical records. I am not sure why that process took a year. Theoretically, the information in medical records should belong to the patient. Augusto had a negative PCR test at the hospital. When Augusto called the trial site on September 14 to inform them he had been hospitalized, they wrote in his clinical trial records that he had bilateral pneumonia that had nothing to do with the "investigational product." That is not what Augusto says he told the clinicians. Remember, his doctor had diagnosed an adverse vaccine reaction. On October 7, at the request of BioNTech, the clinical trial notes were "updated" with a code for suspected COVID-19 disease rather than a code for a vaccine adverse event.

Whoa! First, even if he had developed COVID (and we have no evidence he did), this would have been of clinical interest since we know that immune systems are suppressed after the COVID jabs, especially in the first two weeks. Second, efficacy calculations of the jab depend on a positive PCR test to confirm the diagnosis. Since his case was only "suspected" (by the trial administrators in disagreement with the treating physicians), it did not damage their precious 95 percent efficacy claim. Third, by "sweeping the diagnosis under the rug," as Jack wrote, this case of pericarditis soon after vaccine receipt magically disappeared.

On October 9, Augusto was formally unblinded, and the patient and the investigators learned he was in the vaccine arm of the trial. Theoretically, the trial could be unblinded if the subject's life was in danger; on October 9, this did not seem to be the case for Augusto. It gets worse. The trial's principal investigator, a pediatrician, wrote in Augusto's clinical trial records that he had a severe anxiety attack starting on September 23 and that it was not caused by the vaccine. The investigator wrote that Augusto suspected a conspiracy involving the hospitals. To make matters worse, the mental health diagnosis of anxiety was added to his personal medical records.[842]

Let's think for a moment about this volunteer's predicament. He heroically volunteers for the trial in an act of altruism and concern for his mother. He gets very sick and his heart is inflamed. As a lawyer, he is trained in analytic thinking and has likely seen his share of human malfeasance. Given his experience, I would argue that anxiety is a rational response. A novel technology never tried before, inflammation of his heart sac diagnosed by CT scan three days later, hospital staff telling him they had seen a lot of patients with illness onset after the new jabs—after all that, most rational people would be anxious.

Augusto told Dr. Lyons-Weiler that he wrote to the FDA in late 2020 but did not hear back. He wrote to the European counterpart, the EMA (European Medicines Agency), but did not get an answer. He filed a VAERS report in November 2020 and received a temporary VAERS number in confirmation, but his report never made it into the system as far as he can tell.

Pfizer and FDA Ask for Clinical Trial Data to Stay Secret for Seventy-Five Years

The famous FDA FOIA data release (which the FDA and Pfizer had petitioned the court to delay for seventy-five years), showed all protocol deviations recorded during the trial. Thanks to a judicial ruling, records are being released every month and analyzed by three thousand volunteer scientists and clinicians at Daily Clout.[843] Shockingly, in the ninety days after its mRNA vaccine was rolled out under Emergency Use, Pfizer documented 1,223 deaths and 158,000 adverse reactions.[844] There is an entire book about the revelations the scientists are discovering as they pore through the documents, provocatively titled *The Pfizer Papers: Pfizer's Crimes against Humanity,* written by Daily Clout analysts who volunteered their time and expertise.[845]

Protocol Deviations

Reporting protocol deviations is an important part in the scientific process. Sadly, I screwed up in a clinical trial and had to report a protocol deviation before the paper was published, which is embarrassing to say the least. In one clinical trial about children with autism, I failed to adequately oversee my staff to ensure they got parents to fill out all the study questionnaires—we missed one. I was a busy pediatrician running a general pediatric practice, I was seeing complicated autism patients and lecturing about their problems in multiple conferences, and I had no specific research training. Did I mention the dog ate our study protocol? I still feel bad about it. Apologies to Dr. Dan Rossignol, the PI on the study. But we reported the mistake when we published the paper in the peer-reviewed journal, as ethical scientists do.

Protocol Deviations in Augusto's Case

Here are the three deviations listed for Augusto's case, as reported by Dr. Lyons-Weiler:

- Sept. 12, 2020 Nasal swab not collected for the visit where it is required.
  - o But he did have a relevant negative nasal swab done on Sept 12 at the other hospital which seems to have been deleted as important information; it would raise suspicions about the "suspected COVID" label.
- Sept. 12, 2020 Visit performed outside of protocol specified window.
  - o He had pericarditis and was admitted to a hospital. Seems like a valid deviation.
- Oct. 9, 2020 Blind compromised

I wonder if there are other trial participants like Augusto who did not have the training or resources to make a paper trail of their difficulties reporting their adverse reactions to the agencies tasked with post-marketing adverse event tracking.

Maddie de Garay

You have met Maddie—the teen who developed paralysis and gastroparesis during an mRNA trial—in previous chapters. Her condition was reported as "functional abdominal pain" instead of her very serious neurologic conditions. Pfizer executives surely suspected it would be bad for vaccine uptake if parents learned that one of about a thousand kids in the trial ended up in a wheelchair.

Liability Protection

Vaccines have enjoyed special protection against liability because vaccine makers were getting sued so often due to adverse reactions that it was not profitable for them to continue making vaccines. During the Reagan administration, pharmaceutical companies threatened to stop making vaccines if they were not granted liability protection. Since the story of vaccines as saviors of humanity is so deeply ingrained, they got a waiver. Vaccine manufacturers have no liability for products given to babies on day one of life, infants at multiple well-baby checks, teens entering college, and adults urged to get yearly flu and COVID vaccines.[846]

Ninety-Five Percent Effective?

In May 2021, the previously esteemed medical journal *The Lancet* published a study claiming 95 percent effectiveness for the Pfizer COVID vaccine. You heard that number repeated on newscasts and advertisements urging people to get the shots. Immediately, some eminent statisticians from Queen Mary University of London wrote a letter to *The Lancet* outlining the flaws in the conclusions and how the efficacy claims were inflated. Nearly two years later, after an initial email from *The Lancet* that they would check into the flaws, Fenton and Neil reported that

> On 13 January 2023 we got a response from *The Lancet* saying they had decided against publishing the letter, asserting that any claim questioning the efficacy and safety of the Pfizer vaccine was "misinformation" and that they did not consider the position of SA-P an undeclared conflict of interest or a challenge to the integrity of the data.[847]

FOOD FOR THOUGHT

Have you noticed how many ads on TV are for prescription drugs? Have you heard of vaccine trial fraud allegations before? Did you know Pfizer and the FDA wanted to keep their clinical trial results secret for seventy-five years? Have you read any of the concerning information being discovered in the clinical trial records? Did you know that the United States is one of only two countries that allow marketing of prescription drugs directly to the consumer? How would you rate your level of trust in drug companies?

CHAPTER 33

Are COVID Injections Contaminated with DNA?

The great enemy of truth is very often not the lie—deliberate, contrived and dishonest—but the myth—persistent, persuasive, and unrealistic.
—John F. Kennedy

Clean up your own mess.
Everything I Needed to Know I Learned in Kindergarten
—Robert Fulgham from

Bait and Switch

OK, this is bad news. For those of us schooled in science and cell biology, it is horrifying. We now have evidence from multiple independent scientists that COVID injections are contaminated with DNA.[848] A different manufacturing process was used to scale up production for widespread distribution than was used in the original clinical trials. This is a very big deal. It may help explain why some recipients develop such terrible reactions to the shots. It may help explain why we see unprecedented numbers of adverse events and deaths with these injections. It definitely explains why regulatory agencies should not take pharmaceutical industries at their word when billions of dollars of profits are at stake.

There is a lot to unpack here. The science is complicated and beyond the expertise of a pediatrician like me. Know that independent scientists all over the world are calling for an immediate cease and desist of promotion and distribution of COVID injections for humanitarian reasons. The safety concerns are enormous. Some are calling this crisis "Plasmidgate." This was reported in spring of 2023, but as of this writing, most people have not heard about it.

Pfizer and Moderna Vials Contaminated with DNA Plasmids

A plasmid is a circular ring of DNA. Plasmid DNA is derived from bacterial chromosomes. Plasmids can create resistance to some antibiotics. If plasmids are integrated into the human genome, people could keep producing that modified RNA forever. Plasmids were used in the "Process 2" phase of manufacturing modified RNA injections. The use of plasmids was not revealed to regulatory agencies in Europe or the United States, as we now know from a careful examination of pharmaceutical documents submitted to the FDA. Plasmids and DNA fragments were not included on the list of ingredients submitted to regulatory agencies.[849] Is this scandal finally the turning point in the use of these novel gene products?

Those of you who watch the nightly news "brought to you by Pfizer" and hear Ken Jennings refer to "our friends at Moderna" on *Jeopardy!* might have illusions of altruistic pharmaceutical execs who use their well-earned wealth to bring you TV shows. I imagine that many people working to develop drugs are well-intentioned and honest. In some cases, the reality is much darker. Robert Kennedy Jr. refers to some pharmaceutical companies as corporate serial felons.[850] Their profits are so large that huge fines for breaking the law (on the order of hundreds of millions or even billions of dollars) are built into the budget as a regrettable but necessary expense.[851]

Kevin McKernan, PhD, was the first scientist to expose vials of Pfizer and Moderna COVID injections that were contaminated with DNA plasmids. His work has been replicated by at least half a dozen independent scientists. Let's look at McKernan's credentials. He is a leader in DNA sequencing technology. He led a research team as part of the Human Genome Project. As Director of Research and Development of Life

Technologies, he managed next generation SOLiD sequencing technology and oversaw more than one hundred research collaborations. He was instrumental in progress in human tumor gene sequencing, which has led to more specific cancer therapies. I have heard him lecture. He knows what he is talking about. I find him completely credible.

Like many discoveries in science, Kevin's crucial revelation was the result of serendipity. Dr. McKernan was trying to sort out a sequencing problem in his lab. He obtained bivalent Pfizer and Moderna COVID-19 vials, expecting them to be pure. He reasoned they should be perfect mRNA controls for use in working out his mRNA sequencing challenges. He was shocked to discover that the vials included a lot of DNA. He proved there was plasmid-derived double-stranded DNA in COVID vials from Pfizer and Moderna.[852]

Manufacturing Issues

Let's talk about the use of plasmids in manufacturing. Bacterial plasmids are easily manipulated. They are used as templates in science to replicate messenger RNA. However, and this is absolutely crucial, these ingredients must be removed in the manufacturing process or plasmids will be incorporated into the nucleus of cells. We have evidence that this crucial step was not taken. Why am I not surprised? Remember Robert Fulgham's book *All I Really Need to Know I Learned in Kindergarten*? One of the lessons is "clean up your own mess." Pfizer and Moderna didn't clean up their own mess—Big Pharma flunks kindergarten.

Well, you might think, this sounds really complicated. Maybe Pfizer and Moderna made an honest mistake, but the process is a basic one. Another of my COVID-era heroes, Byram Bridle, explained that plasmid removal is a basic scientific process. Dr. Bridle specializes in vaccinology, immunology, and biology and has extensive publications in the scientific literature. Pfizer and Moderna, with all the taxpayer funding and scientists on staff, should have been expected to achieve this basic scientific metric. During a lecture for the World Council for Health, Dr. Bridle explained, "we use plasmid DNA all the time in our vaccinology lab." He then told the story of supervising a student fresh out of college working on a master's degree. As part of his basic competency, the rookie scientist had to prove he removed all the DNA plasmids by using specialized enzymes. This kid just out of college accomplished what Pfizer and Moderna did not.[853] Dr. Bridle thinks there is no excuse for Pfizer and Moderna not getting rid of these contaminants. I think it is an egregious and costly error. Not costly to Pfizer or Moderna, at least not as I write this. Costly to the millions of people around the world who have had tragic outcomes in the wake of these injections.

It gets worse. It turns out that the contents of vials discussed in pharmaceutical documents submitted to the FDA were not the same contents that were injected into millions of arms around the world. A process labeled "Process 1" did not use E. coli or plasmids, and this is what was approved by the FDA.[854] In what seems to be the "bait and switch" of the century, after the clinical trials were completed, the game moved to production of unprecedented amounts of solution for injection. The tried-and-true bacterial plasmid technique was employed to ramp up production. Now we have evidence that rumors of sloppy manufacturing techniques were true. Let me make it clear: the vaccines that were approved under Emergency Use Authorization were not the same vaccines that were rolled out to the public with much fanfare, at least based on commercially produced vials analyzed so far. Scientists have documented a wide range of DNA contamination. Furthermore, vials with higher DNA concentrations are associated with more adverse events. It seems fair to conclude that at least a subset of the injections was contaminated with variable amounts of DNA.

The Immune System Is Built to Attack Foreign DNA

So why is that such a big deal, you might ask. We all are full of DNA, right? Yes, we are full of our *own* DNA. The problem arises when our immune system, exquisitely trained to know the difference between self and nonself, starts attacking anything identified as foreign. This is not a new problem. Other vaccines have used fetal cells in manufacturing. Many people object to the use of fetal tissue in vaccines on ethical or religious grounds, but there is a theoretical immunologic argument also—injecting foreign DNA can induce autoimmune disease. In another chapter we explored the concerns about autoimmunity raised by qualified scientists even before the rollout of COVID injections. This issue of plasmid DNA contamination takes those concerns far beyond the molecular mimicry due to similarities with human tissues that careful scientists foresaw.[855]

Cell-free DNA concentration in blood is usually 10–30 ng/ml, but the concentration increases with pregnancy, tumor growth, and inflammation. Now we can add receipt of modified RNA injections to the list. Some scientists are reporting concentrations of DNA that surpass regulatory guidelines by as much as a thousand-fold![856] When the body recognizes bacterial plasmids, it reacts by triggering inflammation. Plasmid DNA can last a very long time. Its presence opens the door to prolonged expression of spike protein. How long? We do not know. Long-lived plasmids can lead to chronic inflammation. Chronic inflammation is a risk factor for cancer and many other chronic illnesses.[857,858]

SV40 Promoter Contamination

It gets even worse. WE learned that multiple independent researchers have confirmed the presence of SV40 promoter in vials of mRNA COVID solutions. Pfizer did not mention SV40 promoter in the list of components submitted to the FDA.[859] You may have heard of SV40 contamination in polio vaccines, and how SV40 has been identified in multiple types of tumors. In this case, SV40 *promoters* can be used as an enhancer to drag products into the nucleus of cells. Lipid nanoparticles act as "Trojan horses" to get products into cell nuclei. Since DNA is packaged into lipid nanoparticles, it is "transfection ready," meaning it can be transported into the nucleus. The concentration of the sticky ends of DNA fragments is another factor that governs risk of integration into the nuclei of cells. All these factors raise concerns for mutations and potential inheritable changes.

What Is Transfection and Why Should We Care?

How do I transfect thee? Let me count the ways. Transfection means to infect a cell with free nucleic acids (DNA or RNA, the building blocks of life). Genetic material can be transfected. The lipid nanoparticles used in the mRNA vaccines are excellent candidates for transfecting material because they are designed to cross cell membranes. This quality has been exploited in the past to deliver chemotherapy drugs to the brain.

Polyethylene glycol, or PEG for short, is widely found in cosmetics and various medications. As many as 72 percent of people have antibodies to PEG, which was a concern raised early on by some of us who looked at the proposed ingredients in the developing mRNA technology injections. In September of 2020, prior to the rollouts, RFK Jr. and the Children's Health Defense team wrote to the FDA expressing concerns about PEG causing anaphylaxis in vaccine recipients.[860,861]

PEG can bind DNA. Lipid nanoparticles are PEGylated lipids. Lipid nanoparticles are electrically charged, which means they can attract and attach all kinds of molecules. This process is called adsorption, meaning something sticks to the outside of the molecule. Can PEGylated lipids grab bacterial plasmids? If so, the plasmids will be transported into cells, where they will live long and prosper.

SV40 promoter can carry plasmids into the nucleus. Adenoviruses, used as a vector for the Johnson & Johnson COVID injections, may enable integration into nuclei.[862] Radiation-induced DNA damage promotes integration of foreign plasmid DNA into the cell nucleus.[863] Genome plasticity may enhance plasma integration into nuclei[864] This is a huge concern for children, especially as we consider the implications of transfection into the cells of young, developing brains. Let's review the list: lipid nanoparticles, PEG, SV40 promoter, adenoviruses, radiation, genome plasticity—too many ways to disrupt the nucleus, which guards our "book of life" in the form of human DNA.

Free Circulating DNA Promotes Chronic Inflammation

Remember that tough chapter on the cell danger response? We learned how persnickety mitochondria are. If they go into cell defense mode because they are too hot or too cold, or too acid or too basic, can you imagine how much they will hate free circulating foreign bacterial DNA? A subset of people who received these modified RNA injections could lock down into cell defense mode and be unable to break free into playing offense—making proteins, working through metabolic pathways, conducting routine cellular business, etc. Lipid nanoparticles complexed with plasmid DNA would be expected to lead to chronic activation of the immune system, promoting inflammation that would be very difficult to clear. Phagocytes, designed to pick up debris outside of cells, will be furiously Pac-manning their way around human bodies trying to clear the bacterial plasmid DNA fragments, activating proinflammatory cytokines along the way. Are these potential mechanisms of prolonged vaccine injuries that many of my clinician friends are struggling to treat?

Could This Break the PREP Indemnity Clause Defense for Pharma?

Katie Ashby Koppins is an Australian lawyer who is orchestrating class action suits on behalf of people misled about COVID injection technology. She is eager to commit the movers and shakers to talking on the record.[865] Since the science is now showing that the LNP-modified RNA complex can transfer genetic material in a variety of ways, and since the pharmaceutical agencies did not disclose the presence of plasmid DNA or SV40 promoter to regulatory agencies, there may be an open door for liability for the harms that have been demonstrated so heartbreakingly. Clearly, there was a failure of good regulatory oversight worldwide. Independent scientists, free of corporate or commercial interests, are replicating experiments proving DNA plasmid contamination. Koppins reports that the tests to find contamination in the vials would have cost about ten dollars and taken a few hours.[866] She argues that the regulators should have done those tests and not taken the word of the manufacturers at face value. Good on you, Katie, as my friends in Australia say.

Conclusion

This was a hard chapter to write, not just because understanding and translating the science is difficult but confronting these discoveries is a gut punch. This will be hard for many people to understand unless they have backgrounds in biochemistry and genetics, but here is the fundamental issue: Cell uptake of partial or whole plasmids can potentially lead to alteration of the human genome. Did we just witness mass injection of innocent people with products that could alter gene expression?

FOOD FOR THOUGHT

Who do you trust? Who do you believe? Has the COVID experience fundamentally changed your perspectives on the world and its people? Do you think pharmaceutical companies should be held accountable for egregious errors?

CHAPTER 34

Is Robert F. Kennedy Jr. Wrong about Vaccines?

Few men are willing to brave the disapproval of their peers, the censure of their colleagues, the wrath of their society. Moral courage is a rarer commodity than bravery in battle or great intelligence. Yet it is the one essential, vital quality for those who seek to change a world that yields most painfully to change.

—Robert F. Kennedy

I never considered a difference of opinion in politics, in religion, in philosophy, as cause for withdrawing from a friend.

—Thomas Jefferson

Misquoted or Taken Out of Context?

For those of you who do media interviews or return calls from reporters, are you ever misquoted or taken out of context? For those of you who are doctors, is what you try to convey to a patient always what the patient understands? For those of you who are spouses, do you ever feel that what you thought you said and what your spouse heard are two different things?

When I watch a video or hear a speech from RFK Jr. in person, and then read what *CNN* or *Time* magazine reports, I am usually surprised that we saw the same speech. Remember, how we interpret what we see and hear is subjective, based on our gender, upbringing, political inclinations, and prior positive and traumatic life experiences—and our funding sources.

Childhood Vaccines and Lack of Placebos

RFK Jr. stirred up a hornets' nest when he went on record stating that almost all vaccines for children were licensed after clinical trials that did not include placebos. I watched a mainstream reporter push back against this claim, and I sincerely think she was shocked to even consider the possibility that what Kennedy said was true, given how we have all been conditioned to think vaccines are safe and lifesaving. As an experienced pediatrician, based on extensive review of vaccine clinical trials, CDC committee meetings online and reading the package inserts for vaccines given to children, I can affirm he is telling the truth. I would swear under oath that he is correct: Most childhood vaccines were not tested against true placebos. Are you shocked? I have copied and pasted the definition of placebo directly from the CDC website.

> Placebo: A substance or treatment that has no effect on living beings, usually used as a comparison to vaccine or medicine in clinical trials.[867]

Perfect placebos have classically included saline injections or pills with small amounts of sugar in them (although now that we know the effects of sugar on human metabolism, I'm not sure if a sugar pill qualifies as inert).

Attorney Aaron Siri walks us through one example. The investigation involves RotaTeq, a vaccine developed by Dr. Paul Offit's team. To protect infants from rotavirus infections, RotaTeq is given by mouth to infants, usually in the first 2–6 months. Since infants that age cannot talk about what they taste, a good placebo would have been distilled water drops. Instead, investigators used bioactive ingredients in the vaccine, leaving out only the viral component. There was no way to tell if those extra ingredients caused any identifiable problems!

Aaron Siri wrote about this on Substack on June 25, 2023.[868]

In 2018, on behalf of ICAN, we were investigating the control used in each clinical trial relied upon by the FDA to license each childhood vaccine.

In that review, we found that while the package insert for the RotaTeq vaccine says the control in its clinical trial was a "placebo,"[869] when we read the FDA's clinical trial review for RotaTeq, the ingredients of this so-called "placebo" were redacted.

So, on behalf of ICAN we submitted a Freedom of Information Act Request to the FDA for *"Documents sufficient to identify the ingredients of the 'placebo' in the prelicensure clinical trials identified in Section 6.1 of the package insert for RotaTeq."*[870]

In a response dated June 14, 2018, the FDA provided the requested documents which clearly show that the control was not a placebo. Rather, it included polysorbate-80, sodium citrate, sodium phosphate, and sucrose.[871]

These same four ingredients are also contained in RotaTeq. The only difference between the vaccine and the control is that RotaTeq also included tissue culture medium and rotavirus reassortments. So, bottom line: the control used in the RotaTeq clinical trial was **not a placebo** since it included bioactive ingredients. RFK Jr. was right!

Kennedy knows more about vaccines than many doctors and is willing to go head-to-head with vaccinologists. Dr. Paul Offit said in his article "Should Scientists Debate the Undebatable"—wait for it—"All vaccines are tested in placebo-controlled trials before licensure."[872] The only way that can be deemed "true" is if you define "placebo" to include vaccines, vaccine adjuvants, and vaccines-minus-adjuvants—all of which clearly have an effect on living beings. Two lawyers, Robert F. Kennedy Jr. and Aaron Siri, have publicly offered to debate Drs. Offit and Hotez about this issue. As a former state debate champion who cares a lot about safe vaccines for kids, I might even throw my hat in the ring for a little gender diversity. But Offit and Hotez claim the issue is "not debatable," so we are not likely to get the chance.

Here is a chart compiling the data about the use of placebos in clinical trials:

HHS'S CHILDHOOD SCHEDULE: ONE DAY TO 6 MONTHS OF LIFE			
VACCINE TYPE	TEST GROUP RECEIVED	CONTROL GROUP RECEIVED[15]	PLACEBO CONTROL?
DTaP	Infanrix (GSK)[16]	DTP	NO
	Daptacel (Sanofi)[17]	DT or DTP	NO
Hib	ActHIB (Sanofi)[18]	Hepatitis B Vaccine	NO
	Hiberix (GSK)[19]	ActHIB	NO
	PedvaxHIB (Merck)[20]	Lyophilized PedvaxHIB[21]	NO
Hepatitis B	Engerix-B (GSK)[22]	No control group	NO
	Recombivax HB (Merck)[23]	No control group	NO
Pneumococcal	Prevnar 13 (Pfizer)[24]	Prevnar[25]	NO
Polio	Ipol (Sanofi)[26]	No control group	NO

HHS'S CHILDHOOD SCHEDULE: ONE DAY TO 6 MONTHS OF LIFE			
VACCINE TYPE	TEST GROUP RECEIVED	CONTROL GROUP RECEIVED[15]	PLACEBO CONTROL?
Combination Vaccines	Pediarix (GSK)[27]	ActHIB, Engerix-B, Infanrix, IPV, and OPV	NO
	Pentacel (Sanofi)[28]	HCPDT, PolioVAX, ActHIB, Daptacel, and IPOL	NO

HHS'S CHILDHOOD SCHEDULE: 6 TO 18 MONTHS OF LIFE			
VACCINE TYPE	TEST GROUP RECEIVED	CONTROL GROUP RECEIVED	PLACEBO CONTROL?
Hepatitis A	Havrix (GSK)[29]	Engerix-B	NO
	Vaqta (Merck)[30]	AAHS and Thimerosal	NO
MMR	M-M-R II (Merck)[31]	No control group	NO
Chicken Pox	Varicella (Merck)[32]	Stabilizer and 45mg of Neomycin	NO
Combo Vaccine	ProQuad (Merck)[33]	M-M-R II and Varivax	NO
Flu[34]	Fluarix (IIV4) (GSK)[35]	Prevnar13, Havrix and/or Varivax or unlicensed vaccine	NO
	FluLaval (IIV4) (ID Bio)[36]	Fluzone (IIV4), Fluarix (IIV3) or Havrix	NO
	Fluzone (IIV4) (Sanofi)[37]	Fluzone (IIV3)	NO

HHS'S CHILDHOOD SCHEDULE: 18 MONTHS TO 18 YEARS OF LIFE			
VACCINE TYPE	TEST GROUP RECEIVED	CONTROL GROUP RECEIVED	PLACEBO CONTROL?
Tdap	Boostrix (GSK)[38]	DECAVAC or Adacel	NO
	Adacel (Sanofi)[39]	Td (for adult use)	NO
HPV	Gardasil (Merck)[40]	AAHS or Gardasil carrier solution (Sodium Chloride, L-histidine, Polysorbate 80, Sodium Chloride, and Yeast Protein) (594 subjects)	NO
	Gardasil-9 (Merck)[41]	Gardasil or Placebo (306 subjects that recently received 3 doses of Gardasil)	YES[42]
Mening-ococcal	Menactra (Sanofi)[43]	Menomune	NO
	Menveo (GSK)[44]	Menomune, Boostrix, Menactra, or Mencevax	NO
Combination Vaccines	Kinrix (GSK)[45]	Infanrix and Ipol	NO
	Quadracel (Sanofi)[46]	Daptacel and Ipol	NO
Flu[47]	Afluria (IIV3) (Seqirus)[48]	Fluzone (IIV3)	NO
	Afluria (IIV4) (Seqirus)[49]	Fluarix (IIV4)	NO
	Flucelvax (IIV4) (Seqirus)[50]	Flucelvax (IIV3) or a (Seqirus) investigational vaccine	NO

ICAN has published an 88-page report detailing their investigation.[873]

Thimerosal: Let the Science Speak

When I decided to write this book, I reached out to Mr. Kennedy for advice about how to structure it. He made a sandwich while I talked with him on FaceTime. I told him I wanted to write a very well documented, scientifically sound book about what parents were not hearing from Fauci, the CDC, and the FDA. He warned me that the public wants to read about controversy, not scientific facts that are difficult to integrate into their knowledge base. He gave as an example his book *Thimerosal: Let the Science Speak*, which was intense science. "Nobody read it," he said. Did he mean that literally? Of course not. I had read it and so did more than ten thousand other people. But dense science books are not as likely to be widely read as celebrity memoirs or thrillers.

When I was preparing to testify in the 2006 Omnibus Autism Proceeding, I read a stack of articles about adverse effects of thimerosal that filled three three-inch binders. I had done a research project in my own practice about the prevalence of neurodevelopmental disorders in patients whose mothers received thimerosal-containing Rhogam shots in pregnancy and those who did not. The babies who were exposed to prenatal thimerosal had more neurodevelopmental problems (including autism and attention deficit disorders) than the unexposed babies.[874]

I was already aware that there were significant differences in the pharmacokinetics (how the body processes) of methylmercury (the type in fish) and ethylmercury (the type in thimerosal-containing vaccines). I knew mercury liked to live in fat. I wondered if the mercury-containing Rhogam injections I had gotten during pregnancy contributed to a kidney problem in my son. Thimerosal is 49.6 percent mercury by weight; it breaks down in the body to ethylmercury and thiosalicylate.[875]

Kennedy's staff provided a stack of thimerosal papers over two-feet high for a press conference at the National Press Club, but I did not see any reporters look at the articles. RFK Jr. worked on getting mercury out of fish before he moved to getting mercury out of children's vaccines

Thimerosal Clears Blood Quickly—the Pichichero Study

One of the most widely quoted studies used by the pediatric community to downplay the potential risks to children from ethylmercury in thimerosal-containing vaccines was published by Dr. Pichichero in 2002.[876] The researchers looked at blood, urine, and stool samples of seventy-two infants in three age groups (newborn, two months, and six months) before and after thimerosal-containing vaccines. Half of the mercury in the blood was gone by 3.7 days and blood levels returned to pre-vaccination levels by thirty days. There was increased mercury in the stools after vaccination (which suggests that the babies were able to poop out some of the mercury) but no detectable mercury in urine. The researchers interpreted that to mean ethylmercury was less dangerous than methylmercury. They concluded:

> The blood half-life of intramuscular ethyl mercury from thimerosal in vaccines in infants is substantially shorter than that of oral methyl mercury in adults . . . exposure guidelines based on oral methyl mercury in adults may not be accurate for risk assessments in children who receive thimerosal-containing vaccines.[877]

Fair enough, but where did the rest of the mercury go? The researchers made an unjustified assumption that because the levels declined, mercury was eliminated, without considering the possibility that it was bound to other tissues, like in the nervous system. This is a fatal flaw in the paper.

Mercury Likes Fat

Sadly, mercury goes into the brain, a very fatty tissue. Mercury has an affinity for fatty organs like kidneys and brains. Let me take you through a compelling research study that has not gotten the attention it deserves in the nearly twenty years since it was published. Burbacher and colleagues noted that methylmercury oral exposure guidelines from the US Environmental Protection Agency were being extrapolated to injected ethylmercury exposures and wondered if that was appropriate. They compared the systemic and brain distribution of oral methylmercury with ethylmercury plus controls in forty-one monkeys, with the groups balanced for gender and weight. They exposed baby monkeys at one, two, and three weeks of age and measured blood, total mercury, and inorganic mercury at two, four, and seven days after exposure. They looked at brain levels again at twenty-eight days. There was a twofold difference in the blood concentrations of mercury, showing that levels vary among individuals. They proved that the half-life of ethylmercury from shots was significantly shorter than the half-life of oral methylmercury, making the case that methylmercury guidelines were indeed inappropriate for calculating ethylmercury exposure. The decrease in *organic* Hg in the brain over time was not statistically significant between the two mercury types.[878] However, there's a rub.

Inorganic Mercury in the Brain

The average brain-to-blood concentration ratio was slightly *higher* in thimerosal-exposed monkeys. And, yikes, "a higher percentage of the total Hg in the brain was in the form of *inorganic* Hg for the thimerosal-exposed monkeys (34% vs. 7%)."[879] Furthermore, the average concentration of inorganic mercury did not change during the twenty-eight days. The inorganic mercury was between 21% to 86% of the total mercury in the brain. The inorganic fraction from methylmercury (the fish kind) was only 6–10%. Bottom line: the type of mercury that stayed in the brain for a long time was higher from thimerosal-containing shots than orally ingested fish. This is what the scientists wrote:

> There was a *much higher* proportion of *inorganic* Hg in the brain of *thimerosal monkeys* than in the brains of [methylmercury] monkeys (up to 71% vs. 10%). Absolute inorganic Hg concentrations in the brains of the thimerosal-exposed monkeys were approximately twice that of the [methylmercury] monkeys.[880]

Synthesizing Science

Now, let's tie these results in with the Pichichero study. Burbacher noted that the blood mercury half-life was remarkably similar between his baby macaques and Pichichero's baby humans. This gives me confidence that the monkey studies show concerns that are applicable to human babies. Yes, there is a significant difference in the way fish-derived mercury and shot-derived mercury are processed. But we should not be reassured by the rapid disappearance of mercury from an infant's blood after vaccination! Mercury is extensively distributed all over the body, especially in fatty tissues like brains and kidneys. Only one-twentieth of the body burden of mercury is confined to blood vessels, so measuring thimerosal breakdown products in the blood does not give any information about total body burden. Do most pediatricians know about this? RFK Jr. does. Multiple studies decades ago found that organic mercury cleared from the brain more quickly than inorganic methylmercury and that changes in astrocytes and microglia followed mercury exposure.[881,882,883,884] (Note that the five references I cited are from 1994, 1995, and 1996, so this was not new knowledge when the thimerosal debate began.) The estimated half-life of inorganic Hg in adult brains varied greatly across some regions of the brain, from 227 days to 540 days,[885] which is a long time to have a neurotoxin in a person's brain. In some cases, the inorganic type even **increased after the exposure ceased**.[886,887] That's not good.

Burbacher explains that studies show persistence of inorganic mercury is associated with more microglia in the brain. Microglia are immune cells that have a "mothering" function. I was at an NIH meeting when Diana Vargas presented groundbreaking work from Johns Hopkins that showed that brains of people with autism exhibit a marked activation of microglia and an active neuroinflammatory process.[888] The evidence of microglial activation and neuroinflammation in the brains of children with autism drove a lot of our research agendas and therapeutic strategies during my tenure as Medical Director of the Autism Research Institute.

Burbacher ended his article with the conclusion that "knowledge of the toxicokinetics and developmental toxicity of thimerosal is needed to afford a meaningful assessment of the developmental effects of thimerosal-containing vaccines."[889] A Google Scholar search for "mercury and neurotoxicity" yields 43,700 articles. In 1990, as new mercury-containing vaccines were being added to the childhood schedule, there were published articles showing how mercury travels across the blood-brain barrier.[890]

Since Burbacher published his study, scientists have demonstrated that neurodegeneration from heavy metal toxicity can cause devastating consequences. Genetic, endogenous, and environmental factors leading to oxidative stress, mitochondrial dysfunction, and phosphorylation impairment are the major shared pathways.[891] Much of my work trying to help children with autism involves treating oxidative stress, mitochondrial dysfunction, and phosphorylation abnormalities.

When aluminum and mercury are given together in vaccines, the combination is more toxic than either of them on their own, causing inflammation in human brain cells. Parents, if your child gets a multi-dose flu vaccine and other well-baby vaccines on the same day, they will likely be exposed to both neurotoxins simultaneously, kicking off the synergy.[892]

Remember that Hopkins researchers showed neuroinflammation in the brains of kids with autism.[893] My colleagues and I have been treating neuroinflammation in kids' brains for decades.

Thimerosal toxicity disrupts glutamate signaling and excitotoxicity, both of which are seen in autistic children. Redox homeostasis is also disrupted by mercury and seen in kids with autism. [894] Microglial and astrocyte activation (as shown by Vargas and her colleagues) are immunotoxic effects of mercury. Genetic predispositions make some children more vulnerable to thimerosal toxicity than others. Genes related to the metabolism of glutathione, which is depleted in most kids with autism, are crucial in mercury detoxification.[895]

In 1999, the American Academy of Pediatrics and the US Public Health Service encouraged "all government agencies to work rapidly toward reducing children's exposure to mercury from all sources," including thimerosal-containing vaccines[896] However, multi-dose vials containing thimerosal continued to be shipped to other countries, in a terrible affront to developing countries. Those parents love their children just as much and don't want neurotoxins in their vaccines either.[897]

In 2001, an Institute of Medicine analysis found there was not enough evidence to render an opinion on the relationship between thimerosal and developmental disorders in children and recommended further study.[898]

I attended an Institute of Medicine meeting in 2004 and was shocked at their conclusions *"The committee also concludes that the body of epidemiological evidence favors rejection of a causal relationship between thimerosal-containing vaccines and autism.*[33] As Burbacher wrote, the IOM "appears to have abandoned the earlier recommendation as well as backed away from the American Academy of Pediatrics goal. This approach is difficult to understand . . ."[34] He is being diplomatic; I would say horrifying.

Conspiracy Theories or Knowledge of COVID and Genomics?

It has been widely reported that RFK Jr. spread unfounded conspiracy theories about COVID-19. A videoed dinner table conversation in which he discussed published scientific papers about the racial and ethnic differences in ACE 2 receptors (the target of SARS-CoV-2) was distorted to suggest that he is nothing more than a conspiracy theorist. His concern about bioweapons development was reinforced by spending two and a half years doing research for *The Wuhan Cover-Up: And the Terrifying Bioweapons Arms Race.*[899]

I do not think RFK Jr. is racist. First, Kennedy was exposed to more ethnic diversity during his childhood than most of us in those days. He embraces people from all races, genders, and walks of life. Second, his remarks referred to research about ACE receptor differences in various people,[900] including ethnic groups, which is a legitimate avenue for research. Understanding how individuals respond differently to COVID (or other illnesses) allows doctors to make educated judgments about risks and treatment options. Age stratification data show that the elderly are much more at risk than children, for example. Scientific papers studying possible genetic factors affecting the immune response to COVID-19 emerged starting in 2020. Scientists wondered whether strong expression of the angiotensin-converting enzyme 2 (ACE-2) in various sub-populations could predict more severe disease. One fact that emerged is that men have genes associated with more severe disease. Good to know. A doctor might be wise to be more aggressive with treatment for a seventy-year-old man versus a thirty-year-old woman.

Because the ACE protein has a huge role in how easily the virus gets into cells, genetic factors affecting ACE will impact infection. Endocytosis (moving into the cell) of viral particles is initiated by the receptor-binding domain of the spike protein, specifically the S1 subunit.[901]

When the S1 subunit is cleaved, the virus is released into the cell.[902] Looking at small genetic differences, called single nucleotide polymorphisms or SNPs (pronounced "snips") gives insights about how individual patients process certain illnesses or metabolic challenges. There are genetic susceptibilities to COVID, which should come as no surprise.[903]

Recently, a study in *Nature Genetics* investigated the associations across the whole genome. They found a genetic variant (rs190509934—you can look it up in SNPedia) that downregulated ACE 2 expression by 37 percent, reducing the risk of COVID-19 infection by 40 percent.[904] That, too, is good to know. Other researchers found that ethnic differences in ACE 1 polymorphisms (changes in the genetics of the ACE receptor from person to person) like JT2 seem to explain the difference in mortality between West and East Asia.[905]

Tasnim Beacon et al. wrote about epigenetic regulation of ACE receptors and SARS-CoV-2 in a 2021 article in *Genome*.[906] That same year, Zafer Yildirum et al. found that the Lys26Arg allele, which is found more often in Ashkenazi Jews, can result in more severe disease.[907] Human genome interactions in Ashkenazi Jews and other ethnic groups are also relevant for how they affect metabolization of repurposed drugs used to treat COVID.[908]

Numerous other researchers have investigated how genomic predispositions in people of African descent affect their response to COVID.[909,910,911] Other researchers found a negative correlation between ACE 2 expression and COVID-19 fatality in China.[912] Kennedy understands this important arena of genomics; the reporters do not. Kennedy argues for the importance of individualized patient care. Isn't that what we want from our health-care providers?

Anne Frank, Hitler, and the Lincoln Memorial

This is an example of CNN not only taking Mr. Kennedy's remarks out of context but reporting the topic of his remarks in error. They misreported that "anti-vaccine advocate Robert F. Kennedy Jr. suggested the stubbornly unvaccinated are worse off than Anne Frank. . . . Even in Hitler Germany (sic), you could, you could cross the Alps into Switzerland. You could hide in an attic, like Anne Frank did."[913]

I was there when he made the controversial remarks at the "Defeat the Mandates" march at the Lincoln Memorial because I do not believe in mandating experimental gene therapy products or giving the exact same treatment to everyone despite their individual risks and medical histories.

Mr. Kennedy was pointing out that *if digital surveillance existed* during World War II, as it does now, Anne Frank would not have stayed hidden as long before she was tragically captured and killed. When I see CNN make such a fundamental reporting error about something I heard with my own ears (and they could have easily replayed the recording to get it right), I wonder whether I can trust anything they say about the multitudes of geopolitical subjects I don't know much about. CNN should have done their job and looked at the entire speech.

Criticizing Mr. Kennedy

A widely circulated story in *Vanity Fair* on May 13, 2021, said the following:

> It is difficult to comprehend how a spectacularly educated person (undergrad at Harvard, classes at the London School of Economics, law school at the University of Virginia, and a master's in environmental law from Pace University) feels comfortable promoting the kind of arguments that Kennedy puts forth—ones in opposition to scientific consensus."[914]

Kennedy would be the first to acknowledge that his last name may have facilitated his access to such higher institutions of learning. In a weird synchronicity, we overlapped at UVA when I was a pediatric resident and he was a law student. I would argue that it is precisely his "spectacular education," combined with his childhood training around the dining room table to look at all sides of an issue, as well as his insatiable curiosity that should make people want to understand the arguments he puts forth.

I take issue with the author's allegiance to the concept of "scientific consensus." Science should be ever evolving. There is no question that "scientific consensus" has gotten many things wrong over the years, especially when it comes to medical science. Scientists with integrity welcome discussions with scientists who have different training and fields of expertise. For Kennedy to push back against the manufactured "scientific consensus" during COVID was an act of courage and intelligence.

Time Magazine published the following in 2019:

> No vaccines except some formulations of the flu vaccine contain thimerosal, and the type of mercury it uses is *ethylmercury*, which is cleared from the body *quickly and harmlessly* . . . RFK, Jr. could take a bigger step still if he heeds his family, learns his science and walks away from the anti-vax rhetoric for good. Every day misinformation gets peddled is another day that children will suffer."[915]

You now know that *Time*'s statements about thimerosal and ethylmercury are not true. What does that say about the quality of their reporting?

Smart, Well-Read, but Not Perfect

Rarely have I met someone as open to feedback as Bobby Kennedy Jr. When he discussed the differences in all-cause mortality among the participants in the original Pfizer trial, he correctly pointed out that there were more deaths in the vaccine group than the placebo people. During a Zoom meeting, an epidemiologist, a bench scientist, and a pediatrician (OK, it was me) pointed out that the way he explained the differences did not allow for confidence interval calculations. He did not seem offended at all and accepted our criticism graciously.

Based on reading his books, helping to edit his book on Fauci, listening to many of his podcasts, and talking with him in person, my judgment is that he cares deeply about truth, justice, and children. He has an almost uncanny ability to withstand multiple hostile verbal attacks that would derail me. After his relatives criticized him in the press several years ago, he texted me, "I am used to it; it really does not bother me anymore." He believes in the freedom of citizens to make their own choices about vaccination. He wants to help those who have been injured by vaccines. I know his dad and uncles would be very proud of his courage in standing up for his principles. I bet Kennedy would agree with the Jefferson quote at the top of this chapter, as he nurtures friendships with those with whom he disagrees. In fact, vigorous debates with friend, foe, or family seem to energize him.

FOOD FOR THOUGHT

Were you surprised by the lack of placebos in childhood vaccines documented on the chart? What about the evidence for inorganic mercury accumulation in the brain? Were you aware that these topics are rarely covered honestly in the news? Do you think Kennedy is wrong? Do you think his remarks about COVID predispositions make him racist or antisemitic? What do you think about his fortitude in the face of so much criticism?

CHAPTER 35

What Is Informed Consent and Why Is It Important?

How can I provide informed consent to parents if I do not know what is in COVID vaccines?

—Renata Moon, MD

The consent of the governed is not consent if it is not informed.

—Edward Snowden

True Informed Consent

Informed consent is defined as "permission granted in full knowledge of the possible consequences, typically that which is given by a patient to a doctor for treatment with knowledge of the possible risks and benefits."[916] During COVID, recipients of the vaccines were denied informed consent because the package insert for the vaccines were initially BLANK. You read that right. The boxes with vials of COVID vaccines initially included a folded-up piece of paper the usual size of package inserts but without any information about the contents, mechanism of action, or expected side effects. As mentioned earlier, Renata Moon, MD, a pediatrician from Washington State, testified at a Senate hearing that she could not possibly give parents information to make informed consent decisions when the package inserts for COVID products were "intentionally blank."[917] People who were mandated to get COVID injections to keep their livelihoods were not being offered true informed consent, since coercion was involved.

Putting Informed Consent Front and Center

Dr. Paul Thomas, the pediatrician who lost his medical license after his practice data revealed inconvenient truths about chronic disease and vaccines, believes passionately in all patients' right to fully informed consent. He describes it as the guiding principle of his practice of pediatrics. Five aspects are included in the process:

1. disclosure – the clinician must explain the risks and benefits of the proposed procedure or treatment,
2. understanding – the patient or guardian must comprehend what is explained,
3. voluntariness – there should be no coercion,
4. competence – the patient or guardian must not have physical or mental impairments that interfere with the process of giving consent, and, finally,
5. consent – which can be verbal or written.[918]

Anyone who has had to undergo surgery probably signed a document that laid out the presumed benefits of the proposed procedure and a list of potential complications, which tend to include death. Usually, a nurse or administrator "gets consent" from the patient. In pediatrics, immunizations are the most common intervention for which patients or their parents should learn about the risks and benefits. In some practices, there is an assumption that parents may consent, verbally or in writing, after a discussion that might consist of just "your baby is due for these four shots today." A full discussion of each vaccine, the benefits and potential risks, and the synergistic risks of getting multiple vaccines on the same day followed by answering parental questions would be difficult to fit into the seven- to fifteen-minute office visits typical in insurance-based pediatric practices today.[919] A telephone survey reported in the journal *Pediatrics* revealed that the twenty percent of parents who reported longer than twenty-minute visits for well-child care were more satisfied and had the opportunity to ask questions.[920]

Clinicians administering vaccines to children are required by law to show patients or guardians VIS forms (Vaccine Information Statements) edited and updated by the Centers for Disease Control (CDC). Each VIS form has information about reporting adverse events following vaccination to VAERS (Vaccine Adverse Events Reporting System), but there are no penalties for physicians who fail to report.[921]

Dr. Thomas prominently displayed the CDC Recommended Immunization Schedule in each exam room. Some parents chose to follow it precisely. Others chose to use a modified vaccine schedule, such as his "vaccine-friendly plan." Some chose to delay or decline vaccinations, as is their legal right in Oregon. By allowing for parental fully informed consent, which, by definition, has to include the ability to refuse a recommended treatment, Dr. Thomas obeyed the ethics of medicine, complied with the laws of Oregon, and provided a medical home to those families who had been kicked out of other practices for not being up-to-date on their shots per the CDC recommended schedule. Remember that he got kicked out of practicing pediatrics after publishing his data that unvaccinated or under vaccinated kids had fewer chronic illnesses than fully vaccinated kids.

Coercion Has No Role in True Informed Consent

Ideally, consent should happen in a time frame that allows you to consider your options and decide. Consent for elective surgery, for example, should be obtained when surgery is scheduled, not when staff is ready to wheel you into the operating room. Similarly, consent for immunizations should be obtained before the vaccine is in the syringe.

No type of coercion or bullying or bribing should be a part of the informed consent process, according to the principles of the Nuremburg code.[922] During COVID, I was horrified to hear that peer pressure was used to manipulate children into getting COVID injections. It is unethical to promise pizza parties or field trips in exchange for everyone in the class getting an injection, especially since the shots were approved under emergency use and did not have convincing data about benefits for pediatric patients nor long-term data about side effects. The strategy of bribing children exploited the children's developmentally appropriate desire for approval from their peers, playmates, and teachers.

Children Receiving Vaccines Without Parental Consent

In Washington, DC, the mayor declared that students as young as eleven years old could consent to a COVID injection without the consent of the parent and without the primary care physician's knowledge. Fortunately, that ill-conceived strategy was overturned in court after lots of work by parents and professionals who pointed out that tweens do not have the cognitive maturity to make abstract decisions about medical interventions.[923,924,925]

One pragmatic flaw in the plan to immunize children at school without parental knowledge is this: If the child reacts badly to the vaccine or develops a new medical problem, the parent and doctor will not know the immunization was a preceding event that could have triggered the problem. Immunizations can be antecedents, triggers or mediators of medical conditions. Receipt of vaccines is information that must be considered in medical decision making.

The 1986 Act: Liability Waived for Vaccine Adverse Events.

COVID was traumatic for many people around the globe. Fear of losing loved ones or becoming ill made it more difficult for many people to remember their rights as patients, parents, and citizens. Now that COVID is endemic, meaning here to stay and for us to live with, it is important to think about everyday aspects of informed consent. As parents, we are surrogate decision makers for our children. In pediatrics, informed consent is crucial when parents consider making vaccine decisions for their children. I hope the information in the earlier chapters has given you more information about the risks versus benefits of yearly COVID injections for your child.

Once a vaccine is put on the US recommended childhood immunization schedule, the pharmaceutical industry is protected from liability if adverse effects occur.[11] You cannot sue the manufacturers or your medical providers if your child has a severe disabling reaction. Death *immediately* after vaccines is extremely rare, but it can happen; it did happen with COVID shots.[926]

This pharmaceutical industry protection from being sued[927] contrasts with product liability rules for other companies. For example, Ford was held responsible for Pintos that exploded, Monsanto was held legally liable for lymphomas caused by Roundup, and the tobacco industry settled with victims when it was proven in court that cigarettes cause lung cancer. One could argue that money never truly compensates the victims who became disabled or died. Profitable industries build lawsuit losses and court costs into their budgets as the cost of doing business.[928,929,930,931]

In the United States, if your child has a bad reaction to a vaccine, you must go through the National Vaccine Injury Compensation Program, also known as "vaccine court." I have been there as an expert witness for injured children, and despite the program's billing, the evidentiary bar is very high to get justice and compensation for vaccine injuries. Compensation is considered for injuries on the predetermined vaccine injury table and are limited to times soon after the injections.

Every medical product I prescribe has both known and unknown side effects. Medical professionals must constantly make treatment decisions based on imperfect and incomplete information. Patients and parents must decide whom to trust for medical advice. As my colleague Eugene Dean in Bermuda says, "a doctor is meant to inform our decisions, not make them for us. We need to take responsibility."

Difficult Decisions About Childhood Vaccines

Before your children are conceived or at least before they are born, parents need to have honest discussions about their views on vaccines. Parents frequently disagree about multiple aspects of child rearing, which is the hardest job most people ever have. Here are some crucial vaccine questions to consider:

1. Do we want to give hepatis B vaccine at birth if our baby is not at risk?
2. Do we want to follow the CDC's recommended schedule?

3. Is there merit in modifying the schedule to avoid numerous vaccines (and more aluminum) given at the same time?
4. Should we give vaccines when our child is sick or still on antibiotics?
5. Do sincerely held religious or philosophical beliefs impact our decisions about certain (or all) vaccines?
6. Are unvaccinated or under vaccinated children healthier?
7. What level of risk do vaccine-preventable diseases pose in my country?

Package inserts for almost all vaccines are now on the internet. You can download and print them to see lots of information about their mechanisms of action, potential side effects, the frequency of common side effects and ingredients. In my pediatric office, we had notebooks of vaccine information in each exam room. We did not believe a "one size fits all" schedule is appropriate for vaccines. I did not feel comfortable offering vaccines while the child was acutely ill. I took unexplained gastrointestinal symptoms or evolving neurologic disorders into consideration. I considered the family history and the child's developmental status before making any vaccine recommendations. Even then, the family was always free to disagree with me. My staff did not shame or blame parents for sincerely made decisions. Other pediatricians see the risk-to-benefit ratios differently, and we hear reports of parents feeling pressure from medical providers who are worried about vaccine hesitancy.

The Right to Decline Medical Treatments

The other side of the *informed consent* coin is *informed dissent*. You may have concerns about heavy metals or ingredients in vaccines, as I do. Even doctors like me who try to order the "cleanest" vaccines have limited choices in the marketplace. You may look at the data indicating that unvaccinated children have fewer chronic illnesses and decide to take the risk that your child might get a vaccine-preventable disease. For those of you who believe declining or delaying vaccines is irresponsible, remember that people with free will make decisions all the time that are not in their own best interests. Americans have lots of chronic diseases that are linked to lifestyle choices of eating poor quality food, not exercising enough, and taking other health risks. Doctors still take care of people who develop lifestyle-related diabetes, heart disease, or obesity.

Most parents sincerely try to make good health decisions for their children. Be aware that your doctor may not know the answers to all your questions—we are fallible human beings after all. But I do recommend you find doctors who respect your right to ask questions, especially questions about your children. For most people, love for their children is such a powerful force that they are willing to make huge sacrifices to nurture and raise them. Decisions about COVID vaccines should be well thought out, not reflexive or passive, based on peer pressure or inadequate medical information.

FOOD FOR THOUGHT

Did you think you were adequately informed about the risks and benefits of COVID vaccines? Did you know that vaccine injuries are handled by "compensation programs" and not actual courts? Are you confused about the risks and benefits of childhood immunizations? Did you know that the 1986 Act and the PRPE Act provided liability protection to vaccine makers?

CHAPTER 36

Where Do We Go from Here?

No one saves us but ourselves. . . We ourselves must walk the path.
—Gautama Buddha

You never change things by fighting the existing reality. To change something, build a new model that makes the existing model obsolete.
—Buckminster Fuller

Take a Strong Stand Even If You Are in the Minority

Taking a stand often means stepping into discomfort. My decision not to recommend COVID vaccines for pediatric patients as young as six months challenges the prevailing guidance from the American Academy of Pediatrics and the Centers for Disease Control. It's not an easy position, especially as someone raised in the South with the mantra, "If you can't say something nice, don't say anything at all." Criticizing government agencies and enduring online backlash is a departure from my nature as a people-pleaser. The mean comments sting, so I've learned to avoid reading them.

As I stated in the first chapter, our perspectives shape how we see the world. Now, I'll share my perspective on how parents and grandparents can protect their children from being commodified by pharmaceutical companies and conflicted agencies.

Accept Past Vaccine Decisions with Grace

Many parents chose to vaccinate their children against COVID based on the best available information. Children are resilient, and most may tolerate these interventions without long-term consequences. I hope that's true for your family.

As new evidence of vaccine-related adverse events emerges, some parents may wrestle with guilt. This emotion is common, especially for mothers, who often carry a disproportionate sense of responsibility for their child's well-being. I have had mothers look at me with fear in their eyes and ask if the two glasses of wine they had before they realized they were pregnant might have caused their child's autism. The relief that washes over their face when I reassure them is a joy to watch and a testament to the powerful evolutionary forces that strengthen the incredible urge for mothers to protect their cubs. Fathers have been visibly uncomfortable while working up the courage to ask me if the roughhousing that is instinctive for fathers to do with their toddlers may have hurt the child. Assuming no falls from high heights, the answer is that rough and tumble play is part of normal development. Letting go of guilt is transformative. Parents need space to focus on their children's complex needs rather than dwell on past decisions. Remember, you acted out of love and with the information you had. As Robin Williams's character in *Good Will Hunting* said, "It is not your fault."

All the usual sources of information—your government, your news channels, and doctors and nurses you trust—were likely telling you that COVID vaccines were safe and effective. Many of those thought they were telling you the truth; some lied. Some doctors have come forth to confess that they recommended the vaccines early on and now feel guilty that some of their patients had bad side effects. Many more doctors cannot deal with the emotional burden of realizing they hurt people they were trying to help. I think this is a barrier to pediatricians recognizing vaccine injury in their patients.

Make Vaccine Decisions Based on Individual Risks and Benefits

The first year of the pandemic unfolded like a high stakes drama—with an enemy virus and heroic "vaccines." Fauci characterized the arrival of the vaccines as akin to the calvary coming. Yet, the narrative missed crucial nuance. A "one-size-fits-all" approach to vaccination ignored individual risk factors like age stratification and preexisting conditions, leading to unintended consequences.

Moving forward, demand transparency. If vaccine package inserts are blank, take that as a warning. Decisions should be based on clear risk-benefit analyses tailored to your child. If your child got the vaccine and did well, you taught your child an important lesson about altruism and the value of trying to protect one's tribe. Altruism, while noble, must never replace informed consent.

Explore New Perspectives

The COVID era exposed deep flaws in our systems, including collusions between governments and corporations. Educate yourself about the following:

- The origins and risks of gain-of-function research, which creates more dangerous viruses funded by taxpayer dollars.
- The Department of Defense's oversight of vaccine development under Operation Warp Speed.
- The history of vaccine manufacturing errors and cover-ups to maintain public confidence.
- The push to accelerate vaccine timelines by removing regulatory safeguards.

Seek out diverse viewpoints, even those outside your comfort zone. Sophisticated computer algorithms feed us what we already believe. Check out arguments that are counter to what you already know or believe. Challenge yourself to break free from echo chambers.

Recognize Manipulation by Social Media, Marketing, and Censorship

The pandemic showcased the power of coordinated messaging by governments, pharmaceutical companies, and marketing strategists. Voices like Bobby Kennedy Jr. and Barbara Loe Fisher, who raised legitimate concerns, were deplatformed. This censorship created a chilling effect on open dialogue. The National Vaccine Information Center was thrown off YouTube,

Instagram, Twitter, and Facebook. Fisher testified under oath to Senator Ron Johnson's committee on February 26, 2024, during a censorship campaign condoned by the "highest levels of government. . . . I never imagined when I began this work in 1982 that the day would come when I would not be able to exercise freedom of thought and conscience in the country I love."[932] Between the time I wrote these chapters and sent my footnotes to be formatted for publication, many citations and CDC slides were no longer on the web. This was a chilling experience for me. History is literally being rewritten by removing information documenting harms of COVID products!

Listen to Rebels

During the COVID crisis, I watched as many of my colleagues, often highly respected as academic leaders or expert clinicians, had their Wikipedia pages rewritten to include disparaging names like "anti-vaxxer" or "conspiracy theorist." Question: What is the difference between a conspiracy theory and scientific truth? Answer: the passage of time sometimes—as happened during the COVID era.

History vindicates many who were initially dismissed as conspiracy theorists. Galileo was persecuted for observing the solar system's truth. Ignaz Semmelweis was ostracized for advocating handwashing to prevent infection. Dr. Barry Marshall, who discovered Helicobacter pylori as a causative agent of stomach ulcers, was mocked at my alma mater, the University of Virginia, until he took a vial of the bacteria himself and got a biopsy-proven ulcer. During COVID, dissenting experts faced character assassination but often proved correct over time.

Be willing to question authority. Truth often emerges through resistance.

Fight for Parental Rights

Global organizations like the World Health Organization have proposed pandemic treaties that could erode local sovereignty. Such measures risk centralizing decision-making, stripping parents of their rights. I would be embarrassed and ashamed if the United States and the WHO dictated what African countries must do in the next pandemic, since low mortality numbers across the Africa continent put the United States to shame.

Legislative proposals in the US already seek to bypass parental consent for vaccines administered in schools. Protect your rights by staying informed and advocating for local decision-making. Parental instincts, honed over millennia, remain a cornerstone of our children's well-being.

Keep Your Child Healthy by Getting Back to Basics

Humans need love and relationships, light and dark, shelter and sleep, food and play. Amid the technological marvels of modern medicine, fundamental practices often hold the key to health:

- Nourish your child with whole, unprocessed foods.
- Encourage active play and limit screen time.
- Prioritize sleep to support brain detoxification and learning.
- Foster loving relationships and teach stress management.

These basics build a foundation for lifelong health, happiness, and joy.

Consider Natural Healing Strategies Next Time

Throughout this book I have advocated natural anti-inflammatories like Quercetin and vitamins A & C & D for fighting viruses. While I refrain from giving specific dosages here, refer to the Institute for Functional Medicine approaches,[933] the American Academy of Anti-Aging Medicine Pediatric Modules,[934] and the Medical Academy of Pediatrics and Special Needs website.[935] Be on the lookout for emerging guidelines from the Independent Medical Alliance and the American Academy of Parents.

Faith, Forgiveness, and Fractured Families

The COVID crisis fractured many relationships. The unvaccinated were called stupid and selfish and often disinvited to Thanksgiving and Christmas. Families divided over vaccine decisions often declared the topic off-limits to preserve harmony. Forgiveness is essential to rebuilding these bonds. Fear and stress compromise our ability to think critically and act with compassion.

Rediscovering faith and community may help restore a sense of purpose and direction. The places where people communally worshipped, and worshippers supported one another, were deemed less important and closed when liquor stores and big box altars to consumerism were left open. Consider that this era may have, like a tornado, blown many of us off track. The COVID era challenged many of our core values; it's time to find our true paths again.

Build a Better Model: Take Away What Harms and Give What Heals

The COVID pandemic exposed profound defects in our medical systems. Many disillusioned professionals are now working to create parallel systems rooted in informed consent and collaborative decision-making. These new models aim to:

- Address vaccine injuries and long COVID with dedicated care.
- Focus on cellular-level healing, root causes, and holistic medicine.
- Remove conflicts of interest from medical practice.
- Restore health, not just address symptoms.

Doctors gain trust by admitting uncertainty and learning alongside their patients. Medical gaslighting has no place in true healing. Collaboration, not authoritarianism, should be at the core of the new model. Clinicians who help the vaccine injured will need to know how to treat oxidative stress, mitochondrial dysfunction, detoxification impairments, and much more. But first, we must avoid doing harm.

Parental instincts have been honed over millennia to ensure the survival of the species and must be protected. Let people of goodwill work together to learn from the lessons of COVID and create the future our children deserve.

FOOD FOR THOUGHT

Did the COVID crisis affect any important relationships with your friends or families? Did the COVID experience teach you any valuable insights?

Afterword

This book was born from the events of March 2020, a moment when the world faced unprecedented challenges. By late 2021, mounting evidence suggested that COVID-19 countermeasures were disproportionately harming children—an observation that demanded deeper scrutiny.

As this book goes to press, Mark Zuckerberg has revealed that Facebook faced government pressure to censor speech that diverged from official narratives. Whether his public acknowledgment and promise to cease such practices reflect a genuine change of heart or external influences remains unclear. Meanwhile, Dr. Anthony Fauci has received a presumptive pardon for any potential misconduct dating back to 2014, raising questions about the Biden administration's awareness of underlying issues.

In parallel, mainstream medical outlets have noted a troubling rise in strokes and late-stage cancers among young people, though they remain uncertain about the causes driving this shift. These developments underscore the urgent need for transparency and accountability as we navigate the long-term consequences of the pandemic response.

This book highlights several individuals who have been nominated for key leadership roles in public health. Among them are Jay Bhattacharya, MD, PhD, proposed to lead the National Institutes of Health, and Marty Makary, MD, nominated for the Food and Drug Administration. Dave Weldon, MD, a former House representative from Florida and a longstanding advocate for autism and vaccine safety, has been put forward to head the CDC.

Meanwhile, child health advocates are closely watching the potential confirmation of Bobby Kennedy Jr. as Secretary of Health and Human Services. Despite significant public relations campaigns labeling him an anti-vaccine conspiracy theorist, Kennedy continues to champion critical issues such as healthier food systems, eliminating environmental toxins harmful to children, and ensuring vaccine safety and transparency. His nomination challenges Congress to weigh these concerns against the controversies surrounding his public persona.

Many clinicians lacked the critical information needed to make informed decisions for their patients during the pandemic. For those who now feel betrayed or embarrassed by this reality, they deserve our utmost compassion and understanding.

However, there were others—clinicians who dismissed or ridiculed colleagues successfully treating COVID-19 with repurposed medications, government officials who prioritized allegiance to a rigid narrative over analyzing emerging data, and individuals who shamed or dismissed the vaccine-injured—who must now reflect on their actions. Accountability is essential. These individuals should consider why they acted as they did and take steps to make amends or at least accept responsibility.

As for pharmaceutical executives who profited immensely while denying promised support for vaccine-injured patients, greed and pride are not justifications for neglecting those harmed. This lack of accountability undermines public trust and highlights the need for greater oversight.

Fear-driven messaging, which activates instinctive reactions rather than rational decision-making, played a significant role in the challenges we faced. Mandating new vaccine technologies that failed to prevent transmission and, in many cases, increased susceptibility to COVID-19 was not a sustainable solution. Moving forward, we must act with integrity and foresight, crafting plans that safeguard our children's future and prevent a repetition of the failures witnessed during the COVID crisis.

COVID-19 Deaths Before and After Mass Vaccination Program[1]

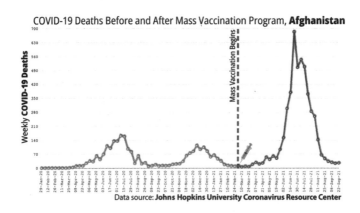

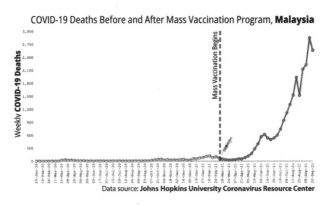

---

1   Appendix charts courtesy of Ed Dowd, *Cause Unknown: The Epidemic of Sudden Death in 2021 and 2022*, pages 129–137.

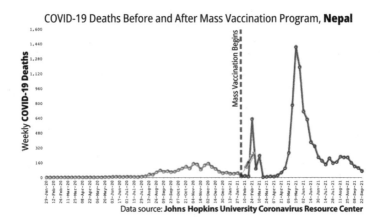

COVID-19 Deaths Before and After Mass Vaccination Program, **Nepal**

Data source: **Johns Hopkins University Coronavirus Resource Center**

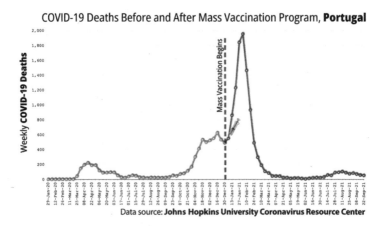

COVID-19 Deaths Before and After Mass Vaccination Program, **Portugal**

Data source: **Johns Hopkins University Coronavirus Resource Center**

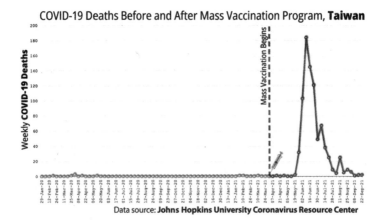

COVID-19 Deaths Before and After Mass Vaccination Program, **Taiwan**

Data source: **Johns Hopkins University Coronavirus Resource Center**

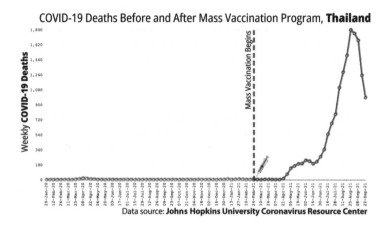

COVID-19 Deaths Before and After Mass Vaccination Program, **Thailand**

Data source: **Johns Hopkins University Coronavirus Resource Center**

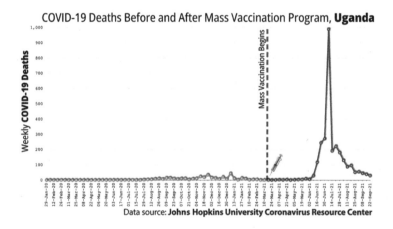

COVID-19 Deaths Before and After Mass Vaccination Program, **Uganda**

Data source: **Johns Hopkins University Coronavirus Resource Center**

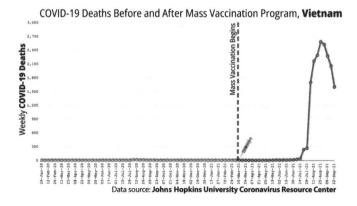

COVID-19 Deaths Before and After Mass Vaccination Program, **Vietnam**

Data source: **Johns Hopkins University Coronavirus Resource Center**

Recommended Reading

Books

Attkisson, Sharyl. *Follow the Science: How Big Pharma Misleads, Obscures, and Prevails*. Harper, 2024.

Bhattacharya, Jay. *Health Economics*. Red Globe Press, 2013.

Children's Health Defense. *Profiles of the Vaccine-Injured: "A Lifetime Price to Pay."* Skyhorse, 2020.

Craig, Clare. *Expired: COVID the Untold Story*. Publishing Aloud, Ltd., 2023.

Hooker, Brian. *Vax-Unvax: Let the Science Speak*. Skyhorse, 2023.

Kennedy Jr, Robert F. *The Real Anthony Fauci: Bill Gates, Big Pharma and the Global War on Democracy and Public Health*. Skyhorse, 2021.

Kennedy Jr, Robert F. *The Wuhan Cover-Up: And the Terrifying Bioweapons Arms Race*. Skyhorse, 2023.

Kheriaty, Aaron. *The New Abnormal: The Rise of the Biomedical Security State*. Regnery, 2022.

C. Klotz, editor. *Canary In a Covid World: How Propaganda and Censorship Changed Our World*. Canary House Publishing House, 2023.

Canary In a (Post) Covid World: Money, Fear and Power. Canary House Publishing House, 2024.

Kory, Pierre. *The War on Ivermectin: The Medicine that saved Millions and could have ended the Pandemic*. Skyhorse, 2023.

Leake, John and McCullough, Peter. *The Courage to face COVID 19: Preventing Hospitalization and Death While Battling the Bio-Pharmaceutical Complex*. Skyhorse, 2022.

Makary, Marty. *Blind Spots: When Medicine Gets It Wrong, and What It Means for Our Health*. Bloomsbury Publishing, 2024.

McCarthy, Ken. *What the Nurses Saw: An Investigation into Systemic Medical Murders That Took Place in Hospitals During the COVID Panic and the Nurses Who Fought Back . . . Their Patients*. Brasscheck Press, 2023.

Mercola, Joseph. *The Truth about COVID: Exposing The Great Reset, Lockdowns, Vaccine Passports, and the New Normal*. Chelsea Green, 2021.

Paul, Rand. *Deception: The Great Covid Cover-Up*. Regnery Publishing, 2023.

O'Toole, Zoey, editor. *Turtles All the Way Down: Vaccine Science and Myth.* The Turtles Team, 2022.

Pover, Caroline. *Worth a Shot?: Secrets of the Clinical Trial Participant Who Inspired a Global Movement—Brianne Dressen's Story.* Skyhorse, 2024.

Thomas, Paul. *Vax Facts: What to Consider Before Vaccinating at All Ages & Stages of Life.* Morgan James Publishing, 2024.

Tyson, Brian and Fareed, George. *Overcoming COVID Darkness: How Two Doctors Successfully Treated 7000 Patients.* Self-published by Tyson and Fareed, 2022.

Winter, Nasha. *Metabolic Approach to Cancer: Integrating Deep Nutrition, the Ketogenic Diet, and Nontoxic Bio-Individualized Therapies.* Chelsea Green, 2017.

Wolf, Naomi. *The Pfizer Papers: Pfizer's Crimes Against Humanity,* War Room Books, 2024.

Substack

Dr's Newsletter: Christopher Exley, PhD
The Truth Barrier: Celia Farber
Jefferey Jaxon on Substack
Pierre Kory's Medical Musings
Human Flourishing: Aaron Kheriaty, MD
Courageous Discourse with Peter McCoulough, MD and John Leake
The Forgotten Side of Medicine: A Midwestern Doctor
Unacceptable Jessica: Jessica Rose, PhD
James Roguski on Substack
Injecting Freedom: Aaron Siri, JD
In Many Words: Ginger Taylor
Dr. Paul's Newsletter: Paul Thomas, MD

Websites

The Autism Community in Action tacanow.org
The Brownstone Institute www. brownstone.org
Children's Health Defense www childrenshealthdefense.org
Daily Clout www/dailyclout.io
Epoch Times www.theepochtimes.com
The Highwire www. thehighwire.com
www.howdovaccinescauseautism.com
Aaron Kheriaty, MD: www.aaronkheriaty.com

Independent Medical Alliance (formerly FLCCC) www.covid19criticalcare.com
The Informed Consent Action Network www.icandecide.org
Millions Against Mandates www.millionsagainstmandates.org
The National Vaccine Information Center www.nvic.org
Physicians for Informed Consent: www.physiciansforinformedconsent.org

The previous list is not comprehensive but may include resources that many physicians and patients have not seen.

Endnotes

Chapter 2

1   Megan O'Driscoll, Gabriel Ribeiro Dos Santos, Lin Wang, et al., "Age-specific mortality and immunity patterns of SARS-CoV-2," *Nature*, 590 no. 7844 (2020): p. 140–145, doi: 10.1038/s41586-020-2918-0.

2   Lauren E. Kushner, Alan R. Schroeder, Joseph Kim, and Roshni Mathew, "'For COVID' or 'With COVID': Classification of SARS-CoV-2 Hospitalizations in Children," *Hospital Pediatrics*, 11 no. 8 (2021): p. e151-e156, doi: 10.1542/hpeds.2021-006001.

3   O'Driscoll, Ribeiro Dos Santos, Wang, et al., "Age-specific mortality and immunity patterns of SARS-CoV-2," *Nature* 590, 140–145 (2021). https://doi.org/10.1038/s41586-020-2918-0.

4   Sunil S. Bhopal, Jayshree Bagaria, Bayanne Olabi, and Raj Bhopal, "Children and young people remain at low risk of COVID-19 mortality," *The Lancet Child & Adolescent Health*, 5 no. 5 (2021): p. e12-e13, doi: 10.1016/s2352-4642(21)00066-3.

5   FAIR Health, Inc., Marty Makary, MD, MPH, "Risk Factors for COVID-19 Mortality among Privately Insured Patients: A Claims Data Analysis," press release, November 19, 2020, https://www.fairhealth.org/article/relationship-between-covid-19-comorbidities-and-mortality-uncovered-by-fair-health-study.

6   Marty Makary, MD, MPH. "The Flimsy Evidence Behind the CDC's Push to Vaccinated Children." Opinion Commentary, *Wall Street Journal*, July 19, 2021. https://www.wsj.com/articles/cdc-covid-19-coronavirus-vaccine-side-effects-hospitalization-kids-11626706868.

7   Clare Smith, David Odd, Rachel Harwood, et al., "Deaths in children and young people in England after SARS-CoV-2 infection during the first

pandemic year," *Nature Medicine*, 28 no. 1 (2021): p. 185–192, doi : 10.1038/s41591-021-01578-1.

8   Anna-Lisa Sorg, Markus Hufnagel, Maren Doenhardt, et al., "Risk for severe outcomes of COVID-19 and PIMS-TS in children with SARS-CoV-2 infection in Germany," *European Journal of Pediatrics*, 181 no. 10 (2022): p. 3635–3643, doi: 10.1007/s00431-022-04587-5.

9   Katherine E. Fleming-Dutra, MD. "COVID-19 epidemiology in children ages 6 months-4 years," https://www.cdc.gov/vaccines/acip/meetings/downloads /slides-2022-06-17-18/02-covid-fleming-dutra-508.pdf.

10   Emma Colton. "'Related to obvious other causes': Gunshot victims included in Washington coronavirus death tally." News, *The Washington Examiner* (Washington DC), May 24, 2020 2020. https://www.washingtonexaminer .com/news/related-to-obvious-other-causes-gunshot-victims-included-in -washington-coronavirus-death-tally.

11   NCHS, "Provisional COVID-19 Death Counts by Age in Years, 2020-2023," https://data.cdc.gov/NCHS/Provisional-COVID-19-Death-Counts-by -Age-in-Years-/3apk-4u4f.

12   Emma Colton. "'Related to obvious other causes': Gunshot victims included in Washington coronavirus death tally." News, *The Washington Examiner* (Washington DC), May 24, 2020 2020. https://www.washingtonexaminer .com/news/related-to-obvious-other-causes-gunshot-victims-included-in -washington-coronavirus-death-tally.

13   R. B. Brown, "Public Health Lessons Learned From Biases in Coronavirus Mortality Overestimation," *Disaster Med Public Health Prep*, 14 no. 3 (2020): p. 364–371, doi: 10.1017/dmp.2020.298.

14   Lindsey Wang, Nathan A. Berger, David C. Kaelber, et al., "Incidence Rates and Clinical Outcomes of SARS-CoV-2 Infection With the Omicron and Delta Variants in Children Younger Than 5 Years in the US," *JAMA Pediatrics*, 176 no. 8 (2022): p. doi: 10.1001/jamapediatrics.2022.0945.

15   Ibid.

16   Benjamin Lee and William V. Raszka, "COVID-19 Transmission and Children: The Child Is Not to Blame," *Pediatrics*, 146 no. 2 (2020): p. doi : 10.1542/peds.2020-004879.

17   David Isaacs, Philip Britton, Annaleise Howard-Jones, et al., "To what extent do children transmit SARS-CoV-2 virus?," *Journal of Paediatrics and Child Health*, 56 no. 6 (2020): p. 978–979, doi: 10.1111/jpc.14937.

18   Yanshan Zhu, Conor J. Bloxham, Katina D. Hulme, et al., "A Meta-analysis on the Role of Children in Severe Acute Respiratory Syndrome Coronavirus 2 in Household Transmission Clusters," *Clinical Infectious Diseases*, 72 no. 12 (2021): p. e1146-e1153, doi: 10.1093/cid/ciaa1825.

19   Jieun Kim, Young June Choe, Jin Lee, et al., "Role of children in household transmission of COVID-19," *Archives of Disease in Childhood*, 106 no. 7 (2021): p. 709–711, doi: 10.1136/archdischild-2020-319910.

20   Benjamin Lee and William V. Raszka, "COVID-19 in Children: Looking Forward, Not Back," *Pediatrics*, 147 no. 1 (2021): p. doi: 10.1542/peds.2020-029736.

21   Ibid.

22   Ibid

23   Kristine Macartney, Helen E. Quinn, Alexis J. Pillsbury, et al., "Transmission of SARS-CoV-2 in Australian educational settings: a prospective cohort study," *The Lancet Child & Adolescent Health*, 4 no. 11 (2020): p. 807–816, doi: 10.1016/s2352-4642(20)30251-0.

24   Kostas Danis, Olivier Epaulard, Thomas Bénet, et al., "Cluster of Coronavirus Disease 2019 (COVID-19) in the French Alps, February 2020," *Clinical Infectious Diseases*, 71 no. 15 (2020): p. 825–832, doi: 10.1093/cid/ciaa424.

25   Jakob P. Armann, Manja Unrath, Carolin Kirsten, et al., "SARS-CoV-2 IgG antibodies in adolescent students and their teachers in Saxony, Germany (SchoolCoviDD19): persistent low seroprevalence and transmission rates between May and October 2020," *medRxiv*, no. (2020): p. doi: 10.1101/2020.07.16.20155143.

26   Laura Heavey, Geraldine Casey, Ciara Kelly, et al., "No evidence of secondary transmission of COVID-19 from children attending school in Ireland, 2020," *Eurosurveillance*, 25 no. 21 (2020): p. doi: 10.2807/1560 -7917.Es.2020.25.21.2000903.

27   Chee Fu Yung, Kai-qian Kam, Karen Donceras Nadua, et al., "Novel Coronavirus 2019 Transmission Risk in Educational Settings," *Clinical Infectious Diseases*, 72 no. 6 (2021): p. 1055–1058, doi: 10.1093/cid/ciaa794.

28   Ruth Link-Gelles, PhD; Amanda L. DellaGrotta, MPH; Caitlin Molin; Ailis Clyne, MD; Kristine Campagna, MED; Tatiana M. Lanzieri, MD; Marisa A. Hast, PhD; Krishna Palipudi, PhD; Emilio Dirlikov, PhD; Utpala Bandy, MD,

"Limited secondary transmission of SARS-CoV-2 in child care programs -
Rhode Island, June 1–July 31, 2020," *MMWR Morb Mortal Wkly Rep*, no. doi:

29   Kevin W. Ng, Nikhil Faulkner, Georgina H. Cornish, et al., "Preexisting and
de novo humoral immunity to SARS-CoV-2 in humans," *Science*, 370 no.
6522 (2020): p. 1339–1343, doi: 10.1126/science.abe1107.

30   Julian Braun, Lucie Loyal, Marco Frentsch, et al., "SARS-CoV-2-reactive T
cells in healthy donors and patients with COVID-19," *Nature*, 587 no. 7833
(2020): p. 270–274, doi: 10.1038/s41586-020-2598-9.

31   Nina Le Bert, Anthony T. Tan, Kamini Kunasegaran, et al., "SARS-CoV-2-
specific T cell immunity in cases of COVID-19 and SARS, and uninfected
controls," *Nature*, 584 no. 7821 (2020): p. 457–462, doi: 10.1038/
s41586-020-2550-z.

32   Ibid.

33   Alexander C. Dowell, Megan S. Butler, Elizabeth Jinks, et al., "Children
develop robust and sustained cross-reactive spike-specific immune responses
to SARS-CoV-2 infection," *Nature Immunology*, 23 no. 1 (2021): p. 40–49,
doi: 10.1038/s41590-021-01089-8.

34   O'Driscoll, Ribeiro Dos Santos, Wang, et al., "Age-specific mortality and
immunity patterns of SARS-CoV-2," *Nature* 590, 140–145 (2021).
https://doi.org/10.1038/s41586-020-2918-0.

35   Ibid.

36   Ibid.

37   A. Saguil, M. Fargo, and S. Grogan, "Diagnosis and management of kawasaki
disease," *Am Fam Physician*, 91 no. 6 (2015): p. 365–71.

38   Centers for Disease Control and Prevention, "About Kawasaki Disease," May
29, 2020 [cited January 9, 2024], https://www.cdc.gov/kawasaki/about.html.

39   CBS NEWS. *Cuomo directs New York hospitals to prioritize COVID-19 testing
for children showing symptoms of mysterious illness*. 2020 May 12. 2020;
Available from: https://www.cbsnews.com/news/cuomo-directs-new-york
-hospitals-to-prioritize-covid-19-testing-for-children-showing-symptoms-of
-mysterious-illness/.

40   Associated Press. "'A pretty scary thing:' Kid illness tied to virus worries
NY." *Chicago Sun Times*, May 12, 2020. https://chicago.suntimes.com
/coronavirus/2020/5/12/21256502/kid-illness-tied-to-virus-worries-new-york
-andrew-cuomo-coronavirus-covid-19.

41   Centers for Disease Control and Prevention, "About Kawasaki Disease," May
      29, 2020 [cited January 9, 2024], https://www.cdc.gov/kawasaki/about.html.

42   N. A. Nakra, D. A. Blumberg, A. Herrera-Guerra, and S. Lakshminrusimha,
      "Multi-System Inflammatory Syndrome in Children (MIS-C) Following
      SARS-CoV-2 Infection: Review of Clinical Presentation, Hypothetical
      Pathogenesis, and Proposed Management," *Children (Basel)*, 7 no. 7 (2020):
      p. doi: 10.3390/children7070069.

43   T. Radia, N. Williams, P. Agrawal, et al., "Multi-system inflammatory
      syndrome in children & adolescents (MIS-C): A systematic review of clinical
      features and presentation," *Paediatr Respir Rev*, 38 no. (2021): p. 51–57, doi
      : 10.1016/j.prrv.2020.08.001.

44   Ibid.

45   Ibid.

46   Ibid.

47   S. Mahmoud, M. El-Kalliny, A. Kotby, et al., "Treatment of MIS-C in
      Children and Adolescents," *Curr Pediatr Rep*, 10 no. 1 (2022): p. 1-10, doi
      : 10.1007/s40124-021-00259-4.

48   Centers for Disease Control and Prevention, "Health Department-Reported
      Cases of Multisystem Inflammatory Syndrome in Children (MIS-C) in the
      United States," COVID Data Tracker [cited January 09, 2024], https://covid
      .cdc.gov/covid-data-tracker/#mis-national-surveillance.

49   B. W. McCrindle, A. H. Rowley, J. W. Newburger, et al., "Diagnosis, Treatment,
      and Long-Term Management of Kawasaki Disease: A Scientific Statement for
      Health Professionals From the American Heart Association," *Circulation*, 135
      no. 17 (2017): p. e927–e999, doi: 10.1161/CIR.0000000000000484.

Chapter 3

50   "Open Letter from FDA to Robert R. Redfield, Md, Director, Centers for Disease
      Control and Prevention." 2020, https://www.fda.gov/media/134919/download.

51   "Molecular Diagnostic Template for Laboratories. Policy for Coronavirus
      Disease-2019 Tests During the Public Health Emergency (Revised)." 2023,
      https://www.fda.gov/media/135659/download.

52   "Your Coronavirus Test Is Positive. Maybe It Shouldn't Be." *The New York
      Times*, 2020, https://www.nytimes.com/2020/08/29/health/coronavirus
      -testing.html.

53   H. Rahman, I. Carter, K. Basile, L. Donovan, S. Kumar, T. Tran, D. Ko, et al. "Interpret with Caution: An Evaluation of the Commercial Ausdiagnostics Versus in-House Developed Assays for the Detection of Sars-Cov-2 Virus." *Journal of Clinical Virology* 127 (2020). https://doi.org/10.1016/j .jcv.2020.104374.

54   Sin Hang Lee. "Testing for Sars-Cov-2 in Cellular Components by Routine Nested Rt-PCR Followed by DNA Sequencing." *Int J Geriatr Rehabil* 2 (2020): 69–96. http://dnalymetest.com/images/IJGeriatRehabLee_on _SARSCoV2_test.pdf.

55   https://portal.census.gov/pulse/data/.

56   https://www.businessinsider.com/ thousands-us-restaurants-closed-coronavirus-pandemic-2020-12.

57   D. Lahr, A. Adams, A. Edges, J. Bletz. "Where Do We Go From Here? The Survival and Recovery of Black-Owned Businesses Post-COVID-19." *Humanity Soc.* 2022 Aug;46(3):460–77. doi: 10.1177/01605976211049243 .PMCID: PMC8784976.

58   https://www.newyorkfed.org/medialibrary/media/smallbusiness/Double Jeopardy_COVID19andBlackOwnedBusinesses.

Chapter 4

59   Dr. Senetra Gupta, Dr. Jay Bhattacharya, Dr. Martin Kulldorff. *The Great Barrington Declaration*, 2020: Great Barrington, MA.

60   Dr. Joseph Mercola. "Fauci, NIH Colluded to Discredit Scientists Behind Great Barrington Declaration, Emails Reveal." *The Defender,* Children's Health Defense, 2022.

61   Nick Gillespie. "Dr. Jay Bhattacharya: How to Avoid 'Absolutely Catastrophic' COVID Mistakes" in *Reason,* 2022, Reason Foundation, Los Angeles, CA.

62   Josie Ensor, "Stanford anti-lockdown professor Jay Bhattacharya secretly blacklisted on Twitter, new leak shows" [cited Dec 28,2022].

63   Gupta, Bhattacharya, Kulldorff, *Great Barrington Declaration*.

64   Ibid.

65   Ibid.

66   Ibid.

Chapter 5

67 Anna Aiello, et al., "Immunosenescence and Its Hallmarks: How to Oppose Aging Strategically? A Review of Potential Options for Therapeutic Intervetion," *Frontiers in Immunology,* 10 no. (2019): p. doi: 10.3389 /fimmu.2019.02247.

68 FLCCC ALLIANCE, "Treatment Protocols," [cited Jan 9, 2024] https ://covid19citicalcare.com/treatment-protocols.

69 Dr. Pierre Kory, Dr. Paul Marik, Dr. Keith Berkowitz, Dr. Flavio Cadegiani, Dr. Suzanne Gazda, Dr. Meryl Nass, Dr. Tina Peers, Dr. Robin Rose, Dr. Yusul (JP) Saleeby, Dr. Mobeen Syed, Dr. Eugene Shippen, Dr. Fred Wagshul, "I-RECOVER: Post-Vaccine Treatment protocol," FLCCC ALLIANCE.

70 F. A. Cadegiani, A. Goren, C. G. Wambier, and J. McCoy, "Early COVID-19 therapy with azithromycin plus nitazoxanide, ivermectin or hydrozychloroquine in outpatient settings significantly improved COVID-19 outcomes compared to known outcomes in untreated patients," *New Microbes New Infect,* 43 no. (2021): p 100915, doi: 10.1016/j .nmni.2021.100915.

71 F. A. Cadegiani, J. McCoy, C. Gustavo Wambier, et al, "Proxalutamide Significantly Accelerated Viral Clearance and Reduces Time to Clinical Remission in Patients with Mild to Moderate COVID-19: Results from a Randomized, Double-Blinded, Placebo-Controlled Trial," *Cureus,* 13 no. 2 (2021): p. e13492, doi: 10.7759/cureus.13492.

72 Cadegiani, Flavio A., John McCoy, Carlos Gustavo Wambier, and Andy Goren. "Early antiandrogen therapy with dutasteride reduces viral shedding, inflammatory responses, and Time-to-Remission in males with COVID-19: a randomized, double-blind, placebo-controlled interventional trial (EAT-DUTA AndroCoV trial–biochemical)." *Cureus* 13, no. 2 (2021).

73 Philippe Gautret, Jean-Christophe Lagier, Philippe Parola, Line Meddeb, Morgane Mailhe, Barbara Doudier, Johan Courjon et al. "Hydroxychloroquine and azithromycin as a treatment of COVID-19: results of an open-label non-randomized clinical trial." *International journal of antimicrobial agents* 56, no. 1 (2020): 105949.

74 Colunga Biancatelli, Ruben Manuel Luciano, Max Berrill, and Paul E. Marik. "The antiviral properties of vitamin C." *Expert review of anti-infective therapy* 18, no. 2 (2020): 99–101.

75   Jose Iglesias, Andrew V. Vassallo, Vishal V. Patel, Jesse B. Sullivan, Joseph Cavanaugh, and Yasmine Elbaga. "Outcomes of metabolic resuscitation using ascorbic acid, thiamine, and glucocorticoids in the early treatment of sepsis: the ORANGES trial." *Chest* 158, no. 1 (2020): 164–173.

76   Lorenz Borsche, Bernd Glauner, and Julian von Mendel. "COVID-19 mortality risk correlates inversely with vitamin D3 status, and a mortality rate close to zero could theoretically be achieved at 50 ng/mL 25 (OH) D3: results of a systematic review and meta-analysis." *Nutrients* 13, no. 10 (2021): 3596.

77   B. Seven, O. Gunduz, A. S. Ozgu-Erdinc, Dilek Sahin, O. Moraloglu Tekin, and H. L. Keskin. "Correlation between 25-hydroxy vitamin D levels and COVID-19 severity in pregnant women: a cross-sectional study." *The Journal of Maternal-Fetal & Neonatal Medicine* 35, no. 25 (2022): 8817–8822.

78   Amare Teshome, Aynishet Adane, Biruk Girma, and Zeleke A. Mekonnen. "The impact of vitamin D level on COVID-19 infection: systematic review and meta-analysis." *Frontiers in public health* 9 (2021): 624559.

79   Doctors 4 Covid Ethics, https://doctors4covidethics.org/

80   Peter I. Parry, Astrid Lefringhausen, Conny Turni, Christopher J. Neil, Robyn Cosford, Nicholas J. Hudson, and Julian Gillespie. "'Spikeopathy': COVID-19 spike protein is pathogenic, from both virus and vaccine mRNA." *Biomedicines* 11, no. 8 (2023): 2287.

81   Ronald B. Brown. "Public health lessons learned from biases in coronavirus mortality overestimation." *Disaster medicine and public health preparedness* 14, no. 3 (2020): 364–371.

82   John Ioannidis, PA. "Infection fatality rate of COVID-19 inferred from seroprevalence data." *Bulletin of the world health organization* 99, no. 1 (2021): 19.

83   John Ioannidis, PA. "Reconciling estimates of global spread and infection fatality rates of COVID-19: an overview of systematic evaluations." *European journal of clinical investigation* 51, no. 5 (2021): e13554.

84   John Ioannidis, PA. "A fiasco in the making? As the coronavirus pandemic takes hold, we are making decisions without reliable data." *Stat* 17 (2020): 1–6.

85   Ibid.

86   Michael Palmer, Sucharit Bhakdi, Margot DesBois, Brian Hooker, David Rasnick, Mary Holland, and JD Catherine Austin Fitts. "mRNA Vaccine Toxicity." 2023, *Doctors for COVID Ethics*, p. 218.

87   Peter A. McCullough, Paul E. Alexander, Robin Armstrong, Cristian Arvinte, Alan F. Bain, Richard P. Bartlett, Robert L. Berkowitz et al. "Multifaceted highly targeted sequential multidrug treatment of early ambulatory high-risk SARS-CoV-2 infection (COVID-19)." *Reviews in cardiovascular medicine* 21, no. 4 (2020): 517.

88   Paul E. Marik, Jose Iglesias, Joseph Varon, and Pierre Kory. "A scoping review of the pathophysiology of COVID-19." *International journal of immunopathology and pharmacology* 35 (2021): 20587384211048026.

89   https://www.who.int/news-room/feature-stories/detail/who-advises-that -ivermectin-only-be-used-to-treat-covid-19-within-clinical-trials.

90   Michael Palmer, M. D., M. D. Sucharit Bhakdi, and M. D. Wolfgang Wodarg. "Expert Report on the Johnson & Johnson COVID-19 Vaccine." https://doctors4covidethics.org/wp-content/uploads/2022/02/jassen -published.pdf

91   Doctors for COVID Ethics, "About Doctors for COVID Ethics," [cited Jan 9, 2024] https://doctors4covidethics.org/about/.

92   Palmer, Bhakdi, DesBois et al., "mRNA Vaccine Toxicity."

93   Nazeeh Hanna, Ari Heffes-Doon, Xinhua Lin, Claudia Manzano De Mejia, Bishoy Botros, Ellen Gurzenda, and Amrita Nayak. "Detection of messenger RNA COVID-19 vaccines in human breast milk." *JAMA pediatrics* 176, no. 12 (2022): 1268–1270.

94   Palmer, Bhakdi, DesBois et al., "mRNA Vaccine Toxicity."

95   Ibid.

96   Doctors for Covid Ethics, "Books," https://doctors4covidethics.org/books/

97   David M. Morens, Jeffery K. Taubenberger, and Anthony S. Fauci. "Rethinking next-generation vaccines for coronaviruses, influenzaviruses, and other respiratory viruses." *Cell host & microbe* 31, no. 1 (2023): 146–157.

Chapter 6

98   Yaseen M. Arabi, George P. Chrousos, and G. Umberto Meduri. "The ten reasons why corticosteroid therapy reduces mortality in severe COVID-19." *Intensive care medicine* 46, no. 11 (2020): 2067-2070.

99   Gretchen L. Sacha, Alyssa Y. Chen, Nicole M. Palm, and Abhijit Duggal. "Evaluation of the initiation timing of hydrocortisone in adult patients with septic shock." *Shock* 55, no. 4 (2021): 488–494.

100   Luis Corral-Gudino, Alberto Bahamonde, Francisco Arnaiz-Revillas, Julia
      Gómez-Barquero, Jesica Abadía-Otero, Carmen García-Ibarbia, Víctor
      Mora et al. "Methylprednisolone in adults hospitalized with COVID-19
      pneumonia: An open-label randomized trial (GLUCOCOVID)." *Wiener
      klinische Wochenschrift* 133 (2021): 303–311.

101   FLCCC Alliance, "Ivermectin,: [cited January 9, 2024], https
      ://covid19criticlcare.com/ivermectin.

102   Pierre Kory, Gianfranco Umberto Meduri, Joseph Varon, Jose Iglesias, and
      Paul E. Marik. "Review of the emerging evidence demonstrating the efficacy
      of ivermectin in the prophylaxis and treatment of COVID-19." *American
      journal of therapeutics* 28, no. 3 (2021): e299–e318.

103   Ibid.

104   Ibid.

105   Sara Diani, Erika Leonardi, Attilio Cavezzi, Simona Ferrari, Oriana Iacono,
      Alice Limoli, Zoe Bouslenko et al. "Sars-Cov-2—the role of natural
      immunity: A narrative review." *Journal of clinical medicine* 11, no. 21 (2022):
      6272.

106   Leon Caly, Julian D. Druce, Mike G. Catton, David A. Jans, and Kylie M.
      Wagstaff. "The FDA-approved drug ivermectin inhibits the replication of
      SARS-CoV-2 in vitro." *Antiviral research* 178 (2020): 104787.

107   Sundy Yang, NY, Sarah C. Atkinson, Chunxiao Wang, Alexander Lee, Marie
      A. Bogoyevitch, Natalie A. Borg, and David A. Jans. "The broad spectrum
      antiviral ivermectin targets the host nuclear transport importin α/β1
      heterodimer." *Antiviral research* 177 (2020): 104760.

108   Steven Lehrer and Peter H. Rheinstein. "Ivermectin docks to the SARS-
      CoV-2 spike receptor-binding domain attached to ACE2." *in vivo* 34, no. 5
      (2020): 3023–3026.

109   Xinxin Ci, Hongyu Li, Qinlei Yu, Xuemei Zhang, Lu Yu, Na Chen, Yu
      Song, and Xuming Deng. "Avermectin exerts anti-inflammatory effect by
      downregulating the nuclear transcription factor kappa-B and mitogen-
      activated protein kinase activation pathway." *Fundamental & clinical
      pharmacology* 23, no. 4 (2009): 449–455.

110   X. Zhang, Y. Song, X. Ci, N. An, Y. Ju, H. Li, X. Wang, C. Han, J. Cui, and
      X. Deng. "Ivermectin inhibits LPS-induced production of inflammatory
      cytokines and improves LPS-induced survival in mice." *Inflammation
      Research* 57 (2008): 524–529.

111 Muhammad Sohail Sajid, Zafar Iqbal, Ghulam Muhammad, Mansur Abdullah Sandhu, Muhammad Nisar Khan, Muhammad Saqib, and Muhammad Umair Iqbal. "Effect of ivermectin on the cellular and humoral immune responses of rabbits." *Life sciences* 80, no. 21 (2007): 1966–1970.

112 Ibid.

113 Lucy Kerr, Fernando Baldi, Raysildo Lobo, Washington Luiz Assagra, Fernando Carlos Proença, Juan J. Chamie, Jennifer A. Hibberd, Pierre Kory, and Flavio A. Cadegiani. "Regular use of ivermectin as prophylaxis for COVID-19 led up to a 92% reduction in COVID-19 mortality rate in a dose-response manner: results of a prospective observational study of a strictly controlled population of 88,012 subjects." *Cureus* 14, no. 8 (2022).

Chapter 7

114 Edward R. Melnick and John P. A. Ioannidis, "Should governments continue lockdown to slow the spread of covid-19?," *BMJ*, no. (2020): p. m1924, doi : 10.1136/bmj.m1924.

115 John P. A. Ioannidis, "Global perspective of COVID-19 epidemiology for a full-cycle pandemic," *Eur J Clin Invest*, 50 no. 12 (2020): p. e13423, doi : 10.1111/eci.13423.

116 John P.A. Ioannidis, Cathrine Axfors, Despina G. Contopoulos-Ioannidis, "Population-level COVID-19 mortality risk for non-elderly individuals overall and for non-elderly individuals without underlying diseases in pandemic epicenters 2020," *Environmental Research*, 188 no. (2020): p. doi : https://doi.org/10.1016/j.envres.2020.109890.

117 Anders Tegnell, "Coronavirus: Anders Tegnell, State Epidemiologist of Sweden, on herd immunity - BBC HARDtalk", BBC HARDtalk. S. Sackur. May 19, 2020; Available from: https://www.youtube.com /watch?v=Biqq34aUJcQ.

118 "LOCKDOWN LUNACY: The Thinking Person's Guide." *Children's Health Defense*, June 04, 2020 https://childrenshealthdefense.org/news /lockdown-lunacy-the-thinking-persons-guide/.

119 Children's Health Defense Team. "A Timeline—Pandemic and Erosion of Freedoms Have Been Decades in the Making." *Children's Health Defense*, May 21, 2020. https://childrenshealthdefense.org/news/a-timeline-pandemic-and -erosion-of-freedoms-have-been-decades-in-the-making/.

120 "Schools MUST go "Back to Normal" in the Fall—A Scientist's Perspective." *Children's Health Defense*, June 25, 2020. https://childrenshealthdefense.org /news/editorial/schools-must-go-back-to-normal-in-the-fall-a-scientists -perspective/.

121 Gabriel Hays. "Fauci admits he knew 'draconian' lockdowns would have 'collateral negative consequences' on schoolchildren." *New York Post*, Sep. 21, 2022. https://nypost.com/2022/09/21/fauci-admits-he-knew-draconian -lockdowns-would-have-collateral-negative-consequences-on-schoolchildren/.

122 Meryl Nass MD. "Fauci Emails: How Top Public Health Officials Spun Tangled Web of Lies Around COVID Origin, Treatments." *Children's Health Defense*, June 4, 2021. https://childrenshealthdefense.org/defender /fauci-emails-top-public-health-officials-lies-covid-origin-treatments/.

123 Children's Health Defense Team. "3 Ways Fauci Emails Expose Trail of Manipulation and Deception." *The Defender*. https://childrenshealthdefense .org/ defender/fauci-kristian-andersen-emails-manipulation-deception/.

124 Rebecca T. Leeb PhD, Rebecca H. Bitsko PhD, Lakshmi Radhakrishnan MPH, et al. *Mental Health–Related Emergency Department Visits Among Children Aged <18 Years During the COVID-19 Pandemic—United States, January 1–October 17, 2020*. 2020. 69, 1675–1680.

125 Ibid.

126 Ryan M. Hill, Katrina Rufino, Sherin Kurian, et al., "Suicide Ideation and Attempts in a Pediatric Emergency Department Before and During COVID-19," *Pediatrics*, 147 no. 3 (2021): p. doi: 10.1542/peds.2020-029280.

127 U.S. Senate, *COVID-19: A Second Opinion*, 117th Congress, 2nd Session, 2022

128 Dan Brown and Elisabetta De Cao, *The impact of unemployment on child maltreatment in the United States*. 2018. No.2018-04.

129 Prateek Kumar Panda, Juhi Gupta, Sayoni Roy Chowdhury, Rishi Kumar, Ankit Kumar Meena, Priyanka Madaan, Indar Kumar Sharawat, Sheffali Gulati, "Psychological and Behavioral Impact of Lockdown and Quarantine Measures for COVID-19 Pandemic on Children, Adolescents and Caregivers: A Systematic Review and Meta-Analysis," *J Trop Pediatr*, 67 no. 1 (2021): p. doi: 10.1093/tropej/fmaa122.

130 Michiel A J Luijten, Maud M van Muilekom, Lorynn Teela, et al., "The impact of lockdown during the COVID-19 pandemic on mental and social

health of children and adolescents," *Qual Life Res*, 30 no. 10 (2021): p. 2795–2804, doi: 10.1007/s11136-021-02861-x.

131   Yunyu Xiao, Timothy T Brown, Lonnie R Snowden, et al., "COVID-19 Policies, Pandemic Disruptions, and Changes in Child Mental Health and Sleep in the United States," *JAMA Netw Open*, 6 no. 3 (2023): p. e232716, doi: 10.1001/jamanetworkopen.2023.2716.

132   Kuldeep Dhama, Shailesh Kumar Patel, Rakesh Kumar, et al., "The role of disinfectants and sanitizers during COVID-19 pandemic: advantages and deleterious effects on humans and the environment," *Environ Sci Pollut Res Int*, 28 no. 26 (2021): p. 34211–34228, doi: 10.1007/s11356-021-14429-w.

133   Francisco Guarner, Raphaëlle Bourdet-Sicard, Per Brandtzaeg, et al., "Mechanisms of Disease: the hygiene hypothesis revisited," *Nature Clinical Practice Gastroenterology & Hepatology*, 3 no. 5 (2006): p. 275–284, doi : 10.1038/ncpgasthep0471.

134   Jean-François Bach, "The hygiene hypothesis in autoimmunity: the role of pathogens and commensals," *Nature Reviews Immunology*, 18 no. 2 (2017): p. 105–120, doi: 10.1038/nri.2017.111.

135   Sridevi Devaraj, Janice Wang-Polagruto, John Polagruto, et al., "High-fat, energy-dense, fast-food–style breakfast results in an increase in oxidative stress in metabolic syndrome," *Metabolism*, 57 no. 6 (2008): p. 867–870, doi : 10.1016/j.metabol.2008.02.016.

136   Jalal Bohlouli, Amir Reza Moravejolahkami, Marjan Ganjali Dashti, et al., "COVID-19 and Fast Foods Consumption: a Review," *International Journal of Food Properties*, 24 no. 1 (2021): p. 203–209, doi : 10.1080/10942912.2021.1873364.

137   Berthold Koletzko, Christina Holzapfel, Ulrike Schneider, and Hans Hauner, "Lifestyle and Body Weight Consequences of the COVID-19 Pandemic in Children: Increasing Disparity," *Annals of Nutrition and Metabolism*, 77 no. 1 (2021): p. 1-3, doi: 10.1159/000514186.

138   Tommaso Celeste Bulfone, Mohsen Malekinejad, George W. Rutherford, and Nooshin Razani, "Outdoor Transmission of SARS-CoV-2 and Other Respiratory Viruses: A Systematic Review," *The Journal of Infectious Diseases*, 223 no. 4 (2021): p. 550–561, doi: 10.1093/infdis/jiaa742.

139 Jonas F Ludvigsson, "Children are unlikely to be the main drivers of the COVID-19 pandemic–a systematic review," *Acta Paediatrica*, 109 no. 8 (2020): p. 1525–1530

140 Edeh Michael Onyema, Chika Nwafor, Faith Obafemi, and Shuvro Sen, "Impact of Coronavirus Pandemic on Education," *Journal of Education and Practice*, no. (2020): p. doi: 10.7176/jep/11-13-12.

141 Sean Cl Deoni, Jennifer Beauchemin, Alexandra Volpe, et al., "The COVID-19 Pandemic and Early Child Cognitive Development: A Comparison of Development in Children Born During the Pandemic and Historical References," *medRxiv*, no. (2022): p. doi: 10.1101/2021.08.10.21261846.

142 Ibid.

143 Ibid.

144 Paul Elias Alexander. "More than 170 Comparative Studies and Articles on Mask Ineffectiveness and Harms." *Brownstone Institute*, December 20, 2021. https://brownstone.org/articles/studies-and-articles-on-mask-ineffectiveness -and-harms/.

145 Manfred Spitzer, "Masked education? The benefits and burdens of wearing face masks in schools during the current Corona pandemic," *Trends in Neuroscience and Education*, 20 no. (2020): p. doi: 10.1016/j .tine.2020.100138.

146 Ibid.

147 S. Vicari, J. S. Reilly, P. Pasqualetti, et al., "Recognition of facial expressions of emotions in school-age children: the intersection of perceptual and semantic categories," *Acta Paediatr*, 89 no. 7 (2000): p. 836–45

148 Ibid.

149 Tim Langdell, "Recognition of faces: an approach to the study of autism," *J Child Psychol Psychiatry*, 19 no. 3 (1978): p. 255-68, doi: 10.1111/j.1469 -7610.1978.tb00468.x.

150 T. A. Walden and T. M. Field, "Discrimination of facial expressions by preschool children," *Child Dev*, 53 no. 5 (1982): p. 1312-9

151 Ze Liu, Jianqun Wang, Xuetong Yang, et al., "Generation of environmental persistent free radicals (EPFRs) enhances ecotoxicological effects of the disposable face mask waste with the COVID-19 pandemic," *Environ Pollut*, 301 no. (2022): p. 119019, doi: 10.1016/j.envpol.2022.119019.

152 Ana M Oliveira, Ana L Patricio Silva, Amadeu M V M Soares, et al., "Current knowledge on the presence, biodegradation, and toxicity of discarded face

masks in the environment," *J Environ Chem Eng*, 11 no. 2 (2023): p. 109308, doi: 10.1016/j.jece.2023.109308.

153 James Roguski, *Mask Charade* https://substack.com/@jamesroguski.

Chapter 8

154 James Lyons-Weiler, and Paul Thomas. "RETRACTED: Relative Incidence of Office Visits and Cumulative Rates of Billed Diagnoses Along the Axis of Vaccination." *International Journal of Environmental Research and Public Health* 17, no. 22 (2020): 8674.

155 Jeremy R. Hammond, *The War on Informed Consent: The Persecution of Dr. Paul Thomas by the Oregon Medical Board* (New York: Skyhorse, 2021), 168.

156 FLCCC Alliance, "Paul E. Marik, MD, FCCM, FCCP," https://covid19criticalcare.com/experts/paul-e-marik/.

157 https://www.youtube.com/ watch?v=0IgbZWUMKmk and vt62y6-covid-19 -a-second-opinion.html starting 4:19:30.

158 Dr. Paul Marik, "'Sham Peer Review': How hospitals exert control over physicians who dare to speak out," February, 22. 2022, https://covid19criticalcare.com/sham-peer-review-how-hospitals-exert-control -over-physicians-who-dare-to-speak-out/.

159 Lawrence R. Huntoon, PhD, "Tactics Characteristic of Sham Peer Review," *Journal of American Physicians and Surgeons*, 14 no. Fall 2009 (2009): p. 3.

160 Marik, "Sham Peer Review."

161 Ibid.

162 Ibid.

163 Ibid.

164 FLCCC Alliance, "Paul E. Marik, MD, FCCM, FCCP," https://covid19criticalcare.com/experts/paul-e-marik/.

Chapter 9

165 Stewart Sell, "How vaccines work: immune effector mechanisms and designer vaccines," *Expert Review of Vaccines*, 18 no. 10 (2019): p. 993–1015, doi : 10.1080/14760584.2019.1674144.

166 Panagis Polykretis, "Role of the antigen presentation process in the immunization mechanism of the genetic vaccines against COVID-19 and the

need for biodistribution evaluations," *Scandinavian Journal of Immunology*, 96 no. 2 (2022): doi: 10.1111/sji.13160.

167 Pfizer, *SARS-CoV-2 mRNA Vaccine (BNT162, PF-07302048) 2.6.4 Summary of pharmacokinetic study*. 2021, PFIZER CONFIDENTIAL. p. 6.

168 EMA, "Assessment report Comirnaty Common name: COVID-19 mRNA vaccine (nucleoside-modified)," https://www.ema.europa.eu/en/documents /assessment-report/comirnaty-epar-public-assessment-report_en.pdf.

169 EMA, "Assessment report COVID-19 vaccine Moderna common name: COVID-19 mRNA vaccine (nucleoside-modified)," https://www.ema.europa .eu/en/documents/assessment-report/spikevax-previously-covid-19-vaccine -moderna-epar-public-assessment-report_en.pdf.

170 Sucharit Bhakdi MD and Michael Palmer, MD, "Elementary, my dear Watson: why mRNA vaccines are a very bad idea," https ://doctors4covidethics.org/elementary-my-dear-watson-why-mrna -vaccines-are-a-very-bad-idea/.

171 Sucharit Bhakdi MD and Michael Palmer, MD, "Long-term persistence of the SARS-CoV-2 spike protein: evidence and implications," December 21, 2021, https://doctors4covidethics.org/long-term persistence-of-the-sars-cov -2-spike-protein-evidence-and-implications-2/.

172 Bhakdi and Palmer, "Elementary, my dear Watson."

173 Bhakdi and Palmer, "Long-term persistence."

174 Swank, Z., Senussi, Y., Manickas-Hill, Z., Yu, X.G., Li, J.Z., Alter, G., Walt, D.R., 2023. Persistent Circulating Severe Acute Respiratory Syndrome Coronavirus 2 Spike Is Associated With Post-acute Coronavirus Disease 2019 Sequelae. Clinical Infectious Diseases 76, e487–e490.. https://doi .org/10.1093/cid/ciac722.

175 Sharyl Attkisson. "CDC changes definition of "vaccines" to fit Covid-19 vaccine limitations." September 8, 2021 https://sharylattkisson. com/2021/09/read-cdc-changes-definition-of-vaccines-to-fit-covid-19 -vaccine-limitations/.

176 @RepThomasMassie. *"Check out @CDCgov's evolving definition of "vaccination." They've been busy at the Ministry of Truth:".* Twitter.com September 8, 2021; Available from: 10:11 AM.

Chapter 10

177 Stephanie Seneff, Greg Nigh, Epidemic NCDs, "Worse Than the Disease? Reviewing Some Possible Unintended Consequences of the mRNA Vaccines Against COVID-19," *International Journal of Vaccine Theory, Practice, and Research*, 2 no. 1 (2021): doi: https://doi.org/10.56098/ijvtpr.v2i1.23.

178 Aristo Vojdani and Datis Kharrazian, "Potential antigenic cross-reactivity between SARS-CoV-2 and human tissue with a possible link to an increase in autoimmune diseases," *Clinical Immunology*, 217 no. (2020): doi: 10.1016/j .clim.2020.108480.

179 Dario Pellegrini, Rika Kawakami, Giulio Guagliumi, et al., "Microthrombi as a Major Cause of Cardiac Injury in COVID-19," *Circulation*, 143 no. 10 (2021): p. 1031–1042, doi: 10.1161/circulationaha.120.051828.

180 SARS-CoV-2 MRNA Vaccine (BNT162, PF-0 7302048)2.6.4. Pfizer-BioNTech, reported to Japan's Pharmaceuticals and Medical Devices Agency (PMDA).

181 https://childrenshealthdefense.org//wp-content/uploads/Whelan-FDA-letter -re-EAU-Pfizer-.pdf.

182 Ibid.

183 Elizabeth M. Rhea, Aric F. Logsdon, Kim M. Hansen, et al., "The S1 protein of SARS-CoV-2 crosses the blood–brain barrier in mice," *Nature Neuroscience*, 24 no. 3 (2020): p. 368–378, doi: 10.1038/s41593-020-00771-8.

184 Tetyana P Buzhdygan, Brandon J DeOre, Abigail Baldwin-Leclair, et al., "The SARS-CoV-2 spike protein alters barrier function in 2D static and 3D microfluidic in-vitro models of the human blood-brain barrier," *Neurobiol Dis*, 146 no. (2020): p. 105131, doi: 10.1016/j.nbd.2020.105131.

185 https://childrenshealthdefense.org//wp-content/uploads/Whelan-FDA-letter -re-EAU-Pfizer-.pdf

186 Ibid.

187 Seneff and Nigh, "Worse Than the Disease?"

188 Cynthia Magro, Gerard J Nuovo, Toni Shaffer, Hamdy Awad, David Suster, Sheridan Mikhail, Bing He, Jean-Jacques Michaille, Benjamin Liechty, Esmerina Tili, "Endothelial cell damage is the central part of COVID-19 and a mouse model induced by injection of the S1 subunit of the spike protein," *Annals of Diagnostic Pathology*, 51 no. April 2021 (2021): doi: https://doi .org/10.1016/j.anndiagpath.2020.151682.

189   Seneff and Nigh, "Worse Than the Disease?"

190   Ibid.

191   Ibid.

192   Daniel Wrapp, Nianshuang Wang, Kizzmekia S Corbett, et al., "Cryo-EM structure of the 2019-nCoV spike in the prefusion conformation," *Science*, 367 no. 6483 (2020): p. 1260–1263, doi: 10.1126/science.abb2507.

193   Danish Idrees and Vijay Kumar, "SARS-CoV-2 spike protein interactions with amyloidogenic proteins: Potential clues to neurodegeneration," *Biochem Biophys Res Commun*, 554 no. (2021): p. 94–98, doi: 10.1016/j.bbrc.2021.03.100.

194   Danish Idrees and Vijay Kumar, "Prion-like Domains in Spike Protein of SARS-CoV-2 Differ across Its Variants and Enable Changes in Affinity to ACE2," *Microorganisms*, 10 no. 2 (2022): doi: 10.3390/microorganisms10020280.

195   J. Bart Classen, "Review of COVID-19 Vaccines and the Risk of Chronic Adverse Events Including Neurological Degeneration," *Journal of Medical - Clinical Research & Reviews*, 5 no. 4 (2021): p. 1–7, doi : 10.33425/2639-944X.1202.

196   Seneff and Nigh, "Worse Than the Disease?"

197   Mohammad N Uddin and Monzurul A Roni, "Challenges of Storage and Stability of mRNA-Based COVID-19 Vaccines," *Vaccines (Basel)*, 9 no. 9 (2021): doi: 10.3390/vaccines9091033.

198   Linde Schoenmaker, Dominik Witzigmann, Jayesh A Kulkarni, et al., "mRNA-lipid nanoparticle COVID-19 vaccines: Structure and stability," *Int J Pharm*, 601 no. (2021): 120586, doi: 10.1016/j.ijpharm.2021.120586.

199   Kapil Bahl, Joe J Senn, Olga Yuzhakov, et al., "Preclinical and Clinical Demonstration of Immunogenicity by mRNA Vaccines against H10N8 and H7N9 Influenza Viruses," *Mol Ther*, 25 no. 6 (2017): p. 1316–1327, doi : 10.1016/j.ymthe.2017.03.035.

200   Vivek P Chavda, Shailvi Soni, Lalitkumar K Vora, et al., "mRNA-Based Vaccines and Therapeutics for COVID-19 and Future Pandemics," *Vaccines (Basel)*, 10 no. 12 (2022): doi: 10.3390/vaccines10122150.

201   Steven Black and Stephen Evans, "Serious adverse events following mRNA vaccination in randomized trials in adults," *Vaccine*, 41 no. 23 (2023): p. 3473–3474, doi: 10.1016/j.vaccine.2023.04.040.

Chapter 11

202 Ian Schwartz. "Fauci: 'The Cavalry Is Coming' With A Vaccine." *RealClear Politics*, November 12, 2020. https://www.realclearpolitics.com /video/2020/11/12/fauci_the_cavalry_is_coming_with_a_vaccine.html.

203 Laura Santhanam, Judy Woodruff, Alison Thoet. "What Dr. Fauci wants you to know about face masks and staying home as virus spreads." *PBS NEWS HOUR*, Apr 3, 2020. https://www.pbs.org/newshour/show/what-dr-fauci -wants-you-to-know-about-face-masks-and-staying-home-as-virus-spreads.

204 Centers for Disease Control and Prevention, "Safety of COVID-19 Vaccines," Nov. 3, 2023 https://www.cdc.gov/coronavirus/2019-ncov/vaccines/safety /safety-of-vaccines.html.

205 Moderna, "Why Spikevax (COVID-19 Vaccine, mRNA)?," December, 2023, https://spikevax.com/healthcare-provider-guide /why-choose-moderna-spikevax-mrna-covid-19-vaccine.

206 U.S. Centers for Disease Control and Prevention, "How I Recommend Vaccination Video Series," November 1, 2021, https://www.cdc.gov/vaccines /howirecommend/index.html.

207 U.S. Centers for Disease Control and Prevention, *COVID-19 Vaccines Are Safe & Effective*. Dec 9, 2022.

208 https://zdoggmd.com/ /podcasting/paul-offit-live/#:~:text=We're%20 talking%20with%20vaccine%20expert%20Dr.%20Paul%20Offit%20 about%20variants.

209 Sri Rezeki Hadinegoro, Jose Luis Arredondo-Garcia, Maria Rosario Capeding, et al., "Efficacy and Long-Term Safety of a Dengue Vaccine in Regions of Endemic Disease," *N Engl J Med*, 373 no. 13 (2015): p. 1195-206, doi : 10.1056/NEJMoa1506223.

210 Ibid.

211 Stephen J. Thomas and In-Kyu Yoon, "A review of Dengvaxia®: development to deployment," *Human Vaccines & Immunotherapeutics*, 15 no. 10 (2019): p. 2295–2314, doi: 10.1080/21645515.2019.1658503.

212 Carol Isoux. "Are Philippine children's deaths linked to dengue vaccine? | South China Morning Post (scmp.com)." *South China Morning Post*, April 21, 2019. https://www.scmp.com/magazines/post-magazine/long-reads /article/3006712/philippines-suspicion-dengue-vaccine-linked.

213 Guzman, Maria G., and Susana Vazquez. "The complexity of antibody-dependent enhancement of dengue virus infection." *Viruses* 2, no. 12 (2010): 2649-2662. https://https://www.nature.com/articles/nri3014].

214 https://www.npr.org/sections/goatsandsoda/2019/05/03/719037789/botched-vaccine-launch-has-deadly-repercussions.

215 U.S. House of Representatives, House Science, Space, and Technology Committee, *House Science, Space, and Technology Committee Hearing on Coronavirus*, 117th Congress, 2nd Session.

216 Julie Steenhuysen. "As pressure for coronavirus vaccine mounts, scientists debate risks of accelerated testing." *Reuters*, March 11, 2020. https://www.reuters.com/article/us-health-coronavirus-vaccines-insight/as-pressure-for-coronavirus-vaccine-mounts-scientists-debate-risks-of-accelerated-testing-idUSKBN20Y1GZ.

217 Nikolai Eroshenko, Taylor Gill, Marianna K. Keaveney, et al., "Implications of antibody-dependent enhancement of infection for SARS-CoV-2 countermeasures," *Nature Biotechnology*, 38 no. 7 (2020): p. 789–791, doi: 10.1038/s41587-020-0577-1.

218 Shan Su, Lanying Du, and Shibo Jiang, "Learning from the past: development of safe and effective COVID-19 vaccines," *Nature Reviews Microbiology*, 19 no. 3 (2020): p. 211–219, doi: 10.1038/s41579-020-00462-y.

Chapter 12

219 Maryland General Assembly, Health and Government Operations Committee, *Maryland House Bill 699 - Proof of Vaccination for Employees and Applicants for Employment - Prohibition (Vaccination by Choice Act)*, 2023 Session, March 6, 2023.

220 Peter Doshi, "Peter Doshi: Pfizer and Moderna's '95% effective' vaccines-let's be cautious and first see the full data," https://blogs.bmj.com/bmj/2020/11/26/peter-doshi-pfizer-and-modernas-95-effective-vaccines-lets-be-cautious-and-first-see-the-full-data/.

221 Ibid.

222 Ibid.

223 Yair Goldberg, Micha Mandel, Yinon M Bar-On, et al., "Waning Immunity after the BNT162b2 Vaccine in Israel," *N Engl J Med*, 385 no. 24 (2021): p. e85, doi: 10.1056/NEJMoa2114228.

224 Daniel M Altmann and Rosemary J Boyton, "Waning immunity to SARS-CoV-2: implications for vaccine booster strategies," *Lancet Respir Med*, 9 no. 12 (2021): p. 1356–1358, doi: 10.1016/S2213-2600(21)00458-6.

225 Einav G Levin, Yaniv Lustig, Carmit Cohen, et al., "Waning Immune Humoral Response to BNT162b2 Covid-19 Vaccine over 6 Months," *N Engl J Med*, 385 no. 24 (2021): p. e84, doi: 10.1056/NEJMoa2114583.

226 Peter Doshi, "Will covid-19 vaccines save lives? Current trials aren't designed to tell us," *BMJ*, no. (2020): p. m4037, doi: 10.1136/bmj.m4037.

227 Ibid.

228 Ibid.

229 Moderna, "A Phase 3, Randomized, Stratified, Observer-Blind, Placebo-Controlled Study to Evaluate the Efficacy, Safety, and Immunogenicity of mRNA-1273 SARS-CoV-2 Vaccine in Adults Aged 18 Years and Older," https://trials.modernatx.com/study/?id=mRNA-1273-P301.

230 RW Frenck Jr., EE Walsh, AR Falsey, et al., "Safety and immunogenicity of two RNA-based Covid-19 vaccine candidates," *New England Journal of Medicine*, 383 no. 2439–50., doi: 0.1056/NEJMoa2027906.

231 Peter Doshi, "Will covid-19 vaccines save lives?"

232 Riley Griffin and John Lauerman, "Pfizer Offers Covid Shots to Health Workers Who Got Placebo," December 16, 2020, https://www.bloomberg.com/news/articles/2020-12-15/moderna-to-offer-covid-shots-to-volunteers-who-got-placebo#xj4y7vzkg.

233 Doshi, "Pfizer and Moderna's '95% effective.'"

234 Ibid.

235 Ibid.

236 Moderna, "A Phase 3, Randomized, Stratified . . . Study"

Chapter 13

237 Emmanuel B. Walter, Kawsar R. Talaat, Charu Sabharwal, Alejandra Gurtman, Stephen Lockhart, Grant C. Paulsen, Elizabeth D. Barnett et al. "Evaluation of the BNT162b2 Covid-19 vaccine in children 5 to 11 years of age." *New England Journal of Medicine* 386, no. 1 (2022): 35-46.

238 https://labeling.pfizer.com/.

239 Ibid.

240 Ibid.

241 Ibid.

242 Clare Craig, *Expired: COVID the Untold Story* (UK: Nielson 2023).

243 Ibid.

244 Evan J. Anderson, C. Buddy Creech, Vladimir Berthaud, Arin Piramzadian, Kimball A. Johnson, Marcus Zervos, Fredric Garner et al. "Evaluation of mRNA-1273 vaccine in children 6 months to 5 years of age." *New England Journal of Medicine* 387, no. 18 (2022): 1673-1687.

245 Ezgi Hacisuleyman, Caryn Hale, Yuhki Saito, et al., "Vaccine Breakthrough Infections with SARS-CoV-2 Variants," *N Engl J Med*, 384 no. 23 (2021): p. 2212-2218, doi: 10.1056/NEJMoa2105000.

246 Norbert Stefan, "Metabolic disorders, COVID-19 and vaccine-breakthrough infections," *Nat Rev Endocrinol*, 18 no. 2 (2022): p. 75-76, doi: 10.1038 /s41574-021-00608-9.

247 https://pa.childrenshealthdefense.org/video/addie-de-garays-story/.

248 "World Medical Association Declaration of Helsinki," *JAMA*, 310 no. 20 (2013): p. 2191, doi: 10.1001/jama.2013.281053.

249 https://phmpt.org/wp-content/uploads/2021/11/5.3.6-postmarketing -experience.pdf.

Chapter 14

250 https://www.fda.gov/drugs/resources-drugs.

251 https://www.fda.gov/vaccines-blood-biologics/guidance-compliance -regulatory-information-biologics.

252 Health Ethics & Governance (HEG), *Expert consultation on the use of placebos in vaccine trials*. World Health Organization.

253 K. Singh and S. Mehta, "The clinical development process for a novel preventive vaccine: An overview," *J Postgrad Med*, 62 no. 1 (2016): p. 4-11, doi: 10.4103/0022-3859.173187.

254 A. Werzberger, B. Mensch, B. Kuter, et al., "A controlled trial of a formalin-inactivated hepatitis A vaccine in healthy children," *N Engl J Med*, 327 no. 7 (1992): p. 453–7, doi: 10.1056/NEJM199208133270702.

255 A. Vojdani, J. B. Pangborn, E. Vojdani, and E. L. Cooper, "Infections, toxic chemicals and dietary peptides binding to lymphocyte receptors and tissue enzymes are major instigators of autoimmunity in autism," *Int J*

Immunopathol Pharmacol, 16 no. 3 (2003): p. 189–99, doi : 10.1177/039463200301600302.

256   S. Havarinasab, L. Lambertsson, J. Qvarnstrom, and P. Hultman, "Dose-response study of thimerosal-induced murine systemic autoimmunity," *Toxicol Appl Pharmacol*, 194 no. 2 (2004): p. 169–79, doi: 10.1016/j .taap.2003.09.006.

257   S. Havarinasab, B. Haggqvist, E. Bjorn, et al., "Immunosuppressive and autoimmune effects of thimerosal in mice," *Toxicol Appl Pharmacol*, 204 no. 2 (2005): p. 109–21, doi: 10.1016/j.taap.2004.08.019.

258   Said Havarinasab and Per Hultman, "Organic mercury compounds and autoimmunity," *Autoimmun Rev*, 4 no. 5 (2005): p. 270–5, doi: 10.1016/j .autrev.2004.12.001.

259   Vijendra K Singh and Wyatt H Rivas, "Detection of antinuclear and antilaminin antibodies in autistic children who received thimerosal-containing vaccines," *J Biomed Sci*, 11 no. 5 (2004): p. 607–10, doi: 10.1007 /BF02256125.

260   Jeff Hanson, Vijendra K. Singh, "Assessment of metallothionein and antibodies to metallothionein in normal and autistic children having exposure to vaccine-derived thimerosal," *Pediatric Allergy and Immunology*, 17 no. 4 291–296, doi: 10.1111/j.1399-3038.2005.00348.x.

261   William Crowe, Philip J Allsopp, Gene E Watson, Pamela J Magee, J J Strain, David J Armstrong, Elizabeth Ball, Emeir M McSorley, "Mercury as an environmental stimulus in the development of autoimmunity – A systematic review," *Autoimmunity Reviews*, 16 no. 1 72–80, doi: 10.1016/j .autrev.2016.09.020.

262   Vera Stejskal, *Allergy and autoimmunity caused by metals: a unifying concept*, in *Vaccines and Autoimmunity*, Yehuda Shoenfeld, Nancy Agmon-Levin, Lucija Tomljenovic, Editor. 2015.

263   Samuel R Goth, Ruth A Chu, Jeffrey P Gregg, et al., "Uncoupling of ATP-mediated calcium signaling and dysregulated interleukin-6 secretion in dendritic cells by nanomolar thimerosal," *Environ Health Perspect*, 114 no. 7 (2006): p. 1083–91, doi: 10.1289/ehp.8881.

264   Per Hultman, *Environmental Factors That Contribute to Autoimmunity*, in *Autoantibodies and Autoimmunity: Molecular Mechanisms in Health and Disease*, P.D.K.M. Pollard, Editor. 2005, Wiley.

265 Lucija Tomljenovic, Christopher A. Shaw, "Mechanisms of aluminum adjuvant toxicity and autoimmunity in pediatric populations," *Lupus*, 21 no. 2 (2012): p. 223–30, doi: 10.1177/0961203311430221.

266 Christopher. A. Shaw and Lucija Tomljenovic, Christopher A. Shaw, "Aluminum in the central nervous system (CNS): toxicity in humans and animals, vaccine adjuvants, and autoimmunity," *Immunol Res*, 56 no. 2-3 (2013): p. 304–16, doi: 10.1007/s12026-013-8403-1.

267 Lucija Tomljenovic, and Christopher A. Shaw, "Aluminum vaccine adjuvants: are they safe?." *Current medicinal chemistry*, 17 no. 18 (2011): p. 2630–2637, doi: 10.2174/092986711795933740.

268 Carlo Perricone, Serena Colafrancesco, Roei D Mazor, et al., "Autoimmune/inflammatory syndrome induced by adjuvants (ASIA) 2013: Unveiling the pathogenic, clinical and diagnostic aspects," *J Autoimmun*, 47 no. (2013): p. 1–16, doi: 10.1016/j.jaut.2013.10.004.

269 Neil Z. Miller, "Aluminum in Childhood Vaccines Is Unsafe," *Journal of American Physicians and Surgeons*, 21 no. 4 (2016): p. 109–116

270 "Yehuda Shoenfeld, Nancy Agmon-Levin, Lucija Tomljenovic, *Vaccines and Autoimmunity*. 2015: Wiley-Blackwell.

271 "Clinical Review for STN 125265/0 Rotarix: Rotavirus Vaccine, Live, Oral," GlasoSmithKline Biologicals, by U.S. Food & Drug Administration. 2008. https://www.fda.gov/media/79986/download.

272 Stan L Block, Timo Vesikari, Michelle G Goveia, et al., "Efficacy, immunogenicity, and safety of a pentavalent human-bovine (WC3) reassortant rotavirus vaccine at the end of shelf life," *Pediatrics*, 119 no. 1 (2007): p. 11–8, doi: 10.1542/peds.2006-2058.

273 International Conference On Harmonisation Of Technical Requirements For Registration Of Pharmaceuticals For Human Use, "ICH Harmonised Tripartite Guideline: Choice Of Control Group And Related Issues In Clinical Trials E10," no.

274 Michelle Roth-Cline, Jason Gerson, Patricia Bright, et al., *Ethical Considerations in Conducting Pediatric Research*, in *Pediatric Clinical Pharmacology*, H.W. Seyberth, A. Rane, and M. Schwab, Editors. 2011, Springer Berlin Heidelberg: Berlin, Heidelberg. p. 219–244.

275 Frederik E Juul, Henriette C Jodal, Ishita Barua, et al., "Mortality in Norway and Sweden during the COVID-19 pandemic," *Scand J Public Health*, 50 no. 1 (2022): p. 38–45, doi: 10.1177/14034948211047137.

276 Jonas Vlachos, Edvin Hertegård, and Helena Svaleryd, *School closures and SARS-CoV-2. Evidence from Sweden's partial school closure.* 2020, Cold Spring Harbor Laboratory.

277 Luis Rajmil, Anders Hjern, Perran Boran, et al., "Impact of lockdown and school closure on children's health and well-being during the first wave of COVID-19: a narrative review," *BMJ Paediatr Open*, 5 no. 1 (2021): p. e001043, doi: 10.1136/bmjpo-2021-001043.

278 Helene Mens, Anders Koch, Manon Chaine, and Aase Bengaard Andersen, "The Hammer vs Mitigation-A comparative retrospective register study of the Swedish and Danish national responses to the COVID-19 pandemic in 2020," *APMIS*, 129 no. 7 (2021): p. 384–392, doi: 10.1111/apm.13133.

279 "World Medical Association Declaration of Helsinki," *JAMA*, 310 no. 20 (2013): p. 2191, doi: 10.1001/jama.2013.281053.

280 Vaccines Committee to Review Adverse Effects of and Medicine Institute of, in *Adverse Effects of Vaccines: Evidence and Causality,* A.F. Kathleen Stratton, Erin Rusch, Ellen Wright Clayton Editor. 2011, National Academies Press (US) Copyright 2012 by the National Academy of Sciences. All rights reserved.: Washington (DC).

281 Ibid.

282 Ibid.

283 Ibid.

284 Dr. Bernadine Healy, "Leading Dr.: Vaccines-Autism Worth Study," CBS NEWS. S. Attkisson. 2008 May 12, 2008; Available from: https://www.cbsnews.com/news/leading-dr-vaccines-autism-worth-study/.

285 Flor M Munoz, Nanette H Bond, Maurizio Maccato, et al., "Safety and immunogenicity of tetanus diphtheria and acellular pertussis (Tdap) immunization during pregnancy in mothers and infants: a randomized clinical trial," *Jama*, 311 no. 17 (2014): p. 1760–9, doi: 10.1001/jama.2014.3633.

286 Doris M Rivera Medina, Alejandra Valencia, Alet de Velasquez, et al., "Safety and immunogenicity of the HPV-16/18 AS04-adjuvanted vaccine: a randomized, controlled trial in adolescent girls," *J Adolesc Health*, 46 no. 5 (2010): p. 414–21, doi: 10.1016/j.jadohealth.2010.02.006.

287 R Lakshman, I Jones, D Walker, et al., "Safety of a new conjugate meningococcal C vaccine in infants," *Arch Dis Child*, 85 no. 5 (2001): p. 391-7, doi: 10.1136/adc.85.5.391.

288 Frederick Varricchio, John Iskander, Frank Destefano, et al., "Understanding vaccine safety information from the Vaccine Adverse Event Reporting System," *The Pediatric infectious disease journal*, 23 no. 4 (2004): p. 287–294, doi: 10.1097/00006454-200404000-00002.

Chapter 15

289 Paul Offit, "Paul Offit, MD," https://www.paul-offit.com/about-paul -offit-md.

290 House - Energy and Commerce; Ways and Means | Senate - Labor and Human Resources, H.R.5546 - National Childhood Vaccine Injury Act of 1986, by Rep. Henry A. Waxman. 1986. https://www.congress.gov /bill/99th-congress/house-bill/5546.

291 Health Resources Services Administration, U.S. Department of Health and Human Services, "National Vaccine Injury Compensation Program," January 2024, https://www.hrsa.gov/vaccine-compensation.

292 H.R.S. Administration, Vaccine Injury Table, by U.S. Department of Health and Human Services. 2024. https://www.hrsa.gov/sites/default/files/hrsa/vicp /vaccine-injury-table-01-03-2022.pdf.

293 Health Resources Services Administration, U.S. Department of Health and Human Services, *National Vaccine Injury Compensation Program Monthly Statistics Report*. 2024. p. 9.

294 House - Energy and Commerce; Ways and Means | Senate - Labor and Human Resources, H.R.5546 - National Childhood Vaccine Injury Act of 1986, by Rep. Henry A. Waxman. 1986. https://www.congress.gov /bill/99th-congress/house-bill/5546.

295 Grant Final Report, Grant ID: R18 HS 017045, Electronic Support for Public Health–Vaccine Adverse Event Reporting System (ESP:VAERS), by Ross Lazarus, MBBS, MPH, MMed, GDCompSci, Michael Klompas, MD, MPH, Steve Bernstein, Harvard Pilgrim Health Care, Inc. 2010. https ://digital.ahrq.gov/sites/default/files/docs/publication/r18hs017045-lazarus -final-report-2011.pdf.

296 U.S. Department of Health and Human Services, Centers for Disease Control and Prevention, ACIP COVID-19 Vaccines Work Group, Overview of myocarditis and pericarditis, by MD Matthew Oster, MPH, CDC

COVID-19 Vaccine Task Force. 2021. https://www.cdc.gov/vaccines/acip
/meetings/downloads/slides-2021-06/02-COVID-Oster-508.pdf.

297   Centers for Disease Control and Prevention U.S. Department of Health and
Human Services, Advisory Committee on Immunization Practices (ACIP),
"COVID-19 Vaccine Safety Technical (VaST) Work Group Reports," May
26, 2022, https://www.cdc.gov/vaccines/acip/work-groups-vast/index.html.

298   Mayme Marshall, Ian D Ferguson, Paul Lewis, et al., "Symptomatic Acute
Myocarditis in 7 Adolescents After Pfizer-BioNTech COVID-19 Vaccination,"
Pediatrics, 148 no. 3 (2021): p. doi: 10.1542/peds.2021-052478.

299   https://medalerts.org/vaersdb/findfield.php?TABLE=ON&GROUP1=AGE&
EVENTS=ON&SYMPTOMS%5B%5D=Myocarditis+(10028606)&SYMP
TOMS%5B%5D=Pericarditis+(10034484)&VAX=COVID19&STATE=NO
TFR.

300   Grant Final Report, Grant ID: R18 HS 017045, Electronic Support for
Public Health–Vaccine Adverse Event Reporting System (ESP:VAERS), by
Ross Lazarus, MBBS, MPH, MMed, GDCompSci, Michael Klompas, MD,
MPH, Steve Bernstein, Harvard Pilgrim Health Care, Inc. 2010. https
://digital.ahrq.gov/sites/default/files/docs/publication/r18hs017045-lazarus
-final-report-2011.pdf.

301   Andrew Bryant, Orcid: 0000-0003-4351-8865, Theresa A Lawrie, et al.,
"Ivermectin for Prevention and Treatment of COVID-19 Infection: A
Systematic Review, Meta-analysis, and Trial Sequential Analysis to Inform
Clinical Guidelines," *Am J Ther*, 28 no. 4 (2021): p. e434-e460, doi
: 10.1097/mjt.0000000000001402.

302   Peter Kasperowicz. "White House wants simultaneous COVID, flu shots:
'This is why God gave us two arms'." *FOX NEWS*, September 6, 2022 2022.
https://www.foxnews.com/politics/this-is-why-god-gave-us-two-arms-white
-house-suggests-simultaneous-covid-flu-shots.

303   Centers for Disease Control and Prevention U.S. Department of Health and
Human Services, Advisory Committee on Immunization Practices (ACIP),
"ACIP Presentation Slides: June 23-25, 2021 Meeting," https://www.cdc.gov
/vaccines/acip/meetings/slides-2021-06.html.

304   Committee on Government Reform House Of Representatives, *Faca: Conflicts
of Interest And Vaccine Development—Preserving the Integrity of the Process*,
106th, 2nd, 2000.

Chapter 16

305 Dana Goldner and Joel E Lavine, "Nonalcoholic Fatty Liver Disease in Children: Unique Considerations and Challenges," *Gastroenterology*, 158 no. 7 (2020): p. 1967–1983 e1, doi: 10.1053/j.gastro.2020.01.048.

306 Matthew J. Maenner PhD, Zachary Warren, PhD, Ashley Robinson Williams, PhD, et. al, "Prevalence and Characteristics of Autism Spectrum Disorder Among Children Aged 8 Years — Autism and Developmental Disabilities Monitoring Network, 11 Sites, United States, 2020," *Morbidity and Mortality Weekly Report (MMWR), Surveillance Summaries* no. 1–14, http://dx.doi.org/10.15585/mmwr.ss7202a1.

307 Paul Shattock and Paul Whiteley, "Biochemical aspects in autism spectrum disorders: updating the opioid-excess theory and presenting new opportunities for biomedical intervention," *Expert Opin Ther Targets*, 6 no. 2 (2002): p. 175–83, doi: 10.1517/14728222.6.2.175.

308 Dag Tveiten, Adrian Finvold, Marthe Andersson, and Karl L. Reichelt, "Peptides and Exorphins in the Autism Spectrum," *Open Journal of Psychiatry*, 04 no. 03 (2014): p. 275–287, doi: 10.4236/ojpsych.2014.43034.

309 Beata Jarmołowska, Marta Bukało, Ewa Fiedorowicz, et al., "Role of Milk-Derived Opioid Peptides and Proline Dipeptidyl Peptidase-4 in Autism Spectrum Disorders," *Nutrients*, 11 no. 1 (2019): p. 87, doi: 10.3390/nu11010087.

310 Diana Di Liberto, Antonella D'Anneo, Daniela Carlisi, et al. *Brain Opioid Activity and Oxidative Injury: Different Molecular Scenarios Connecting Celiac Disease and Autistic Spectrum Disorder.* Brain sciences, 2020. **10**, E437 DOI : 10.3390/brainsci10070437.

311 Daniel Rossignol and Richard Frye, "Folate receptor alpha autoimmunity and cerebral folate deficiency in autism spectrum disorders," *Journal of Pediatric Biochemistry*, 02 no. (2016): p. 263–271, doi: 10.1055/s-0036-1586422.

312 Daniel A. Rossignol and Richard E. Frye, "Cerebral Folate Deficiency, Folate Receptor Alpha Autoantibodies and Leucovorin (Folinic Acid) Treatment in Autism Spectrum Disorders: A Systematic Review and Meta-Analysis," *Journal of Personalized Medicine*, 11 no. 11 (2021): p. 1141, doi: 10.3390/jpm11111141.

313 P. Moretti, T. Sahoo, K. Hyland, et al., "Cerebral folate deficiency with developmental delay, autism, and response to folinic acid," *Neurology*, 64 no. 6 (2005): p. 1088–90, doi: 10.1212/01.WNL.0000154641.08211.B7.

314 Nancy O'Hara MD, *Demystifying PANS/PANDAS: A Functional Medicine Desktop Reference on Basal Ganglia Encephalitis*. 2022. https://www.drohara.com/demystifying-pans-pandas-book/.

315 Henrietta Leonard and Susan Swedo, "Paediatric autoimmune neuropsychiatric disorders associated with streptococcal infection (PANDAS)." *The international journal of neuropsychopharmacology / official scientific journal of the Collegium Internationale Neuropsychopharmacologicum (CINP)*, 4 no. (2001): p. 191–8, doi: 10.1017/S1461145701002371.

316 Susan. E. Swedo, "Pediatric autoimmune neuropsychiatric disorders associated with streptococcal infections (PANDAS)," *Molecular Psychiatry*, 7 no. 2 (2002): p. S24-S25, doi: 10.1038/sj.mp.4001170.

317 L. A. Snider and S. E. Swedo, "PANDAS: current status and directions for research," *Mol Psychiatry*, 9 no. 10 (2004): p. 900-7, doi: 10.1038/sj.mp.4001542.

318 Susan E Swedo and Paul J Grant, "Annotation: PANDAS: a model for human autoimmune disease," *J Child Psychol Psychiatry*, 46 no. 3 (2005): p. 227–34, doi: 10.1111/j.1469-7610.2004.00386.x.

319 AAP Committee on Infectious Diseases, *Red Book (2018): Report of the Committee on Infectious Diseases (31st Edition)*, ed. M. David W. Kimberlin, FAAP, M. Michael T. Brady, FAAP, and M. Mary Ann Jackson, FAAP. 2018.

320 American Academy of Pediatrics; Committee on Infectious Diseases, *Red Book: 2021–2024 Report of the Committee on Infectious Diseases (32nd Edition)*, ed. M. David W. Kimberlin, FAAP, et al. 2021.

321 Elizabeth A. Mumper, "A Call for Action: Recognizing and Treating Medical Problems of Children with Autism," *American Chinese Journal of Medicine and Science*, 5 no. (2012): p. 180, doi: 10.7156/V5I3P180.

322 Elizabeth A. Mumper, "Can Awareness of Medical Pathophysiology in Autism Lead to Primary Care Autism Prevention Strategies," *North American journal of medicine & science*, 6 no. (2013): https://www.najms.com/index.php/najms/article/view/191

323 Ed Dowd, *'Cause Unknown': The Epidemic of Sudden Deaths in 2021 & 2022*. (New York: Skyhorse, 2023).

324 Eugenio Paglino, Dielle J Lundberg, Zhenwei Zhou, et al., "Monthly excess mortality across counties in the United States during the COVID-19 pandemic, March 2020 to February 2022," *Sci Adv*, 9 no. 25 (2023): p. eadf9742, doi: 10.1126/sciadv.adf9742.

325   NBC 10 Philadelphia, "CHOP Vaccine Expert: Wait for More Info Before Getting New Vaccine Booster". R. Conners. 2023 August 29, 2022; Available from: https://www.youtube.com/watch?v=S4DnG5Ar1eE.

Chapter 17

326   https://casetext.com//case/bruesewitz-v-wyeth-llc.

327   Ross Lazarus, Michael Klompas, and Steve Bernstein. "Electronic Support for Public Health–Vaccine Adverse Event Reporting System (ESP: VAERS)." *Grant. Final Report, Grant ID: R18 HS* 17045 (2011).

328   Jessica Rose, "Causation: Let Me Count the Ways," Unacceptable Jessica Substack, January 21, 2022, https://jessicar.substack.com/p/causation-prove -me-wrong?utm_source=%2Fsearch%2FVAERS%2520mortality%2520CO VID%2520shots&utm_medium=reader2.

329   https://odysee.com/$/embed/Highwire-These-Patients-Deserve-to-be-Heard- VAERS-Whistleblower.

330   https://twitter.com/CNNSitRoom/status/1423422301882748929.

331   Jessica Rose, "Died Suddenly," Unacceptable Jessica Substack, December 11, 2022, https://jessicar.substack.com/p/died-suddenly-i-pulled-all-the -adverse?utm_source=%2Fsearch%2FCOVID%2520adverse%2520events%2 520compared%2520to%2520other&utm_medium=reader2.

332   https://www.cdc.gov/coronavirus/2019-ncov/vaccines/safety/vsafe.html.

333   Ibid.

334   https://substack.com//@aaronsiri.

335   https://icandecide.org/v-safe-data/.

336   Ibid.

337   Ibid.

338   Ibid.

339   Ibid.

340   Ibid.

341   https://web.archive.org/web/20230821200440/https:/vsafe.cdc.gov/covid/en/.

342   https://brownstone.org/articles/cdc-refusing-new-covid-vaccine-adverse -event-reports/.

343   https://www.nhtsa.gov/vehicle/1985/FORD/BRONCO%252520II.

344 https://brownstone.org/articles/cdc-refusing-new-covid-vaccine-adverse
-event-reports/.

345 https://openvaers.com/covid-data.

346 Ibid.

347 https://openvaers.com/covid-data/myo-pericarditis.

348 BS Hooker, Miller NZ. Analysis of health outcomes in vaccinated
and unvaccinated children: Developmental delays, asthma, ear
infections and gastrointestinal disorders. *SAGE Open Med.* 2020 May
27;8:2050312120925344. doi: 10.1177/2050312120925344. PMID:
32537156; PMCID: PMC7268563.

349 National Academies of Sciences, Engineering, and Medicine. 1994. "Adverse
Events Associated with Childhood Vaccines: Evidence Bearing on Causality."
Washington, DC: The National Academies Press. https://doi
.org/10.17226/2138.

350 Geir Bjørklund, Massimiliano Peana, Maryam Dadar, Salvatore Chirumbolo,
Jan Aaseth, and Natália Martins. "Mercury-induced autoimmunity:
drifting from micro to macro concerns on autoimmune disorders." *Clinical
immunology* 213 (2020): 108352.

351 Geir Bjørklund, Aleksandra Buha Đorđević, Halla Hamdan, David R.
Wallace, and Massimiliano Peana. "Metal-induced autoimmunity in
neurological disorders: A review of current understanding and future
directions." *Autoimmunity Reviews* (2023): 103509.

352 Kevin Roe. "Autism spectrum disorder initiation by inflammation-facilitated
neurotoxin transport." *Neurochemical Research* 47, no. 5 (2022): 1150–1165.

353 Gerwyn Morris, B. K. Puri, R. E. Frye, and Michael Maes. "The putative
role of environmental mercury in the pathogenesis and pathophysiology of
autism spectrum disorders and subtypes." *Molecular neurobiology* 55 (2018):
4834–4856.

354 JB Handley. *How to End the Autism Epidemic* (Vermont: Chelsea Green,
2018).

355 Unacceptable Jessica | Jessica Rose | Substack.

356 Patrick Provost. "The blind spot in COVID-19 vaccination policies: Underreported adverse events." *International Journal of Vaccine Theory, Practice, and
Research* 3, no. 1 (2023): 707–726.

Chapter 18

357 Peter A. McCullough, John Leake, *The Courage to Face Covid-19: Preventing Hospitalization and Death While Battling the Bio-Pharmaceutical Complex* (New York: Counterplay Books, 2022), 317.

358 Shruti Gupta, Salim S Hayek, Wei Wang, et al., "Factors Associated With Death in Critically Ill Patients With Coronavirus Disease 2019 in the US," *JAMA Intern Med*, 180 no. 11 (2020): p. 1436–1447, doi: 10.1001 /jamainternmed.2020.3596.

359 Peter A. McCullough, John Eidt, Janani Rangaswami, et al., "Urgent need for individual mobile phone and institutional reporting of at home, hospitalized, and intensive care unit cases of SARS-CoV-2 (COVID-19) infection," *Reviews in Cardiovascular Medicine*, 21 no. 1 (2020): p. 1, doi: 10.31083/j .rcm.2020.01.42.

360 Ali Aldujeli, Anas Hamadeh, Kasparas Briedis, et al., "Delays in Presentation in Patients With Acute Myocardial Infarction During the COVID-19 Pandemic," *Cardiology Research*, 11 no. 6 (2020): p. 386–391, doi: 10.14740 /cr1175.

361 Ibid.

362 Ibid.

363 Kevin Bryan Lo, Ruchika Bhargav, Grace Salacup, et al., "Angiotensin converting enzyme inhibitors and angiotensin II receptor blockers and outcomes in patients with COVID-19: a systematic review and meta-analysis," *Expert Rev Cardiovasc Ther*, 18 no. 12 (2020): p. 919–930, doi : 10.1080/14779072.2020.1826308.

364 Ibid.

365 Namrata Singhania, Saurabh Bansal, Divya P. Nimmatoori, et al., "Current Overview on Hypercoagulability in COVID-19," *American Journal of Cardiovascular Drugs*, 20 no. 5 (2020): p. 393–403, doi: 10.1007 /s40256-020-00431-z.

366 Peter A. McCullough, Ronan J. Kelly, Gaetano Ruocco, Edgar Lerma, James Tumlin, Kevin R. Wheelan, Nevin Katz et al. "Pathophysiological basis and rationale for early outpatient treatment of SARS-CoV-2 (COVID-19) infection." *The American journal of medicine* 134, no. 1 (2021): 16–22.

367 O. Elsaid, P. A. McCullough, K. M. Tecson, et al., "Ventricular Fibrillation Storm in Coronavirus 2019," *Am J Cardiol*, 135 no. (2020): p. 177–180, doi : 10.1016/j.amjcard.2020.08.033.

368 Amir Kazory, Claudio Ronco, and Peter A McCullough, "SARS-CoV-2 (COVID-19) and intravascular volume management strategies in the critically ill," *Proc (Bayl Univ Med Cent)*, 0 no. 0 (2020): p. 1–6, doi : 10.1080/08998280.2020.1754700.

369 Jennifer E Flythe, Magdalene M Assimon, Matthew J Tugman, et al., "Characteristics and Outcomes of Individuals With Pre-existing Kidney Disease and COVID-19 Admitted to Intensive Care Units in the United States," *Am J Kidney Dis*, 77 no. 2 (2021): p. 190–203.e1, doi: 10.1053/j .ajkd.2020.09.003.

370 Ibid.

371 Brian C Procter, Casey Ross, Vanessa Pickard, et al., "Clinical outcomes after early ambulatory multidrug therapy for high-risk SARS-CoV-2 (COVID-19) infection," *Rev Cardiovasc Med*, 21 no. 4 (2020): p. 611–614, doi: 10.31083/j.rcm.2020.04.260.

372 Peter McCullough MD, "Peter McCullough, MD, Substack.," [cited September 19, 2023], https://substack.com/@petermcculloughmd.

373 Joseph A. Ladapo, John E. McKinnon, Peter A. McCullough, and Harvey A. Risch, *Randomized Controlled Trials of Early Ambulatory Hydroxychloroquine in the Prevention of COVID-19 Infection, Hospitalization, and Death: Meta-Analysis.* 2020, Cold Spring Harbor Laboratory. https://www.wbur.org /hereandnow/2023/11/08/cdc-director-vaccines.

374 Ibid.

375 Marc K. Halushka and Richard S. Vander Heide, "Myocarditis is rare in COVID-19 autopsies: cardiovascular findings across 277 postmortem examinations," *Cardiovascular Pathology*, 50 no. (2021): p. 107300, doi : https://doi.org/10.1016/j.carpath.2020.107300.

376 Sharon E Fox, Lacey Falgout, and Richard S Vander Heide, "COVID-19 myocarditis: quantitative analysis of the inflammatory infiltrate and a proposed mechanism," *Cardiovasc Pathol*, 54 no. (2021): p. 107361, doi : 10.1016/j.carpath.2021.107361

377 Jacqui Wise, "Covid-19: Should we be worried about reports of myocarditis and pericarditis after mRNA vaccines?," *Bmj*, 373 no. (2021): p. n1635, doi : 10.1136/bmj.n1635.

378 David K Shay, Tom T Shimabukuro, and Frank DeStefano, "Myocarditis Occurring After Immunization With mRNA-Based COVID-19 Vaccines," *JAMA Cardiol*, 6 no. 10 (2021): p. 1115–1117, doi: 10.1001 /jamacardio.2021.2821.

379 Julia W Gargano, Megan Wallace, Stephen C Hadler, et al., "Use of mRNA COVID-19 Vaccine After Reports of Myocarditis Among Vaccine Recipients: Update from the Advisory Committee on Immunization Practices - United States, June 2021," *MMWR Morb Mortal Wkly Rep*, 70 no. 27 (2021): p. 977-982, doi: 10.15585/mmwr.mm7027e2

380 Carolyn M. Rosner, Leonard D. Genovese, Behnam N. Tehrani, et al., "Myocarditis Temporally Associated with COVID-19 Vaccination," *Circulation*, no. (2021): p. doi:

381 Ann Marie Navar, Elizabeth McNally, Clyde W Yancy, et al., "Temporal Associations Between Immunization With the COVID-19 mRNA Vaccines and Myocarditis: The Vaccine Safety Surveillance System Is Working," *JAMA Cardiol*, 6 no. 10 (2021): p. 1117–1118, doi: 10.1001/ jamacardio.2021.2853.

382 Alagarraju Muthukumar, Madhusudhanan Narasimhan, Quan-Zhen Li, et al., "In-Depth Evaluation of a Case of Presumed Myocarditis After the Second Dose of COVID-19 mRNA Vaccine," *Circulation*, 144 no. 6 (2021): p. 487–498, doi: 10.1161/circulationaha.121.056038.

383 Jay R. Montgomery, Margaret A. K. Ryan, Renata J M Engler, et al., "Myocarditis Following Immunization With mRNA COVID-19 Vaccines in Members of the US Military," *JAMA cardiology*, 6 no. 10 (2021): p. 1202–1206, doi:

384 Dror Mevorach, Emilia Anis, Noa Cedar, et al., "Myocarditis After BNT162b2 COVID-19 Third Booster Vaccine in Israel," *Circulation*, 146 no. 10 (2022): p. 802–804, doi: 10.1161/circulationaha.122.060961.

385 James R Gill, Randy Tashjian, and Emily Duncanson, "Autopsy Histopathologic Cardiac Findings in 2 Adolescents Following the Second COVID-19 Vaccine Dose," *Arch Pathol Lab Med*, 146 no. 8 (2022): p. 925–929, doi: 10.5858/arpa.2021-0435-SA.

386 Ibid.

387 Ibid.

388 Ibid.

389 Ibid.

390 Ibid.

391 Nicholas G Kounis, Virginia Mplani, and Ioanna Koniari, "Autopsy Histopathologic Cardiac Findings in 2 Adolescents Following the Second COVID-19 Vaccine Dose: Cytokine Storm, Hypersensitivity, or Something Else," *Arch Pathol Lab Med*, 146 no. 8 (2022): p. 924, doi: 10.5858 /arpa.2022-0102-LE.

392 Ibid.

393 https://childrenshealthdefense.org/ defender/are-covid-vaccines-safe/.

394 Kounis, Mplani, and Koniari, "Autopsy Histopathologic Cardiac Findings."

395 Christopher D. Paddock MD, MPHTM, James R. Gill, Randy Tashjian, Emily Duncanson, "Autopsy Histopathologic Cardiac Findings in 2 Adolescents Following the Second COVID-19 Vaccine Dose/In Reply," *Archives of Pathology & Laboratory Medicine* 146 no. 8 (2022): p. 921–923

396 Centers for Disease Control and Prevention U.S. Department of Health and Human Services, "Risk of COVID-19-Related Mortality," Nov. 16, 2022 [cited September 28, 2023], https://web.archive.org/web/20230922084824 /https://www.cdc.gov/coronavirus/2019-ncov/science/data-review/risk.html.

397 Clement Kwong-Man Yu, Sabrina Tsao, Carol Wing-Kei Ng, et al., "Cardiovascular Assessment up to One Year After COVID-19 Vaccine-Associated Myocarditis," *Circulation*, 148 no. 5 (2023): p. 436–439, doi: 10.1161/circulationaha.123.064772.

398 Jessica Rose and Peter A McCullough, "A report on myocarditis adverse events in the US Vaccine Adverse Events Reporting System (VAERS) in association with COVID-19 injectable biological products," *Curr Probl Cardiol*, no. (2021): p. 101011, doi: 10.1016/j.cpcardiol.2021.101011.

399 Ibid.

400 Mao Hu, Hui Lee Wong, Yuhui Feng, et al., "Safety of the BNT162b2 COVID-19 Vaccine in Children Aged 5 to 17 Years," *JAMA Pediatr*, 177 no. 7 (2023): p. 710–717, doi: 10.1001/jamapediatrics.2023.1440.

401 Gill, Tashjian, and Duncanson, "Autopsy Histopathologic Cardiac Findings," p. 925–929.

402   Jessica Rose, "Causation: Let Me Count the Ways," Unacceptable Jessica
      Substack, January 21, 2022, https://jessicar.substack.com/p/causation-prove
      -me-wrong?utm_source=%2Fsearch%2FVAERS%2520mortality%2520CO
      VID%2520shots&utm_medium=reader2.

403   Natacha Buergin, Pedro Lopez-Ayala, Julia R. Hirsiger, et al., "Sex-specific
      differences in myocardial injuryincidence after COVID-19 mRNA-
      1273booster vaccination," European Journal of Heart Failure, 25 no. 10
      (2023): p. 1871–1881, doi: 10.1002/ejhf.2978.

404   https://twitter.com/karma44921039/status/1749440474035712409?t=ve1M
      0TjRich-fgSIeBkYPA&s=04.

Chapter 19

405   Aristo Vojdani, Elizabeth Mumper, et al., "Low natural killer cell cytotoxic
      activity in autism: the role of glutathione, IL-2 and IL-15," J Neuroimmunol,
      205 no. 1-2 (2008): p. 148–54, doi: 10.1016/j.jneuroim.2008.09.005.

406   Aristo Vojdani and Datis Kharrazian, "Potential antigenic cross-reactivity
      between SARS-CoV-2 and human tissue with a possible link to an increase
      in autoimmune diseases," Clin Immunol, 217 no. (2020): p. 108480, doi:
      10.1016/j.clim.2020.108480.

407   Kim Mack Rosenberg, Mary Holland, Eileen Iorio, The HPV Vaccine On
      Trial: Seeking Justice For A Generation Betrayed (New York: Skyhorse, 2018),
      544.

408   Christopher Exley, "Aluminium-based adjuvants should not be used as
      placebos in clinical trials," Vaccine, 29 no. 50 (2011): p. 9289, doi: 10.1016/j
      .vaccine.2011.08.062.

409   https://www.fda.gov/files/vaccines%20blood%20&%20biologics/published
      /Package-Insert—Gardasil.pdf.

410   Vojdani and Kharrazian, "Potential antigenic cross-reactivity."

411   Ibid.

412   Y. Shoenfeld and A. Aron-Maor, "Vaccination and autoimmunity-'vaccinosis':
      a dangerous liaison?," J Autoimmun, 14 no. 1 (2000): p. 1-10, doi: 10.1006
      /jaut.1999.0346.

413   Yehuda Shoenfeld, Nancy Agmon-Levin, Lucija Tomljenovic, Vaccines and
      Autoimmunity. 2015: Wiley-Blackwell.

414　Darja Kanduc and Yehuda Shoenfeld, "Molecular mimicry between SARS-
CoV-2 spike glycoprotein and mammalian proteomes: implications for the
vaccine," *Immunol Res*, 68 no. 5 (2020): p. 310–313, doi: 10.1007
/s12026-020-09152-6.

415　Segal Yahel, and Yehuda Shoenfeld. "Vaccine-induced autoimmunity: the
role of molecular mimicry and immune crossreaction." *Cellular & molecular
immunology*, 15 no. 6 (2018): 586-594.

416　Shoenfeld, Agmon-Levin, and Tomljenovic, *Vaccines and Autoimmunity*
(Hoboken: Wiley-Blackwell, 2015).

417　Kanduc and Shoenfeld, "Molecular mimicry between SARS-CoV-2."

418　Agnieszka Razim, Katarzyna Pacyga, Małgorzata Aptekorz, et al., "Epitopes
identified in GAPDH from Clostridium difficile recognized as common
antigens with potential autoimmunizing properties," *Scientific Reports*, 8 no. 1
(2018): p. doi: 10.1038/s41598-018-32193-9.

419　Aristo Vojdani and Datis Kharrazian, "Potential antigenic cross-reactivity
between SARS-CoV-2 and human tissue with a possible link to an increase in
autoimmune diseases," *Clin Immunol*, 217 no. (2020): p. 108480, doi
: 10.1016/j.clim.2020.108480.

420　Karolina Akinosoglou, Ilektra Tzivaki, and Markos Marangos, "Covid-
19 vaccine and autoimmunity: Awakening the sleeping dragon," *Clinical
Immunology*, 226 no. (2021): p. doi: 10.1016/j.clim.2021.108721.

421　Rossella Talotta, "Do COVID-19 RNA-based vaccines put at risk of
immune-mediated diseases? In reply to "potential antigenic cross-reactivity
between SARS-CoV-2 and human tissue with a possible link to an increase in
autoimmune diseases"," *Clinical Immunology*, 224 no. (2021): p. doi
: 10.1016/j.clim.2021.108665.

422　Michele Ghielmetti, Helen Dorothea Schaufelberger, Giorgina Mieli-Vergani,
et al., "Acute autoimmune-like hepatitis with atypical anti-mitochondrial
antibody after mRNA COVID-19 vaccination: A novel clinical entity?,"
Journal of Autoimmunity, 123 no. (2021): p. doi: 10.1016/j
.jaut.2021.102706.

423　Ibid.

424　Han Zheng, Ting Zhang, Yiyao Xu, et al., "Autoimmune hepatitis after
COVID-19 vaccination," *Frontiers in Immunology*, 13 no. (2022): p. doi
: 10.3389/fimmu.2022.1035073.

425   Isabel Garrido, Susana Lopes, Manuel Sobrinho Simões, et al., "Autoimmune hepatitis after COVID-19 vaccine – more than a coincidence," *Journal of Autoimmunity*, 125 no. (2021): p. doi: 10.1016/j.jaut.2021.102741.

426   Cathy McShane, Clifford Kiat, Jonathan Rigby, and Órla Crosbie, "The mRNA COVID-19 vaccine – A rare trigger of autoimmune hepatitis?," *Journal of Hepatology*, 75 no. 5 (2021): p. 1252–1254, doi: 10.1016/j.jhep.2021.06.044.

427   Kenneth W. Chow, Nguyen V. Pham, Britney M. Ibrahim, et al., "Autoimmune Hepatitis-Like Syndrome Following COVID-19 Vaccination: A Systematic Review of the Literature," *Digestive Diseases and Sciences*, 67 no. 9 (2022): p. 4574-4580, doi: 10.1007/s10620-022-07504-w.

428   Fatma Elrashdy, Murtaza M. Tambuwala, Sk Sarif Hassan, et al., "Autoimmunity roots of the thrombotic events after COVID-19 vaccination," *Autoimmunity Reviews*, 20 no. 11 (2021): p. doi: 10.1016/j.autrev.2021.102941.

429   Yue Chen, Zhiwei Xu, Peng Wang, et al., "New-onset autoimmune phenomena post-COVID-19 vaccination," *Immunology*, 165 no. 4 (2022): p. 386–401, doi: 10.1111/imm.13443.

430   Ibid.

431   Aaron Lerner, Yehuda Shoenfeld, and Torsten Matthias, "Adverse effects of gluten ingestion and advantages of gluten withdrawal in nonceliac autoimmune disease," *Nutrition Reviews*, 75 no. 12 (2017): p. 1046-1058, doi: 10.1093/nutrit/nux054.

432   Aaron Lerner, Patricia Jeremias, and Torsten Matthias, "The World Incidence and Prevalence of Autoimmune Diseases is Increasing," *International Journal of Celiac Disease*, 3 no. 4 (2016): p. 151–155, doi: 10.12691/ijcd-3-4-8.

Chapter 20

433   Sucheep Piyasirisilp and Thiravat Hemachudha, "Neurological adverse events associated with vaccination," *Current opinion in neurology*, 15 no. 3 (2002): p. 333–338, doi: 10.1097/00019052-200206000-00018.

434   Ravindra Kumar Garg and Vimal Kumar Paliwal, "Spectrum of neurological complications following COVID-19 vaccination," *Neurol Sci*, 43 no. 1 (2022): p. 3–40, doi: 10.1007/s10072-021-05662-9.

435 Berkeley Lovelace Jr. "CDC says these are the most common side effects people report after getting Covid vaccine." *CNBC*, Feb 19 2021. https://www .cnbc.com/2021/02/19/cdc-says-these-are-the-most-common-side-effects -people-report-after-getting-covid-vaccine.html.

436 Garg and Paliwal, "Spectrum of neurological complications."

437 Miguel García-Grimshaw, Santa Elizabeth Ceballos-Liceaga, Laura E. Hernández-Vanegas, et al., "Neurologic adverse events among 704,003 first-dose recipients of the BNT162b2 mRNA COVID-19 vaccine in Mexico: A nationwide descriptive study," *Clinical Immunology*, 229 no. (2021): p. 108786, doi: https://doi.org/10.1016/j.clim.2021.108786.

438 Si Ho Kim, Yu Mi Wi, Su Yeon Yun, et al., "Adverse Events in Healthcare Workers after the First Dose of ChAdOx1 nCoV-19 or BNT162b2 mRNA COVID-19 Vaccination: a Single Center Experience," *J Korean Med Sci*, 36 no. 14 (2021): p. e107, doi: 10.3346/jkms.2021.36.e107.

439 Yun Woo Lee, So Yun Lim, Ji Hyang Lee, et al. "Adverse Reactions of the Second Dose of the BNT162b2 mRNA COVID-19 Vaccine in Healthcare Workers in Korea." *Journal of Korean medical science*, 2021. **36**, e153 DOI: 10.3346/jkms.2021.36.e153.

440 Yasser Aladdin and Bader Shirah, "New-onset refractory status epilepticus following the ChAdOx1 nCoV-19 vaccine," *Journal of neuroimmunology*, 357 no. (2021): p. 577629, doi: 10.1016/j.jneuroim.2021.577629.

441 B. D. Liu, C. Ugolini, and P. Jha, "Two Cases of Post-Moderna COVID-19 Vaccine Encephalopathy Associated With Nonconvulsive Status Epilepticus," *Cureus*, 13 no. 7 (2021): p. e16172, doi: 10.7759/cureus.16172

442 Luca Baldelli, Giulia Amore, Angelica Montini, et al., "Hyperacute reversible encephalopathy related to cytokine storm following COVID-19 vaccine," *Journal of neuroimmunology*, 358 no. (2021): p. 577661, doi: 10.1016/j .jneuroim.2021.577661.

443 Isaac Núñez, Miguel García-Grimshaw, Carlos Yoel Castillo Valencia, Daniel Eduardo Aguilera Callejas, Mónica Libertad Moya Alfaro, María del Mar Saniger-Alba, Alonso Gutiérrez-Romero et al. "Seizures following COVID-19 vaccination in Mexico: a nationwide observational study." *Epilepsia* 63, no. 10 (2022): e144-e149.

444 Ali Rafati, Melika Jameie, Mobina Amanollahi, Mana Jameie, Yeganeh Pasebani, Delaram Sakhaei, Saba Ilkhani et al. "Association of seizure with

COVID-19 vaccines in persons with epilepsy: A systematic review and meta-analysis." *Journal of Medical Virology* 95, no. 9 (2023): e29118.

445 Ali Rafati, Melika Jameie, Mobina Amanollahi, Yeganeh Pasebani, Mana Jameie, Ali Kabiri, Sara Montazeri Namin et al. "Association of New-Onset Seizures With SARS-CoV-2 Vaccines: A Systematic Review and Meta-Analysis of Randomized Clinical Trials." *JAMA neurology* (2024).

446 Toshiaki Iba, Jerrold H Levy, and Theodore E Warkentin, "Recognizing Vaccine-Induced Immune Thrombotic Thrombocytopenia," *Crit Care Med*, 50 no. 1 (2022): p. e80–e86, doi: 10.1097/ccm.0000000000005211.

447 Heidi Ledford, "COVID vaccines and blood clots: five key questions," *Nature*, 592 no. 7855 (2021): p. 495–496, doi: 10.1038 /d41586-021-00998-w.

448 Dennis McGonagle, Gabriele De Marco, and Charles Bridgewood, "Mechanisms of Immunothrombosis in Vaccine-Induced Thrombotic Thrombocytopenia (VITT) Compared to Natural SARS-CoV-2 Infection," *Journal of Autoimmunity*, 121 no. (2021): p. 102662, doi: https://doi .org/10.1016/j.jaut.2021.102662.

449 Talal Al-Mayhani, Sadia Saber, Matthew J Stubbs, et al., "Ischaemic stroke as a presenting feature of ChAdOx1 nCoV-19 vaccine-induced immune thrombotic thrombocytopenia," *J Neurol Neurosurg Psychiatry*, 92 no. 11 (2021): p. 1247–1248, doi: 10.1136/jnnp-2021-326984.

450 Rolf Ankerlund Blauenfeldt, Søren Risom Kristensen, Siw Leiknes Ernstsen, et al., "Thrombocytopenia with acute ischemic stroke and bleeding in a patient newly vaccinated with an adenoviral vector-based COVID-19 vaccine," *Journal of Thrombosis and Haemostasis*, 19 no. 7 (2021): p. 1771–1775, doi: 10.1111/jth.15347.

451 Puja R Mehta, Sean Apap Mangion, Matthew Benger, et al., "Cerebral venous sinus thrombosis and thrombocytopenia after COVID-19 vaccination - A report of two UK cases," *Brain Behav Immun*, 95 no. (2021): p. 514–517, doi: 10.1016/j.bbi.2021.04.006.

452 Anton Pottegard, Lars Christian Lund, Oystein Karlstad, et al., "Arterial events, venous thromboembolism, thrombocytopenia, and bleeding after vaccination with Oxford-AstraZeneca ChAdOx1-S in Denmark and Norway: population based cohort study," *BMJ*, 373 no. (2021): p. n1114, doi : 10.1136/bmj.n1114.

453 Katarzyna Krzywicka, Mirjam R Heldner, Mayte Sánchez van Kammen, et al., "Post-SARS-CoV-2-vaccination cerebral venous sinus thrombosis: an analysis of cases notified to the European Medicines Agency," *Eur J Neurol*, 28 no. 11 (2021): p. 3656–3662, doi: 10.1111/ene.15029.

454 Alberto Vogrig, Francesco Janes, Gian Gigli, et al., "Acute disseminated encephalomyelitis after SARS-CoV-2 vaccination," *Clinical Neurology and Neurosurgery*, 208 no. (2021): p. 106839, doi: 10.1016/j .clineuro.2021.106839.

455 Paolo Pellegrino, Carla Carnovale, Valentina Perrone, et al., "Acute Disseminated Encephalomyelitis Onset: Evaluation Based on Vaccine Adverse Events Reporting Systems," *PLoS ONE*, 8 no. 10 (2013): p. e77766, doi : 10.1371/journal.pone.0077766.

456 Gabriel Torrealba-Acosta, Jennifer C Martin, Yve Huttenbach, et al., "Acute encephalitis, myoclonus and Sweet syndrome after mRNA-1273 vaccine," *BMJ Case Rep*, 14 no. 7 (2021): p. doi: 10.1136/bcr-2021-243173.

457 Liming Cao and Lijie Ren, "Acute disseminated encephalomyelitis after severe acute respiratory syndrome coronavirus 2 vaccination: a case report," *Acta Neurol Belg*, 122 no. 3 (2022): p. 793–795, doi: 10.1007 /s13760-021-01608-2.

458 Frederic Zuhorn, Tilmann Graf, Randolf Klingebiel, et al., "Postvaccinal Encephalitis after ChAdOx1 nCov-19," *Ann Neurol*, 90 no. 3 (2021): p. 506–511, doi: 10.1002/ana.26182.

459 Hardeep Singh Malhotra, Priyanka Gupta, Vikas Prabhu, et al., "COVID-19 vaccination-associated myelitis," *QJM*, 114 no. 8 (2021): p. 591–593, doi: 10.1093/qjmed/hcab069.

460 William Fitzsimmons and Christopher Nance, *Sudden Onset of Myelitis after COVID-19 Vaccination: An Under-Recognized Severe Rare Adverse Event*. 2021, SSRN.

461 Claudia Pagenkopf and Martin Südmeyer, "A case of longitudinally extensive transverse myelitis following vaccination against Covid-19," *J Neuroimmunol*, 358 no. (2021): p. 577606, doi: 10.1016/j.jneuroim.2021.577606.

462 Gustavo C Román, Fernando Gracia, Antonio Torres, et al., "Acute Transverse Myelitis (ATM):Clinical Review of 43 Patients With COVID-19-Associated ATM and 3 Post-Vaccination ATM Serious Adverse Events With the

ChAdOx1 nCoV-19 Vaccine (AZD1222)," *Front Immunol*, 12 no. (2021): p. 653786, doi: 10.3389/fimmu.2021.653786.

463   Asaf Shemer, Eran Pras, and Idan Hecht, "Peripheral Facial Nerve Palsy Following BNT162b2 (COVID-19) Vaccination," *Isr Med Assoc J*, 23 no. 3 (2021): p. 143–144, doi:

464   C. Martin-Villares, A. Vazquez-Feito, M. J. Gonzalez-Gimeno, and B. de la Nogal-Fernandez, "Bell's palsy following a single dose of mRNA SARS-CoV-2 vaccine: a case report," *J Neurol*, 269 no. 1 (2022): p. 47–48, doi : 10.1007/s00415-021-10617-3.

465   G Gomez de Terreros Caro, S Gil Diaz, M Perez Ale, and M L Martinez Gimeno, "Bell's palsy following COVID-19 vaccination: a case report," *Neurologia (Engl Ed)*, 36 no. 7 (2021): p. 567–568, doi: 10.1016/j .nrleng.2021.04.002.

466   Mark Obermann, Maliqe Krasniqi, Nadja Ewers, et al., "Bell's palsy following COVID-19 vaccination with high CSF antibody response," *Neurol Sci*, 42 no. 11 (2021): p. 4397–4399, doi: 10.1007/s10072-021-05496-5.

467   Asaf Shemer, Eran Pras, Adi Einan-Lifshitz, et al., "Association of COVID-19 Vaccination and Facial Nerve Palsy: A Case-Control Study," *JAMA Otolaryngol Head Neck Surg*, 147 no. 8 (2021): p. 739–743, doi: 10.1001 /jamaoto.2021.1259.

468   Al Ozonoff, Etsuro Nanishi, and Ofer Levy, "Bell's palsy and SARS-CoV-2 vaccines-an unfolding story - Authors' reply," *Lancet Infect Dis*, 21 no. 9 (2021): p. 1211–1212, doi: 10.1016/S1473-3099(21)00323-6.

469   Haris Iftikhar, Syeda Mishkaat U Noor, Maarij Masood, and Khalid Bashir, "Bell's Palsy After 24 Hours of mRNA-1273 SARS-CoV-2 Vaccine," *Cureus*, 13 no. 6 (2021): p. e15935, doi: 10.7759/cureus.15935.

470   Daniela Parrino, Andrea Frosolini, Chiara Gallo, et al., "Tinnitus following COVID-19 vaccination: report of three cases," *Int J Audiol*, 61 no. 6 (2022): p. 526–529, doi: 10.1080/14992027.2021.1931969.

471   Helena Wichova, Mia E Miller, and M Jennifer Derebery, "Otologic Manifestations After COVID-19 Vaccination: The House Ear Clinic Experience," *Otol Neurotol*, 42 no. 9 (2021): p. e1213–e1218, doi: 10.1097 /MAO.0000000000003275.

472   Luca Spiro Santovito and Graziano Pinna, "Acute reduction of visual acuity and visual field after Pfizer-BioNTech COVID-19 vaccine 2nd dose: a case

report," *Inflamm Res*, 70 no. 9 (2021): p. 931–933, doi: 10.1007 /s00011-021-01476-9.

473   Owen Dyer, "Covid-19: Regulators warn that rare Guillain-Barré cases may link to J&J and AstraZeneca vaccines," *Bmj*, 374 no. (2021): p. n1786, doi : 10.1136/bmj.n1786.

474   Sadia Waheed, Angel Bayas, Fawzi Hindi, et al., "Neurological Complications of COVID-19: Guillain-Barre Syndrome Following Pfizer COVID-19 Vaccine," *Cureus*, 13 no. 2 (2021): p. e13426, doi: 10.7759/cureus.13426.

475   Christopher Martin Allen, Shelby Ramsamy, Alexander William Tarr, et al., "Guillain-Barre Syndrome Variant Occurring after SARS-CoV-2 Vaccination," *Ann Neurol*, 90 no. 2 (2021): p. 315–318, doi: 10.1002 /ana.26144.

476   Tanveer Hasan, Mustafizur Khan, Farhin Khan, and Ghanim Hamza, "Case of Guillain-Barre syndrome following COVID-19 vaccine," *BMJ Case Rep*, 14 no. 6 (2021): p. doi: 10.1136/bcr-2021-243629.

477   https://www.suzannegazdamd.com/.

478   Suzanne Gazda, "Small fiber neuropathy after COVID infection and vaccination.," October 11, 2023, https://www.suzannegazdamd .com/blog—long-covid/small-fiber-neuropathy-after-covid-infection -and-vaccination.

479   Zoe Swank, Yasmeen Senussi, Zachary Manickas-Hill, et al., "Persistent Circulating Severe Acute Respiratory Syndrome Coronavirus 2 Spike Is Associated With Post-acute Coronavirus Disease 2019 Sequelae," *Clin Infect Dis*, 76 no. 3 (2023): p. e487–e490, doi: 10.1093/cid/ciac722.

480   Sandhya Bansal, Sudhir Perincheri, Timothy Fleming, et al., "Cutting Edge: Circulating Exosomes with COVID Spike Protein Are Induced by BNT162b2 (Pfizer-BioNTech) Vaccination prior to Development of Antibodies: A Novel Mechanism for Immune Activation by mRNA Vaccines," *J Immunol*, 207 no. 10 (2021): p. 2405–2410, doi: 10.4049 /jimmunol.2100637.

481   Vaughn Craddock, Aatish Mahajan, Leslie Spikes, et al., "Persistent circulation of soluble and extracellular vesicle-linked Spike protein in individuals with postacute sequelae of COVID-19," *J Med Virol*, 95 no. 2 (2023): p. e28568, doi: 10.1002/jmv.28568.

482 Andrea Gawaz, Michael Schindler, Elena Hagelauer, et al., "SARS-CoV-2-
     Induced Vasculitic Skin Lesions Are Associated with Massive Spike Protein
     Depositions in Autophagosomes," *J Invest Dermatol*, 144 no. 2 (2024):
     p. 369-377.e4, doi: 10.1016/j.jid.2023.07.018.

483 Tudor Emanuel Fertig, Leona Chitoiu, Daciana Silvia Marta, et al.,
     "Vaccine mRNA Can Be Detected in Blood at 15 Days Post-Vaccination,"
     Biomedicines, 10 no. 7 (2022): p. doi: 10.3390/biomedicines10071538.

484 Michael Mörz, "A Case Report: Multifocal Necrotizing Encephalitis and
     Myocarditis after BNT162b2 mRNA Vaccination against COVID-19,"
     Vaccines (Basel), 10 no. 10 (2022): p. doi: 10.3390/vaccines10101651.

485 Zhouyi Rong, Hongcheng Mai, Saketh Kapoor, et al., *SARS-CoV-2 Spike
     Protein Accumulation in the Skull-Meninges-Brain Axis: Potential Implications
     for Long-Term Neurological Complications in post-COVID-19*. 2023, Cold
     Spring Harbor Laboratory.

486 Suzanne Gazda, "Small fiber neuropathy after COVID infection and
     vaccination.," October 11, 2023, https://www.suzannegazdamd
     .com/blog—long-covid/small-fiber-neuropathy-after-covid-infection
     -and-vaccination.

487 Stephanie Seneff and Greg Nigh, "Worse Than the Disease? Reviewing Some
     Possible Unintended Consequences of the mRNA Vaccines Against COVID-
     19," *International Journal of Vaccine Theory, Practice, and Research*, 2 no. 1 doi:

488 React19, "React19 Research: Persistent Symptoms Survey #2 " [cited May 27,
     2022], https://www.react19.org/science-and-research/lit-reviews-and-surveys
     /react19-research-persistent-symptoms-survey-2.

489 Anthony Flamier, Punam Bisht, Alexsia Richards, et al., "Human iPS cell-
     derived sensory neurons can be infected by SARS-CoV-2," *iScience*, 26 no. 9
     (2023): p. 107690, doi: 10.1016/j.isci.2023.107690.

490 Devon E McMahon, Erin Amerson, Misha Rosenbach, et al., "Cutaneous
     reactions reported after Moderna and Pfizer COVID-19 vaccination: A
     registry-based study of 414 cases," *J Am Acad Dermatol*, 85 no. 1 (2021):
     p. 46–55, doi: 10.1016/j.jaad.2021.03.092.

491 Pooja Arora, Kabir Sardana, Sinu Rose Mathachan, and Purnima Malhotra,
     "Herpes zoster after inactivated COVID-19 vaccine: A cutaneous adverse
     effect of the vaccine," *J Cosmet Dermatol*, 20 no. 11 (2021): p. 3389–3390,
     doi: 10.1111/jocd.14268.

492 Iñigo Lladó, Alberto Fernández-Bernáldez, and Pedro Rodríguez-Jiménez, "Varicella zoster virus reactivation and mRNA vaccines as a trigger," *JAAD Case Rep*, 15 no. (2021): p. 62–63, doi: 10.1016/j.jdcr.2021.07.011.

493 Pedro Rodríguez-Jiménez, Pablo Chicharro, Luisa-Martos Cabrera, et al., "Varicella-zoster virus reactivation after SARS-CoV-2 BNT162b2 mRNA vaccination: Report of 5 cases," *JAAD Case Rep*, 12 no. (2021): p. 58–59, doi : 10.1016/j.jdcr.2021.04.014.

494 Edward Eid, Lina Abdullah, Mazen Kurban, and Ossama Abbas, "Herpes zoster emergence following mRNA COVID-19 vaccine," *Journal of medical virology*, 93 no. 9 (2021): p. 5231–5232, doi: 10.1002/jmv.27036.

495 M. Alpalhão and P. Filipe, "Herpes Zoster following SARS-CoV-2 vaccination - a series of four cases," *J Eur Acad Dermatol Venereol*, 35 no. 11 (2021): p. e750–e752, doi: 10.1111/jdv.17555.

496 A. Tan, K. M. Stepien, and S. T. K. Narayana, "Carnitine palmitoyltransferase II deficiency and post-COVID vaccination rhabdomyolysis," *QJM : monthly journal of the Association of Physicians*, 114 no. 8 (2021): p. 596–597, doi : 10.1093/qjmed/hcab077.

497 Mahmoud Nassar, Howard Chung, Yarl Dhayaparan, et al., "COVID-19 vaccine induced rhabdomyolysis: Case report with literature review," *Diabetes Metab Syndr*, 15 no. 4 (2021): p. 102170, doi: 10.1016/j.dsx.2021.06.007.

498 D. J. Theodorou, S. J. Theodorou, A. Axiotis, et al., "COVID-19 vaccine-related myositis," *Qjm*, 114 no. 6 (2021): p. 424–425, doi: 10.1093/qjmed /hcab043.

Chapter 21

499 Fernando P Polack, Stephen J Thomas, Nicholas Kitchin, et al., "Safety and Efficacy of the BNT162b2 mRNA Covid-19 Vaccine," *N Engl J Med*, 383 no. 27 (2020): p. 2603–2615, doi: 10.1056/NEJMoa2034577.

500 Lindsey R Baden, Hana M El Sahly, Brandon Essink, et al., "Efficacy and Safety of the mRNA-1273 SARS-CoV-2 Vaccine," *N Engl J Med*, 384 no. 5 (2021): p. 403–416, doi: 10.1056/NEJMoa2035389.

501 Pfizer, "Pfizer and BioNTech to Supply the U.S. with 100 Million Additional Doses of COVID-19 Vaccine " press release, December 3, 2020,

https://www.pfizer.com/news/press-release/press-release-detail
/pfizer-and-biontech-supply-us-100-million-additional-doses.

502 Maggie Thorp JD and Jim Thorp MD. "FOIA Reveals Troubling
Relationship between HHS/CDC & the American College of Obstetricians
and Gynecologists." May 7, 2023. https://www.americaoutloud.news/foia
-reveals-troubling-relationship-between-hhs-cdc-the-american-college-of
-obstetricians-and-gynecologists/.

503 ClinicalTrials.gov, To Evaluate the Safety, Tolerability, and Immunogenicity
of BNT162b2 Against COVID-19 in Healthy Pregnant Women 18 Years
of Age and Older, by U.S. Department of Health and Human Services.
NCT04754594, 2023. https://clinicaltrials.gov/study/NCT04754594?term
=NCT04754594&rank=1.

504 Maryanne Demasi, PhD. "EXCLUSIVE: Whatever happened to Pfizer's
covid vaccine trial in pregnant women?," February, 23 2023' https://blog
.maryannedemasi.com/p/exclusive-whatever-happened-to-pfizers.

505 Highlights of Prescribing Information: COMIRNATY® (COVID-19 Vaccine,
mRNA) suspension for injection, for intramuscular use 2023-2024 Formula
Initial U.S. Approval: 2021, by Biontech Pfizer. https://www.fda
.gov/media/151707/download?attachment (September 2023).

506 https://static.modernatx.com/pm/6cef78f8-8dad-4fc9-83d5-
d2fbb7cff867/5efa7d9d-05e8-46b5-945a-637c2867bd00/5efa7d9d-05e8
-46b5-945a-637c2867bd00_viewable_rendition__v.pdf.

507 "COVID-19 Vaccines While Pregnant or Breastfeeding," by Centers for
Disease Control and Prevention U.S. Department of Health and Human
Services. https://web.archive.org/web/20230428174410/https://www.cdc.gov
/coronavirus/2019-ncov/vaccines/recommendations/pregnancy.html (October
20,2022).

508 Bernadine Healy, "Women's health, public welfare," *JAMA*, 266 no. 4 (1991):
p. 566–8, doi:

509 Vaccine Safety Research Foundation, "Pediatrician/Professor Renata Moon
MD -Senator Johnson's Covid-19 vaccine roundtable," no. doi: https
://twitter.com/SenRonJohnson/status/1600634770035122176?lang=https
://www.youtube.com/watch?v=LiP3AF5btDQ

510 Jessica Rose, "The Under Reporting Factor in VAERS," November 16, 2022,
https://jessicar.substack.com/p/the-under-reporting-factor-in-vaers.

511 National Vaccine Information Center, "Search the VAERS Database," https ://www.medalerts.org/vaersdb/index.php.

512 Eliana Romero, Shawn Fry, and Brian Hooker, "Safety of mRNA Vaccines Administered During the First Twenty-Four Months of the International COVID-19 Vaccination Program," *International Journal of Vaccine Theory, Practice, and Research*, 3 no. 1 (2023): p. 891–910, doi: 10.56098/ijvtpr .v3i1.70.

513 Gov.UK, "Coronavirus vaccine - summary of Yellow Card reporting," no. (2023): p. doi:

514 James Thorp, Claire Rogers, Michael Deskevich, et al., *COVID-19 Vaccines: The Impact on Pregnancy Outcomes and Menstrual Function*. 2022, Preprints.org.

515 "VAERS interface at CDC WONDER: Data as of April 7, 2023, retrieved using search terms "Infertility" and "Spontaneous Abortion" for "COVID19 Vaccine" and "All Vaccine Products" in "All Territories.","," https://wonder.cdc .gov/vaers.html.

516 Annamaria Mascolo, Gabriella di Mauro, Federica Fraenza, et al., "Maternal, fetal and neonatal outcomes among pregnant women receiving COVID-19 vaccination: The preg-co-vax study," *Front Immunol*, 13 no. (2022): p. 965171, doi: 10.3389/fimmu.2022.965171.

517 Katharine M N Lee, Eleanor J Junkins, Chongliang Luo, et al., "Investigating trends in those who experience menstrual bleeding changes after SARS-CoV-2 vaccination," *Sci Adv*, 8 no. 28 (2022): p. eabm7201, doi: 10.1126 /sciadv.abm7201.

518 Tiffany Parotto, James A. Thorp, Brian Hooker, Paul J. Mills, Jill Newman, Leonard Murphy, Warren Geick et al. "COVID-19 and the surge in Decidual Cast Shedding." *The Gazette of Medical Sciences* 3 (2022): 107–117.

519 Malini DeSilva, Jacob Haapala, Gabriela Vazquez-Benitez, et al., "Evaluation of Acute Adverse Events after Covid-19 Vaccination during Pregnancy," *N Engl J Med*, 387 no. 2 (2022): p. 187–189, doi: 10.1056 /NEJMc2205276.

520 Manish Sadarangani, Phyumar Soe, Hennady P Shulha, et al., "Safety of COVID-19 vaccines in pregnancy: a Canadian National Vaccine Safety (CANVAS) network cohort study," *Lancet Infect Dis*, 22 no. 11 (2022): p. 1553-1564, doi: 10.1016/s1473-3099(22)00426-1.

521   Aharon Dick, Joshua I Rosenbloom, Einat Gutman-Ido, et al., "Safety of SARS-CoV-2 vaccination during pregnancy- obstetric outcomes from a large cohort study," *BMC Pregnancy Childbirth*, 22 no. 1 (2022): p. 166, doi: 10.1186/s12884-022-04505-5.

522   https://www.bing.com/search?q=united+states+births+per+year.

523   https://www.statista.com//statistics/1287702/number-of-births-in-israel/.

524   Naomi Wolf, "The Covenant of Death," August 29, 2023, https://dailyclout.io/the-covenant-of-death/.

525   Naomi Wolf, "Ob/Gyn Warns of Possible Links Between mRNA Injections, Miscarriages and Malformations," September 12, 2022, https://dailyclout.io/obgyn-warns-mrna-injections-and-miscarriages-malformations-may-be-linked/

526   Amy Kelly. "Report 69: BOMBSHELL – Pfizer and FDA Knew in Early 2021 That Pfizer mRNA COVID "Vaccine" Caused Dire Fetal and Infant Risks, Including Death. They Began an Aggressive Campaign to Vaccinate Pregnant Women Anyway." April 29, 2023. https://dailyclout.io/bombshell-pfizer-and-the-fda-knew-in-early-2021-that-the-pfizer-mrna-covid-vaccine-caused-dire-fetal-and-infant-risks-they-began-an-aggressive-campaign-to-vaccinate-pregnant-women-anyway/.

527   Wolf, "Ob/Gyn Warns of Possible Links."

528   Ibid.

529   Kelly. "Report 69: BOMBSHELL."

530   Peter I. Parry, Astrid Lefringhausen, Conny Turni, Christopher J. Neil, Robyn Cosford, Nicholas J. Hudson, and Julian Gillespie. 2023. "'Spikeopathy': COVID-19 Spike Protein Is Pathogenic, from Both Virus and Vaccine mRNA" *Biomedicines* 11, no. 8: 2287. https://doi.org/10.3390/biomedicines11082287

531   SCHC-OSHA Alliance GHS/HazCom Information Sheet Workgroup, "Hazard Communication Information Sheet reflecting the US OSHA Implementation of the Globally Harmonized System of Classification and Labelling of Chemicals (GHS): Reproductive Toxicity," https://www.schc.org/assets/docs/ghs_info_sheets/schc_osha_reproductive_toxicity_4-4-16.pdf.

532   Lydia Pilar Suárez M.D., M.Sc., Marian Chávez Guardado M. D., Marta Barranquero Gómez B.Sc., M.Sc., Zaira Salvador B.Sc., M.Sc., Sandra Fernández B.A., M.A., "What Are Antisperm Antibodies? – Causes & Treatment," *Invitra*, no. (2022): p. doi:

533 Itai Gat, Alon Kedem, Michal Dviri, et al., "Covid-19 vaccination BNT162b2 temporarily impairs semen concentration and total motile count among semen donors," *Andrology*, 10 no. 6 (2022): p. 1016–1022, doi : 10.1111/andr.13209.

Chapter 22

534 Rosa Maria Pascale, Diego Francesco Calvisi, Maria Maddalena Simile, et al., "The Warburg Effect 97 Years after Its Discovery," *Cancers (Basel)*, 12 no. 10 (2020): p. doi: 10.3390/cancers12102819.

535 Zhao Chen, Weiqin Lu, Celia Garcia-Prieto, and Peng Huang, "The Warburg effect and its cancer therapeutic implications," *J Bioenerg Biomembr*, 39 no. 3 (2007): p. 267–74, doi: 10.1007/s10863-007-9086-x.

536 Steven J Bensinger and Heather R Christofk, "New aspects of the Warburg effect in cancer cell biology," *Semin Cell Dev Biol*, 23 no. 4 (2012): p. 352-61, doi: 10.1016/j.semcdb.2012.02.003.

537 Maria V Deligiorgi, Charis Liapi, and Dimitrios T Trafalis, "How Far Are We from Prescribing Fasting as Anticancer Medicine?," *Int J Mol Sci*, 21 no. 23 (2020): p. doi: 10.3390/ijms21239175.

538 Eleah Stringer, Julian J Lum, and Nicol Macpherson, "Intermittent Fasting in Cancer: a Role in Survivorship?," *Curr Nutr Rep*, 11 no. 3 (2022): p. 500–507, doi: 10.1007/s13668-022-00425-0.

539 Mehmet Emin Adin, Edvin Isufi, Michal Kulon, and Darko Pucar, "Association of COVID-19 mRNA Vaccine With Ipsilateral Axillary Lymph Node Reactivity on Imaging," *JAMA Oncol*, 7 no. 8 (2021): p. 1241-1242, doi: 10.1001/jamaoncol.2021.1794.

540 Nishi Mehta, Rachel Marcus Sales, Kemi Babagbemi, et al., "Unilateral axillary adenopathy in the setting of COVID-19 vaccine: Follow-up," *Clin Imaging*, 80 no. (2021): p. 83–87, doi: 10.1016/j.clinimag.2021.06.037.

541 Tin Van Huynh, Lekha Rethi, Ting-Wei Lee, et al., "Spike Protein Impairs Mitochondrial Function in Human Cardiomyocytes: Mechanisms Underlying Cardiac Injury in COVID-19," *Cells*, 12 no. 6 (2023): p. doi: 10.3390 /cells12060877.

542 Chandan Bhowal, Sayak Ghosh, Debapriya Ghatak, and Rudranil De, "Pathophysiological involvement of host mitochondria in SARS-CoV-2 infection that causes COVID-19: a comprehensive evidential insight," *Mol*

Cell Biochem, 478 no. 6 (2023): p. 1325–1343, doi: 10.1007/s11010 -022-04593-z.

543 Chantal A Pileggi, Gaganvir Parmar, Hussein Elkhatib, et al., "The SARS-CoV-2 spike glycoprotein interacts with MAO-B and impairs mitochondrial energetics," *Curr Res Neurobiol*, 5 no. (2023): p. 100112, doi: 10.1016/j .crneur.2023.100112.

544 Meredith Mihalopoulos, Navneet Dogra, Nihal Mohamed, et al., "COVID-19 and Kidney Disease: Molecular Determinants and Clinical Implications in Renal Cancer," *Eur Urol Focus*, 6 no. 5 (2020): p. 1086–1096, doi: 10.1016/j .euf.2020.06.002.

545 Khaled S Allemailem, Ahmad Almatroudi, Faris Alrumaihi, et al., "Single nucleotide polymorphisms (SNPs) in prostate cancer: its implications in diagnostics and therapeutics," *Am J Transl Res*, 13 no. 4 (2021): p. 3868–3889.

546 K. J. O'Byrne and A. G. Dalgleish, "Chronic immune activation and inflammation as the cause of malignancy," *Br J Cancer*, 85 no. 4 (2001): p. 473-83, doi: 10.1054/bjoc.2001.1943.

547 Gareth Iacobucci, "Covid-19: Fourth vaccine doses-who needs them and why?," *Bmj*, 376 no. (2022): p. o30, doi: 10.1136/bmj.o30.

548 Ibid.

549 Gangning Liang and Daniel J Weisenberger, "DNA methylation aberrancies as a guide for surveillance and treatment of human cancers," *Epigenetics*, 12 no. 6 (2017): p. 416–432, doi: 10.1080/15592294.2017.1311434.

550 Tomoyuki Yamaguchi and Shimon Sakaguchi, "Regulatory T cells in immune surveillance and treatment of cancer," *Semin Cancer Biol*, 16 no. 2 (2006): p. 115–23, doi: 10.1016/j.semcancer.2005.11.005.

551 Chao Gu, Matthew Wiest, Wei Zhang, et al., "Cancer Cells Promote Immune Regulatory Function of Macrophages by Upregulating Scavenger Receptor MARCO Expression," *J Immunol*, 211 no. 1 (2023): p. 57-70, doi: 10.4049 /jimmunol.2300029.

552 Hui Jiang and Ya-Fang Mei, "Retraction: Jiang, H.; Mei, Y.-F. SARS-CoV-2 Spike Impairs DNA Damage Repair and Inhibits V(D) J Recombination In Vitro. Viruses 2021, 13, 2056," *Viruses*, 14 no. 5 (2022): p. doi: 10.3390 /v14051011.

553 Markus Alden, Francisko Olofsson Falla, Daowei Yang, et al., "Intracellular Reverse Transcription of Pfizer BioNTech COVID-19 mRNA Vaccine BNT162b2 In Vitro in Human Liver Cell Line," *Curr Issues Mol Biol*, 44 no. 3 (2022): p. 1115-1126, doi: 10.3390/cimb44030073.

554 Rhoda Wilson. "Pfizer/BioNTech mRNA Incorporates into Human DNA In as Little as Six Hours, A New Study Finds." February 27, 2022. https ://expose-news.com/2022/02/27/mrna-incorporates-into-dna-in-six-hours/

555 Markus Alden, Francisko Olofsson Falla, Daowei Yang, et al., "Intracellular Reverse Transcription of Pfizer BioNTech COVID-19 mRNA Vaccine BNT162b2 In Vitro in Human Liver Cell Line," *Curr Issues Mol Biol*, 44 no. 3 (2022): p. 1115–1126, doi: 10.3390/cimb44030073.

556 https://www.statista.com/statistics/1104709/coronavirus-deaths-worldwide -per-million-inhabitants/.

557 Markus Alden, Francisko Olofsson Falla, Daowei Yang, et al., "Intracellular Reverse Transcription of Pfizer BioNTech COVID-19 mRNA Vaccine BNT162b2 In Vitro in Human Liver Cell Line," *Curr Issues Mol Biol*, 44 no. 3 (2022): p. 1115–1126, doi: 10.3390/cimb44030073.

558 Ibid.

559 Ibid.

560 Ibid.

561 Ibid.

562 Krittaya Mekritthikrai, Peera Jaru-Ampornpan, Piyawat Komolmit, and Kessarin Thanapirom, "Autoimmune Hepatitis Triggered by COVID-19 Vaccine: The First Case From Inactivated Vaccine," *ACG Case Rep J*, 9 no. 7 (2022): p. e00811, doi: 10.14309/crj.0000000000000811.

563 Han Zheng, Ting Zhang, Yiyao Xu, et al., "Autoimmune hepatitis after COVID-19 vaccination," *Front Immunol*, 13 no. (2022): p. 1035073, doi : 10.3389/fimmu.2022.1035073.

564 Kenneth W Chow, Nguyen V Pham, Britney M Ibrahim, et al., "Autoimmune Hepatitis-Like Syndrome Following COVID-19 Vaccination: A Systematic Review of the Literature," *Dig Dis Sci*, 67 no. 9 (2022): p. 4574–4580, doi: 10.1007/s10620-022-07504-w.

565 Liguo Zhang, Alexsia Richards, M Inmaculada Barrasa, et al., "Reverse-transcribed SARS-CoV-2 RNA can integrate into the genome of cultured

human cells and can be expressed in patient-derived tissues," *Proc Natl Acad Sci U S A*, 118 no. 21 (2021): p. doi: 10.1073/pnas.2105968118.

566 Ibid.

567 Robert W. Chandler, MD, MBA. "Report 61: Histopathology Series Part 3 – Ute Krüger, MD, Breast Cancer Specialist, Reveals Increase in Cancers and Occurrences of "Turbo Cancers" Following Genetic Therapy "Vaccines"." *Daily Clout.* https://dailyclout.io/report-61-ute-kruger-md-breast-cancer -specialist-reveals-increase-in-cancers-and-occurrences-of-turbo-cancers -following-genetic-therapy-vaccines/.

568 Ibid.

569 Ibid.

570 Vladimir N Uversky, Elrashdy M Redwan, William Makis, and Alberto Rubio-Casillas, "IgG4 Antibodies Induced by Repeated Vaccination May Generate Immune Tolerance to the SARS-CoV-2 Spike Protein," *Vaccines (Basel)*, 11 no. 5 (2023): p. doi: 10.3390/vaccines11050991.

571 Chandler, "Report 61: Histopathology Series Part 3."

572 Jiping Liu, Junbang Wang, Jinfang Xu, et al., "Comprehensive investigations revealed consistent pathophysiological alterations after vaccination with COVID-19 vaccines," *Cell Discov*, 7 no. 1 (2021): p. 99, doi: 10.1038 /s41421-021-00329-3.

573 Stephanie Seneff, Greg Nigh, Anthony M Kyriakopoulos, and Peter A McCullough, "Innate immune suppression by SARS-CoV-2 mRNA vaccinations: The role of G-quadruplexes, exosomes, and MicroRNAs," *Food Chem Toxicol*, 164 no. (2022): p. 113008, doi: 10.1016/j.fct.2022.113008.

574 Nishant Singh and Anuradha Bharara Singh, "S2 subunit of SARS-nCoV-2 interacts with tumor suppressor protein p53 and BRCA: an in silico study," *Transl Oncol*, 13 no. 10 (2020): p. 100814, doi: 10.1016/j .tranon.2020.100814.

575 Ibid.

576 The World Council for Health. *Urgent Expert Hearing on Reports of DNA Contamination in mRNA Vaccines.* 2023.

577 Ibid.

578 Ibid.

579 Ibid.

580 https://worldcouncilforhealth.org/multimedia/urgent-hearing-dna
     -contamination-mrna-vaccines/.

581 Jess Higgins Kelley and Nasha Winters, *A Metabolic Approach to cancer.*
     ISBN 1603586865

582 Paul E. Marik, MD, FCCM, FCCP, *Cancer Care: The role of repurposed drugs
     and metabolic interventions in treating cancer.* FLCCC Alliance. 2023. 227.

583 Peter Nordström, Marcel Ballin, and Anna Nordström, "Risk of infection,
     hospitalisation, and death up to 9 months after a second dose of COVID-19
     vaccine: a retrospective, total population cohort study in Sweden," *Lancet,*
     399 no. 10327 (2022): p. 814–823, doi: 10.1016/s0140-6736(22)00089-7.

Chapter 23

584 Ed Dowd, *Cause Unknown: The Epidemic of Sudden Deaths in 2021 and 2022*
     (New York: Skyhorse, 2022).

585 "Israeli People Committee's Report Find Catastrophic Side Effects Of Pfizer
     Vaccine To Every System In Human Body," *Great Game India,* https
     ://greatgameindia.com/israel-report-pfizer-vaccine-side-effects/.

586 Ibid.

587 Doug Bailey, "'Excess mortality' continuing surge causes concerns," October
     26, 2023, https://insurancenewsnet.com/innarticle/excess-mortality
     -continuing-surge-causes-concerns.

588 Dowd, *Cause Unknown.*

589 Centers for Disease Control and Prevention U.S. Department of Health and
     Human Services, "Excess Deaths Associated with COVID-19, Provisional
     Death Counts for COVID-19," https://www.cdc.gov/nchs/nvss/vsrr/covid19
     /excess_deaths.htm.

590 FSA Patrick Hurley, MAAA, FSA Mike Krohn, CERA, MAAA, FSA Tony
     LaSala, MAAA, et al., *Group Life COVID-19 Mortality Survey Report.* 2023,
     SAO Research Institute.

591 Aaron Smith, "Carriers Face Disturbing Surge in Non-Covid Deaths,"
     September 29, 2023, https://www.linkedin.com/posts/life-annuity-specialist
     _carriers-face-disturbing-surge-in-non-covid-activity-71135682
     68920967169-MpqM/.

592 Centers for Disease Control and Prevention, U.S. Department of Health and Human Services, "Deaths by Week and State: Provisional Death Counts for COVID-19," January 25, 2024, https://www.cdc.gov/nchs/nvss/vsrr/covid19/index.htm.

593 Organization For Economic Co-operation and Development, "Mortality, by week: Excess deaths by week, 2020–2023," https://stats.oecd.org/index.aspx?queryid=104676.

594 Doug Bailey, "'Excess mortality' continuing surge causes concerns," October 26, 2023, https://insurancenewsnet.com/innarticle/excess-mortality-continuing-surge-causes-concerns.

595 Smith, "Carriers Face Disturbing Surge."

596 @Dr. Califf_FDA. X November 30, 2023.

597 Organization For Economic Co-operation and Development, "Mortality, by week."

598 Smith, "Carriers Face Disturbing Surge."

599 FSA Patrick Hurley, MAAA,, FSA Mike Krohn, CERA, MAAA, FSA Tony LaSala, MAAA, et al., *Group Life COVID-19 Mortality Survey Report*. 2023, SAO Research Institute.

600 Centers for Disease Control, "Excess Deaths Associated with COVID-19."

601 FSA Patrick Hurley, MAAA,, FSA Mike Krohn, CERA, MAAA, FSA Tony LaSala, MAAA, et al., *Group Life COVID-19 Mortality Survey Report*. 2023, SAO Research Institute.

602 Centers for Disease Control, "Deaths by Week and State."

603 Drug Overdose Death Rates, by National Institute on Drug Abuse National Institute of Health. 2023. https://nida.nih.gov/research-topics/trends-statistics/overdose-death-rates

604 Institute of Faculty of Actuaries, "CMI says 2022 had the worst second half for mortality since 2010," January 17, 2023, https://actuaries.org.uk/news-and-media-releases/news-articles/2023/jan/17-january-23-cmi-says-2022-had-the-worst-second-half-for-mortality-since-2010/.

605 Jennifer Clarke. "Covid inquiry: What is it investigating and how does it work?" *BBC*, January 16, 2024. https://www.bbc.com/news/explainers-57085964.

606 USA Facts, "US Coronavirus vaccine tracker: Each state has a different plan—and different challenges—in distributing vaccines. Learn more about

who is getting vaccinated by parsing the data by age, sex and race," https://usafacts.org/visualizations/covid-vaccine-tracker-states/.

607   Ronora Stryker, ASA, MAAA, Senior Practice Research Actuary, SOA Research Institute, et al., *Impact of COVID-19 on Future U.S. Mortality.* 2023, SOA Research Institute.

608   Whit Cornman, "Life Insurance Benefits During COVID Highest On Record," press release, https://www.acli.com/posting/nr21-060.

609   Allison Bell, "Insurers Should Test for Post-COVID Mortality Risk, Group Says," November 20, 2023, https://www.thinkadvisor.com/2023/11/20/insurers-should-test-for-post-covid-mortality-risk-group-says/.

610   Marine Baudin, Joseph Hickey, and Jérémie Mercier, *COVID-19 vaccine-associated mortality in the Southern Hemisphere.* 2023.

611   Ibid.

612   Denis Rancourt PhD, "Probable causal association between India's extraordinary April-July 2021 excess-mortality event and the vaccine rollout," in *Correlation Research in the Public Interest.*

613   Marine Baudin and Jérémie Mercier, "Nature of the COVID-era public health disaster in the USA, from all-cause mortality and socio-geo-economic and climatic data." 2021. https://www.researchgate.net/publication/355574895_Nature_of_the_COVID-era_public_health_disaster_in_the_USA_from_all-cause_mortality_and_socio-geo-economic_and_climatic_data.

614   Marine Baudin and Jérémie Mercier, "Analysis of all-cause mortality by week in Canada 2010- 2021, by province, age and sex: There was no COVID-19 pandemic, and there is strong evidence of response- caused deaths in the most elderly and in young males." 2021. https://denisrancourt.ca/entries.php?id=104&name=2021_08_06_analysis_of_all_cause_mortality_by_week_in_canada_2010_2021_by_province_age_and_sex_there_was_no_covid_19_pandemic_and_there_is_strong_evidence_of_response_caused_deaths_in_the_most_elderly_and_in_young_males.

615   Marine Baudin and Jérémie Mercier, "Probable causal association between Australia's new regime of high all-cause mortality and its COVID-19 vaccine rollout". 2022. https://denisrancourt.ca/entries.php?id=125&name=2022_12_20_probable_causal_association_between_australiarsquos_new_regime_of_high_all_cause_mortality_and_its_covid_19_vaccine_rollout.

616 Marine Baudin and Jérémie Mercier, "COVID-Period Mass Vaccination Campaign and Public Health Disaster in the USA From age/state-resolved all-cause mortality by time, age-resolved vaccine delivery by time, and socio-geo-economic data." 2022. https://www.researchgate.net/publication/362427136_COVID-Period_Mass_Vaccination_Campaign_and_Public_Health_Disaster_in_the_USA_From_agestate-resolved_all-cause_mortality_by_time_age-resolved_vaccine_delivery_by_time_and_socio-geo-economic_data.

617 John P. Ioannidis, "Exposure-wide epidemiology: revisiting Bradford Hill," *Stat Med*, 35 no. 11 (2016): p. 1749–62, doi: 10.1002/sim.6825.

618 Denis G. Rancourt, Marine Baudin, Joseph Hickey, and Jérémie Mercier, "COVID-19 vaccine-associated mortality in the Southern Hemisphere," *Correlation Research in the Public Interest*, no. (2023): p. doi: 10.13140/RG.2.2.24720.79366.

619 Ibid.

620 Ibid.

621 Ibid.

622 Ibid.

Chapter 24

623 Phinance Technologies, " US Disability Data, Part 1 - Overview of the Data, Bureau of Labor Statistics (BLS)," https://phinancetechnologies.com/HumanityProjects/US%20Disabilities%20-%20Part1.htm.

624 Ibid

625 Megan Redshaw. "Army Cuts 60,000 Unvaccinated Guard and Reserve Soldiers From Training and Pay as COVID Vaccine Mandate Deadline Passes." *The Defender*, July 7, 2022. https://childrenshealthdefense.org/defender/army-national-guard-reserve-soldiers-covid-vaccine-mandate/.

626 Michael Nevradakis, PhD. "Exclusive: Pilots Injured by COVID Vaccines Speak Out: 'I Will Probably Never Fly Again'." *The Defender*, May 6, 2022. https://childrenshealthdefense.org/defender/pilots-injured-covid-vaccines-speak/.

627 Alexandra Jones. "The restaurant labor shortage: how we got here and a 2023 update." *Open Table*, 2023. https://restaurant.opentable.com/resources/restaurant-labor-shortage/.

628 Ed Dowd, *"Cause Unknown": The Epidemic of Sudden Deaths in 2021 & 2022* (New York: Skyhorse, 2022).

629 Ibid.

630 Ibid.

631 Ibid.

632 Joseph Fraiman, Juan Erviti, Mark Jones, et al., "Serious adverse events of special interest following mRNA COVID-19 vaccination in randomized trials in adults," *Vaccine*, 40 no. 40 (2022): p. 5798-5805, doi: 10.1016/j .vaccine.2022.08.036.

633 https://live.childrenshealthdefense.org/chd-tv/browse-all/chd-bus-collection/.

Chapter 25

634 Christopher Madias, Barry J Maron, Alawi A Alsheikh-Ali, et al., "Commotio cordis," *Indian Pacing Electrophysiol J*, 7 no. 4 (2007): p. 235–45, doi:

635 Del Bigtree, "Why Are Healthy Athletes Collapsing?," in *The Highwire*. 2022. p. 3:47.

636 Del Bigtree, "Healthy Athletes Are Still Inexplicably Collapsing," in *The Highwire*.

637 Reuters Fact Check. "No evidence COVID-19 vaccines are linked to athletes collapsing or dying." *Reuters*, January 23, 2023. https://www.reuters.com /article/idUSL1N3451N2/.

638 Ibid.

639 Dan Evon, "Does Video Show Athletes Fainting Due to COVID-19 Vaccine?," November 30, 2021, https://www.snopes.com/fact-check /athletes-fainting-covid-19-vaccine/.

640 Harshal R. Patil, James H. O'Keefe, Carl J. Lavie, et al., "Cardiovascular damage resulting from chronic excessive endurance exercise," *Missouri medicine*, 109 no. 4 (2012): p. 312–321, doi:

641 Ibid.

642 Justin E Trivax, Barry A Franklin, James A Goldstein, et al., "Acute cardiac effects of marathon running," *J Appl Physiol (1985)*, 108 no. 5 (2010): p. 1148–53, doi: 10.1152/japplphysiol.01151.2009.

643 J. Gladden, C. Wernecke, S. Rector, et al., "Pilot safety study: The use of Vasper™, a novel blood flow restriction exercise in healthy adults," *Journal of Exercise Physiology* online, 19 no. (2016): p. 99–105.

644   Peter A. McCullough and Claudio Ronco, *Chapter 44 - Acute Kidney Injury in Heart Failure*, in *Critical Care Nephrology (Third Edition)*, C. Ronco, et al., Editors. 2019, Elsevier: Philadelphia. p. 257–263.e1.

645   Panagis Polykretis and Peter McCullough, "Rational harm-benefit assessments by age group are required for continued COVID-19 vaccination," *Scandinavian Journal of Immunology*, no. (2022): p. doi: 10.1111/sji.13242.

646   Joseph Fraiman, Juan Erviti, Mark Jones, et al., "Serious adverse events of special interest following mRNA COVID-19 vaccination in randomized trials in adults," *Vaccine*, 40 no. 40 (2022): p. 5798–5805, doi: 10.1016/j.vaccine.2022.08.036.

647   Kevin Bardosh, Allison Krug, Euzebiusz Jamrozik, et al., "COVID-19 vaccine boosters for young adults: a risk benefit assessment and ethical analysis of mandate policies at universities," *Journal of medical ethics*, no. (2022): p. medethics-2022-108449, doi: 10.1136/jme-2022-108449.

648   Ibid.

649   Karin Bille, David Figueiras, Patrick Schamasch, et al., "Sudden cardiac death in athletes: the Lausanne Recommendations," *Eur J Cardiovasc Prev Rehabil*, 13 no. 6 (2006): p. 859-75, doi: 10.1097/01.hjr.0000238397.50341.4a.

650   Nadya Swart. "Normalisation of sudden death surge among athletes demonstrates the extent of our societal pathology." *Biz News*, January 16, 2023. https://www.biznews.com/health/2023/01/16/sudden-death-athletes.

651   "OpenVAERS," https://openvaers.com/.

652   https://goodsciencing.com/ /covid/athletes-suffer-cardiac-arrest-die-after-covid-shot/

653   Suyanee Mansanguan, Prakaykaew Charunwatthana, Watcharapong Piyaphanee, et al., "Cardiovascular Manifestation of the BNT162b2 mRNA COVID-19 Vaccine in Adolescents," *Trop Med Infect Dis*, 7 no. 8 (2022): p. doi: 10.3390/tropicalmed7080196.

654   Suyanee Mansanguan, Prakaykaew Charunwatthana, Watcharapong Piyaphanee, et al., "Cardiovascular Manifestation of the BNT162b2 mRNA COVID-19 Vaccine in Adolescents," *Trop Med Infect Dis*, 7 no. 8 (2022): p. doi: 10.3390/tropicalmed7080196.

655   Ortal Tuvali, Sagi Tshori, Estela Derazne, et al., "The Incidence of Myocarditis and Pericarditis in Post COVID-19 Unvaccinated Patients-A

Large Population-Based Study," *J Clin Med*, 11 no. 8 (2022): p. doi: 10.3390 /jcm11082219.

656 "Report: Vaccination rate remains at roughly 93 percent, after roster cuts." *NBC*, September 3, 2021. https://www.nbcsports.com/nfl/profootballtalk /rumor-mill/news/report-vaccination-rate-remains-at-roughly -93-percent-after-roster-cuts.

657 Ibid.

658 Ibid.

659 Monte Poole. "NFL, NBA, MLB players avoiding vaccines jeopardize wins." *NBC*, September 2, 2021. https://www.nbcsportsbayarea.com /nba/golden-state-warriors/nfl-nba-mlb-players-avoiding-vaccines -jeopardize-wins/1163211/.

Chapter 26

660 Alexandre O Gerard, Audrey Laurain, Audrey Fresse, et al., "Remdesivir and Acute Renal Failure: A Potential Safety Signal from Disproportionality Analysis of the WHO Safety Database," *Clin Pharmacol Ther*, 109 no. 4 (2021): p. 1021–1024, doi: 10.1002/cpt.2145.

661 Meagan L Adamsick, Ronak G Gandhi, Monique R Bidell, et al., "Remdesivir in Patients with Acute or Chronic Kidney Disease and COVID-19," *J Am Soc Nephrol*, 31 no. 7 (2020): p. 1384–1386, doi: 10.1681/ASN.2020050589.

662 Sabue Mulangu, Lori E Dodd, Richard T Jr Davey, et al., "A Randomized, Controlled Trial of Ebola Virus Disease Therapeutics," *N Engl J Med*, 381 no. 24 (2019): p. 2293–2303, doi: 10.1056/NEJMoa1910993.

663 Mabrouk Bahloul, Sana Kharrat, Malek Hafdhi, et al., "Impact of prone position on outcomes of COVID-19 patients with spontaneous breathing," *Acute Crit Care*, 36 no. 3 (2021): p. 208–214, doi: 10.4266/acc.2021.00500.

664 Robert K Naviaux, "Mitochondrial and metabolic features of salugenesis and the healing cycle," *Mitochondrion*, 70 no. (2023): p. 131–163, doi: 10.1016/j .mito.2023.04.003.

665 Robert K Naviaux, "Perspective: Cell danger response Biology-The new science that connects environmental health with mitochondria and the rising tide of chronic illness," *Mitochondrion*, 51 no. (2020): p. 40–45, doi : 10.1016/j.mito.2019.12.005.

666   Robert K Naviaux, "Metabolic features and regulation of the healing cycle-A new model for chronic disease pathogenesis and treatment," *Mitochondrion*, 46 no. (2019): p. 278–297, doi: 10.1016/j.mito.2018.08.001.

667   Lynn Sagan, "On the origin of mitosing cells," *J Theor Biol*, 14 no. 3 (1967): p. 255–74, doi: 10.1016/0022-5193(67)90079-3.

668   Lynn Sagan, "On the origin of mitosing cells. 1967," *J NIH Res*, 5 no. 3 (1993): p. 65–72, https://pubmed.ncbi.nlm.nih.gov/11541390/.

669   Michael W Gray, "Lynn Margulis and the endosymbiont hypothesis: 50 years later," *Mol Biol Cell*, 28 no. 10 (2017): p. 1285–1287, doi: 10.1091/mbc.E16-07-0509.

670   Antonio Ayala, Jose L Venero, Josefina Cano, and Alberto Machado, "Mitochondrial toxins and neurodegenerative diseases," *Front Biosci*, 12 no. (2007): p. 986–1007, doi: 10.2741/2119.

671   Peter Kovacic, Robert S Pozos, Ratnasamy Somanathan, et al., "Mechanism of mitochondrial uncouplers, inhibitors, and toxins: focus on electron transfer, free radicals and structure-activity relationships." *Current Medicinal Chemistry 12, no. 22 (2005): 2601–2623.*

672   Kunwadee Noonong, Moragot Chatatikun, Sirirat Surinkaew, et al., "Mitochondrial oxidative stress, mitochondrial ROS storms in long COVID pathogenesis," *Front Immunol*, 14 no. (2023): p. 1275001, doi: 10.3389/fimmu.2023.1275001.

673   Elena A Belyaeva, Tatyana V Sokolova, Larisa V Emelyanova, and Irina O Zakharova, "Mitochondrial electron transport chain in heavy metal-induced neurotoxicity: effects of cadmium, mercury, and copper," *Scientific World Journal*, 2012 no. (2012): p. 136063, doi: 10.1100/2012/136063.

674   Sergey Korotkov, "Mitochondrial Oxidative Stress Is the General Reason for Apoptosis Induced by Different-Valence Heavy Metals in Cells and Mitochondria," *International Journal of Molecular Sciences*, 24 no. (2023): p. 14459, doi: 10.3390/ijms241914459.

675   Qiuyu Sun, Ying Li, Lijun Shi, et al., "Heavy metals induced mitochondrial dysfunction in animals: Molecular mechanism of toxicity," *Toxicology*, 469 no. (2022): p. 153136, doi: 10.1016/j.tox.2022.153136.

676   Mariana P Cervantes-Silva, Richard G Carroll, Mieszko M Wilk, et al., "The circadian clock influences T cell responses to vaccination by regulating

dendritic cell antigen processing," *Nat Commun*, 13 no. 1 (2022): p. 7217, doi: 10.1038/s41467-022-34897-z.

677   Robert K Naviaux, "Perspective: Cell danger response Biology-The new science that connects environmental health with mitochondria and the rising tide of chronic illness," *Mitochondrion*, 51 no. (2020): p. 40–45, doi : 10.1016/j.mito.2019.12.005.

678   Michelle Beidelschies, Marilyn Alejandro-Rodriguez, Xinge Ji, et al., "Association of the Functional Medicine Model of Care With Patient-Reported Health-Related Quality-of-Life Outcomes," *JAMA Network Open*, 2 no. 10 (2019): p. e1914017, doi: 10.1001/jamanetworkopen.2019.14017.

679   Nabin K Shrestha, Priyanka Shrestha, Patrick C Burke, et al., "Coronavirus Disease 2019 Vaccine Boosting in Previously Infected or Vaccinated Individuals," *Clinical Infectious Diseases*, 75 no. 12 (2022): p. 2169–2177, doi : 10.1093/cid/ciac327.

Chapter 27

680   https://aaronsiri.substack.com/p/instead-of-fdas-requested-500-pages.

681   Public Health and Medical Professionals for Transparency, et al ., V. Food and Drug Administration. 23, United States District Court for the Northern District of Texas Fort Worth Division.

682   "The Daily Clout," https://dailyclout.io.

683   Sunayna Kanjilal, "A Look at Tech Tycoon Larry Ellison's Journey From Creating Oracle to His $108 Billion Net Worth," August 23, 2023, https ://marketrealist.com/what-is-larry-ellisons-net-worth.

684   Aaron Siri, "V-Safe Part 1: After 464 Days, CDC Finally Coughed up Covid-19 Vaccine Safety Data Showing 7.7% of People Reported Needing Medical Care," November 23, 2022, https://aaronsiri.substack .com/p/v-safe-part-1-after-464-days-cdc. Part 2: https://aaronsiri.substack .com/p/v-safe-part-2-what-is-v-safe-what?utm_source=publication-search.

685   Ibid.

686   Ibid.

687   C.f.D.C.a.P. U.S. Department of Health and Human Services, CDC COVID-19 Vaccine Task Force, Vaccine Safety Team, "CDC post-authorization/post-licensure safety monitoring of COVID-19 vaccines," by

Tom Shimabukuro, MD, MPH, MBA. 2020. https://www.cdc.gov/vaccines /acip/meetings/downloads/slides-2020-10/COVID-Shimabukuro-508.pdf.

688 C.f.D.C.a.P. u.S. Department of Health and Human Services, Advisory Committee on Immunization Practices (ACIP), "ACIP Presentation Slides: October 2020 Meeting," by Dr. A Cohn, Dr. J Romero, Dr. M Freedman, et al. https://www.cdc.gov/vaccines/acip/meetings/slides-2020-10.html.

689 Lisa A Jackson, Evan J Anderson, Nadine G Rouphael, et al., "An mRNA Vaccine against SARS-CoV-2 - Preliminary Report," *N Engl J Med*, 383 no. 20 (2020): p. 1920–1931, doi: 10.1056/NEJMoa2022483.

690 Grace M. Lee, José R. Romero, and Beth P. Bell, "Postapproval Vaccine Safety Surveillance for COVID-19 Vaccines in the US," *JAMA*, 324 no. 19 (2020): p. 1937–1938, doi: 10.1001/jama.2020.19692.

691 Siri, "V-Safe Part 1."

692 ICAN, "V-Safe Data," https://icandecide.org/%20v-safe-data/.

693 Paul D. Thacker, in *The DisInformation Chronicle*.

694 Ori Xabi. "The Ministry of Health writes (Article 4): It is not known to him of a single person in Israel who died of Corona at the age of 18 to 49 (which means 50 less one day) who did not have any background diseases." Facebook May 22,2023; Available from: https://www.facebook.com/ori.xabi/posts /pfbid0DdXCCPAJexKZRvmCyWf85an2RXGcpNWWYoYFjZ93mh BS3LZUr4j81KAvPGa4d2VJl.

695 Retsef Levi. "The Israel MOH admits in response to FOIA request by Ori Shabi that there is no known COV death of under 50 person with no comorbidity!" Twitter; Available from: https://twitter.com/RetsefL /status/1660663833524879365.

696 Division of Viral Diseases National Center for Immunization and Respiratory Diseases (NCIRD), "Getting Your COVID-19 Vaccine," January 23, 2024 [cited February, 18, 2024].

697 Emily Crane. "CDC withholding COVID data over fears of misinterpretation." *The New York Post*, Febuary 22, 2022. https://nypost .com/2022/02/22/cdc-withholding-covid-data-over -fears-of-misinterpretation/.

698 Maryland General Assembly, Health and Government Operations Committee of the Maryland State Assembly, *Dr. Kory Testifies on MD Legislation to End Vaccine Mandates*, 2023, https://rumble.com/

v2c7p0o-dr.-pierre-kory-testifies-on-maryland-legislation-to-end-vaccine-mandates.html.

699 Ibid.

700 Naomi Wolf. ""The Covenant of Death"." *Daily Clout*, August 29, 2023. https://dailyclout.io/the-covenant-of-death/.

701 Ibid.

702 Ibid.

703 Naomi Wolf, The WarRoom/DailyClout Pfizer Documents Analysts, Amy Kelly, Stephen K Bannon, *The Pfizer Papers: Pfizer's Crimes Against Humanity* (New York: Skyhorse, 2024), 312.

Chapter 28

704 Nabin K. Shrestha, Patrick C. Burke, Amy S. Nowacki, James F. Simon, Amanda Hagen, and Steven M. Gordon. "Effectiveness of the coronavirus disease 2019 bivalent vaccine." In *Open Forum Infectious Diseases*, vol. 10, no. 6, p. ofad209. US: Oxford University Press, 2023.

705 Nabin K Shrestha, Patrick C Burke, Amy S Nowacki, et al., "Effectiveness of the Coronavirus Disease 2019 Bivalent Vaccine," *Open Forum Infect Dis*, 10 no. 6 (2023): p. ofad209, doi: 10.1093/ofid/ofad209.

706 Walter Fierz and Brigitte Walz, "Antibody Dependent Enhancement Due to Original Antigenic Sin and the Development of SARS," *Front Immunol*, 11 no. (2020): p. 1120, doi: 10.3389/fimmu.2020.01120.

707 Eric L Brown and Heather T Essigmann, "Original Antigenic Sin: the Downside of Immunological Memory and Implications for COVID-19," *mSphere*, 6 no. 2 (2021): p. doi: 10.1128/mSphere.00056-21.

708 Maryam Noori, Seyed Aria Nejadghaderi, and Nima Rezaei, "'Original antigenic sin': A potential threat beyond the development of booster vaccination against novel SARS-CoV-2 variants," *Infect Control Hosp Epidemiol*, 43 no. 8 (2022): p. 1091–1092, doi: 10.1017/ice.2021.199.

709 FDA News, "FDA Open to Less Than 50 Percent Efficacy in COVID Vaccines for Smallest Kids," May 10, 2022, https://www.fdanews.com /articles/207755-fda-open-to-less-than-50-percent-efficacy-in-covid-vaccines -for-smallest-kids?v=preview.

710   Vanessa Piechotta, Waldemar Siemens, Iris Thielemann, et al., "Safety and effectiveness of vaccines against COVID-19 in children aged 5-11 years: a systematic review and meta-analysis," *Lancet Child Adolesc Health*, 7 no. 6 (2023): p. 379–391, doi: 10.1016/s2352-4642(23)00078-0.

711   Daniel Hungerford and Nigel A. Cunliffe, "Real world effectiveness of covid-19 vaccines," *BMJ*, no. (2021): p. n2034, doi: 10.1136/bmj.n2034.

712   Otavio T Ranzani, Matt D T Hitchings, Murilo Dorion, et al., "Effectiveness of the CoronaVac vaccine in older adults during a gamma variant associated epidemic of covid-19 in Brazil: test negative case-control study," *BMJ*, 374 no. (2021): p. n2015, doi: 10.1136/bmj.n2015.

713   Hannah Chung, Siyi He, Sharifa Nasreen, et al., "Effectiveness of BNT162b2 and mRNA-1273 covid-19 vaccines against symptomatic SARS-CoV-2 infection and severe covid-19 outcomes in Ontario, Canada: test negative design study," *BMJ*, 374 no. (2021): p. n1943, doi: 10.1136/bmj.n1943.

714   Daniel Hungerford and Nigel A. Cunliffe, "Real world effectiveness of covid-19 vaccines," *BMJ*, no. (2021): p. n2034, doi: 10.1136/bmj.n2034.

715   Jamie Lopez Bernal, Nick Andrews, Charlotte Gower, et al., "Effectiveness of the Pfizer-BioNTech and Oxford-AstraZeneca vaccines on covid-19 related symptoms, hospital admissions, and mortality in older adults in England: test negative case-control study," *BMJ*, 373 no. (2021): p. n1088, doi: 10.1136/bmj.n1088.

716   Ibid.

717   Ibid.

718   Lisa Lundberg-Morris, Susannah Leach, Yiyi Xu, et al., "Covid-19 vaccine effectiveness against post-covid-19 condition among 589 722 individuals in Sweden: population based cohort study," *BMJ*, no. (2023): p. e076990, doi: 10.1136/bmj-2023-076990.

719   Zachary Stieber. "CDC Never Saw Raw Data Underpinning Key Study." *Epoch Times*, November 9, 2023. https://www.theepochtimes.com/article/cdc-never-saw-raw-data-underpinning-key-study-5524800.

720   Catherine H. Bozio, PhD, MD Shaun J. Grannis, PhD Allison L. Naleway, et al., "Laboratory-Confirmed COVID-19 Among Adults Hospitalized with COVID-19–Like Illness with Infection Induced or mRNA Vaccine-Induced SARS-CoV-2 Immunity — Nine States, January September 2021," *Morbidity and Mortality Weekly Report (MMWR)*, 70 no. 44 1539–1544.

721   Stieber. "CDC Never Saw Raw Data".

722 Ibid.

723 Epidemiology Task Force CDC COVID-19 Response, "Rates of COVID-19 Cases or Deaths by Age Group and Vaccination Status." 2023.

724 Zachary Stieber. "Exclusive: Millions of Breakthrough COVID-19 Cases in 2021, Files Show." *The Epoch Times*, November 11, 2023. https://www .theepochtimes.com/health/exclusive-millions-of-breakthrough-covid-19 -cases-in-2021-files-show-5523013.

725 Jason Lemon. "Video of Biden Saying Vaccinations Prevent COVID Resurfaces After Infection." *Newsweek*, July 21, 2022. https://www.newsweek .com/joe-biden-2021-video-saying-vaccinations-prevent-covid-resurfaces -1726900.

726 Ibid.

727 Ibid.

728 https://www.businessinsider.com/ cdc-director-data-vaccinated-people-do-not-carry-covid-19-2021-3E

729 Zachary Stieber. "Over 10,000 COVID-19 Infections Recorded in Americans Who Received a Vaccine: CDC." *The Epoch Times*, May 26, 2021. https ://www.theepochtimes.com/article/over-10000-infections-recorded-in -americans-who-received-a-covid-19-vaccine-cdc-3831559.

730 Ibid.

731 Shrestha, Burke, Nowacki, et al., "Effectiveness of the Coronavirus Disease."

732 Nabin K Shrestha, Priyanka Shrestha, Patrick C Burke, et al., "Coronavirus Disease 2019 Vaccine Boosting in Previously Infected or Vaccinated Individuals," *Clin Infect Dis*, 75 no. 12 (2022): p. 2169–2177, doi: 10.1093 /cid/ciac327.

733 Elias Eythorsson, Hrafnhildur Linnet Runolfsdottir, Ragnar Freyr Ingvarsson, et al., "Rate of SARS-CoV-2 Reinfection During an Omicron Wave in Iceland," *JAMA Netw Open*, 5 no. 8 (2022): p. e2225320, doi: 10.1001 /jamanetworkopen.2022.25320.

734 Hiam Chemaitelly, Houssein H. Ayoub, Patrick Tang, et al., *COVID-19 primary series and booster vaccination and potential for immune imprinting.* 2022, Cold Spring Harbor Laboratory.

735 Shrestha, Shrestha, Burke, et al., "Coronavirus Disease 2019."

736 Hiam Chemaitelly, Houssein H. Ayoub, Patrick Tang, et al., *COVID-19 primary series and booster vaccination and potential for immune imprinting.* 2022, Cold Spring Harbor Lab.

737 Pascal Irrgang, Juliane Gerling, Katharina Kocher, et al., "Class switch toward noninflammatory, spike-specific IgG4 antibodies after repeated SARS-CoV-2 mRNA vaccination," *Sci Immunol*, 8 no. 79 (2023): p. eade2798, doi: 10.1126/sciimmunol.ade2798.

738 Ibid.

739 Shrestha, Burke, Nowacki, et al., "Effectiveness of the Coronavirus Disease."

740 Peter McCullough, MD, MPH, "Breaking–Pfizer XBB.1.5 Monovalent Vaccine Tested in 20 Mice, No Control Group, and No Humans," Courageous Discourse September 13, 2023, https://petermcculloughmd .substack.com/p/breaking-pfizer-xbb15-monovalent?utm_source =profile&utm_medium=reader2.

741 Katharina Roltgen, Sandra C A Nielsen, Oscar Silva, et al., "Immune imprinting, breadth of variant recognition, and germinal center response in human SARS-CoV-2 infection and vaccination," *Cell*, 185 no. 6 (2022): p. 1025–1040 e14, doi: 10.1016/j.cell.2022.01.018.

742 S V Subramanian and Akhil Kumar, "Increases in COVID-19 are unrelated to levels of vaccination across 68 countries and 2947 counties in the United States," *Eur J Epidemiol*, 36 no. 12 (2021): p. 1237–1240, doi: 10.1007 /s10654-021-00808-7.

743 Ibid.

Chapter 29

744 A Wilder-Smith and D O Freedman, "Isolation, quarantine, social distancing and community containment: pivotal role for old-style public health measures in the novel coronavirus (2019-nCoV) outbreak," *J Travel Med*, 27 no. 2 (2020): p. doi: 10.1093/jtm/taaa020.

745 Thomas Hale, Noam Angrist, Beatriz Kira, Anna Petherick, Toby Phillips, and Samuel Webster, *Variation in government responses to COVID-19*. 2020, Blavatnik School of Government, University of Oxford.

746 Sandra Leon and Lluis Orriols, "Attributing responsibility in devolved contexts. Experimental evidence from the UK," *Electoral Studies*, 59 no. (2019): p. doi: 10.1016/j.electstud.2019.01.001.

747 Rajeev Goel, Ummad Mazhar, Michael Nelson, and Rati Ram, "Different forms of decentralization and their impact on government performance:

Micro-level evidence from 113 countries," *Economic Modelling*, 62 no. (2017): p. doi: 10.1016/j.econmod.2016.12.010.

748   Gil Luria, Ram A. Cnaan, and Amnon Boehm, "National Culture and Prosocial Behaviors: Results From 66 Countries," *Nonprofit and Voluntary Sector Quarterly*, 44 no. 5 (2014): p. 1041-1065, doi : 10.1177/0899764014554456.

749   Katarina Nygren and Anna Olofsson, "Managing the Covid-19 pandemic through individual responsibility: the consequences of a world risk society and enhanced ethopolitics," *Journal of Risk Research*, 23 no. (2020): p. 1–5, doi : 10.1080/13669877.2020.1756382.

750   Bo Yan, Xiaomin Zhang, Long Wu, et al., "Why Do Countries Respond Differently to COVID-19? A Comparative Study of Sweden, China, France, and Japan," *The American Review of Public Administration*, 50 no. 6–7 (2020): p. 762-769, doi: 10.1177/0275074020942445.

751   Anders Tegnell, "The Swedish public health response to COVID-19," *APMIS*, 129 no. 7 (2021): p. 320-323, doi: 10.1111/apm.13112.

752   Ibid.

753   Ibid.

754   Johns Hopkins Coroavirus Resource Center, "Ongoing Johns Hopkins COVID-19 Resources," https://coronavirus.jhu.edu/.

755   Jonas F Ludvigsson, Lars Engerstrom, Charlotta Nordenhall, and Emma Larsson, "Open Schools, Covid-19, and Child and Teacher Morbidity in Sweden," *N Engl J Med*, 384 no. 7 (2021): p. 669–671, doi: 10.1056 /NEJMc2026670.

756   Ibid.

757   "Thousands march in Melbourne on another weekend of COVID protests against vaccine mandates." November 26, 2021. https://www.abc.net.au /news/2021-11-27/melbourne-protests-thousands-mandate-pandemic -march-cbd/100656014.

758   Dr Ah Kahn Syed, *Genetically Modified Organons*, in *Arkmedic's blog*.

759   T.G. Administration, Nonclinical Evaluation Report, BNT162b2 [mRNA] COVID-19 vaccine (COMIRNATYTM), by Department of Health Australian Government. PM-2020-05461-1-2, 2021. Pfizer Australia Pty Ltd, https://www.tga.gov.au/sites/default/files/foi-2389-06.pdf.

760   Ibid.

761   Ian Brighthope, in *Ian Brighthope's Substack*.

762 Kenneth R. Drysdale, Heather DiGregorio, Dr. Bernard Massie, and Janice Kaikkonen. "N.C. Inquiry, Final Report: Inquiry into the Appropriateness and Efficacy of the COVID-19 Response in Canada," 2023. https ://nationalcitizensinquiry.ca/wp-content/uploads/2023/11/NCI-FINAL -REPORT-FULL-Inquiry-into-the-Appropriateness-and-Efficacy-of-the -COVID-19-Response-in-Canada-November-28-2023.pdf

763 Jack Phillips. "'Freedom Convoy' Protests Are Spreading Around World as Truckers Lead Fight Against Mandates." *The Epoch Times*, February 7, 2022. https://www.theepochtimes.com/article/freedom-convoy-protests-are -spreading-throughout-world-as-truckers-fight-against-mandates-4261451.

764 Ibid.

765 Drysdale, DiGregorio, Massie, and Kaikkonen, "N.C. Inquiry, Final Report."

766 Eugene Dean, "Bermuda Freedom Fighters", Friday Roundtable. B.H. Polly Tommey, Ph.D., P.E., Elizabeth Mumper, M.D., FAAP, Aimee Villella McBride, Meryl Nass, M.D. August 4, 2023; Available from: https://live .childrenshealthdefense.org/chd-tv/shows/friday-roundtable /bermuda-freedom-fighters-friday-roundtable/.

767 Preeti Vankar, "Health and health systems ranking of countries worldwide in 2023, by health index score," October 11, 2023, https://www.statista.com /statistics/1290168/health-index-of-countries-worldwide-by-health- index-score/.

768 Evan D. Gumas Munira Z. Gunja, Reginald D. Williams II, "U.S. Health Care from a Global Perspective, 2022: Accelerating Spending, Worsening Outcomes," January 31, 2023.

769 Ibid.

770 Ibid.

771 Arnav Shah Eric C. Schneider, Michelle M. Doty, Roosa Tikkanen, Katharine Fields, Reginald D. Williams II, "Mirror, Mirror 2021: Reflecting Poorly, Health Care in the U.S. Compared to Other High-Income Countries," August 4, 2021, https://www.commonwealthfund.org/publications/fund -reports/2021/aug/mirror-mirror-2021-reflecting-poorly.

772 Ed Dowd, *Cause Unknown: The Epidemic of Sudden Deaths in 2021 and 2022* (New York: Skyhorse, 2022).

Chapter 30

773 Edward L. Bernays, *Propaganda*. 1928. 159.

774 Vera Sharav. "Propaganda & Its Insidious Tactics of Persuasion —
     Then & Now." *Children's Health Defense*, January 4, 2024. https://live
     .childrenshealthdefense.org/chd-tv/events/propaganda-and-its-insidious
     -tactics-of-persuasion-then-and-now/propaganda-insidious-tactics
     -of-persuasion-then-and-now/.

775 Peter Palm. "The Role Of J.P. Morgan In Providing Loans To England And
     France In World War I; The Souring Of These Loans As It Became Apparent
     That Germany Would Win; The Betrayal Of A British Ship And The Sacrifice
     American Passengers As A Strategem To Bring America." *Seeking Alpha*,
     November 16, 2015. https://seekingalpha.com
     /instablog/25783813-peter-palms/4550806-role-of-j-p-morgan-in-providing
     -loans-to-england-and-france-in-world-war-i-souring-of-loans.

776 Victor Lebow, "Price Competition in 1955," *Journal of Retailing*, (1955)

777 Kerryn Higgs. "How the world embraced consumerism." *BBC*, January 20,
     2021. https://www.bbc.com/future/article/20210120-how-the-world
     -became-consumerist.

778 Denis Rancourt, PhD, https://denisrancourt.ca

779 Joseph Guzman. "Fauci: 'The cavalry is coming' with coronavirus vaccine,
     but public health measures are still needed." *The Hill*, November 12, 2020.
     https://thehill.com/changing-america/well-being/prevention-cures
     /525714-fauci-the-cavalry-is-coming-with-coronavirus/.

780 ClinicalTrials.gov, COVID-19 Vaccine Messaging, Part 1, by Yale University.
     ClinicalTrials.gov ID NCT04460703, 2022. https://clinicaltrials.gov/study
     /NCT04460703

781 Dan Ariely, *Predictably Irrational, Revised and Expanded Edition: The Hidden
     Forces That Shape Our Decisions* (New York: Harper Perennial, 2010).

782 Richard Thaler, *Misbehaving: The Making of Behavioral Economics* (New York:
     W. W. Norton & Company, 2015).

783 Richard Thaler, *Nudge: The Final Edition* (New York: Penguin Books, 2021).

784 https://www.britannica.com/ science/Milgram-experiment

785 Ibid.

786 Margaret Heffernan, *Willful Blindness: Why We Ignore the Obvious at Our Peril*
     (New York: Bloomsbury, 2012).

787 Margaret Heffernan. *Dare to Disagree*. June 2012; Available from: https ://www.ted.com/talks/margaret_heffernan_dare_to_disagree.

Chapter 31

788 Elizabeth Mumper, "Can awareness of medical pathophysiology in autism lead to primary care autism prevention strategies," *N. Am. J Med. Sci.*, 6 no. (2013): p. 134–144, doi: 10.7156/najms.2013.0603134.

789 Brian S Hooker and Neil Z Miller, "Analysis of health outcomes in vaccinated and unvaccinated children: Developmental delays, asthma, ear infections and gastrointestinal disorders," *SAGE Open Med*, 8 no. (2020): p. 2050312120925344, doi: 10.1177/2050312120925344.

790 Elizabeth Mumper, M.D., FAAP. "Black in the Age of Autism: A Tale of Two Boys." *The Defender*, May 20, 2020 https://childrenshealthdefense.org/news /black-in-the-age-of-autism-a-tale-of-two-boys/.

791 Aaron Siri. "Federal Judge Orders Biden Administration to Stop Social Media Censorship." *Injecting Freedom* (Substack.com), July 5, 2023. https://aaronsiri .substack.com/p/federal-judge-orders-biden-administration.

792 Ibid.

793 Ibid.

794 Ibid.

795 Ibid.

796 CCDH Center for Countering Digital Hate, http://counterhate.com.

797 Matt Taibbi. "THE CENSORSHIP-INDUSTRIAL COMPLEX." *The Twitter Files*, April 12, 2023. https://twitterfiles.substack.com/p /the-censorship-industrial-complex.

798 Hans Mahncke Jeff Carlson. "New Emails Reveal Evidence of Government Efforts to Suppress Free Speech." *The Epoch Times*, December 21, 2021. https://www.theepochtimes.com/opinion/new-emails-reveal-evidence-of -government-efforts-to-suppress-free-speech-4171310

799 Katie Spence. "Fired Over Ivermectin, ER Doctor Fights for Medical Freedom." *The Epoch Times*, February, 20, 2023. https://www.theepochtimes .com/article/fired-over-ivermectin-er-doctor-fights-for-medical -freedom-5066238.

800 "Pierre Kory: A War Is Still Being Waged Against Doctors Who Question COVID Orthodoxy." *Daily Caller News Foundation*, January 21, 2023.

https://dailycaller.com/2023/01/21/opinion-a-war-is-being-waged-against -doctors-who-question-covid-orthodoxy-pierre-kory/.

801 *SB-815 Healing arts.* https://www.mbc.ca.gov/About/Laws/SB815.aspx.

802 Paul Sacca. "NIH Director Francis Collins told Anthony Fauci there needs to be a 'quick and devastating' takedown of anti-lockdown declaration by 'fringe' Harvard, Stanford, Oxford epidemiologists: Emails." *The Blaze*, December 18, 2021. https://www.theblaze.com/news/fauci-email-francis -collins-great-barrington-declaration.

803 Peter McCullough MD. "YouTube Is Wiping Safety Content on COVID-19 Vaccines." *Courageous Discourse*, September 19, 023.https ://petermcculloughmd.substack.com/p/youtube-is-wiping-safety-content.

804 Press Briefing by White House COVID-19 Response Team and Public Health Officials, by The White House. https://www.whitehouse.gov/briefing-room /press-briefings/2021/04/27/press-briefing-by-white-house-covid-19 -response-team-and-public-health-officials-32/.

805 Zachary Stieber. "EXCLUSIVE: CDC Finds Hundreds of Safety Signals for Pfizer and Moderna COVID-19 Vaccines." *The Epoch Times*, January 3, 2023. https://www.theepochtimes.com/health/exclusive-cdc-finds-hundreds-of -safety-signals-for-pfizer-and-moderna-covid-19-vaccines-4956733.

806 Ibid.

807 Press Briefing by White House COVID-19 Response Team and Public Health Officials, by The White House. https://www.whitehouse.gov/briefing-room/ press-briefings/2021/04/27/press-briefing-by-white-house-covid-19 -response-team-and-public-health-officials-32/

808 Peter McCullough MD. "YouTube Is Wiping Safety Content on COVID-19 Vaccines." *Courageous Discourse*, September 19, 2023.https ://petermcculloughmd.substack.com/p/youtube-is-wiping-safety-content.

809 Yee Man Margaret Ng, Katherine Hoffmann Pham, and Miguel Luengo-Oroz, "Exploring YouTube's Recommendation System in the Context of COVID-19 Vaccines: Computational and Comparative Analysis of Video Trajectories," *J Med Internet Res*, 25 no. (2023): p. e49061, doi: 10.2196/49061.

810 Forbes Breaking News. *Jim Jordan Resumes Attacks On Dr. Fauci Over COVID-19 Origins, Mask Guidance.* July 28, 2021; Available from: https ://www.youtube.com/watch?v=sj5P65YzphM.

811   PBS NewsHour. *Widely cited hydroxychloroquine study is 'flawed', Fauci tells hearing.* Available from: https://www.youtube.com/watch?v=RkNC5OQD 2UE&t=136s.

812   Lori Weintz. "The Duplicitous Dr. Fauci and His Backpedaling." *Brownstone Institute*, October 30, 2022. https://brownstone.org/articles /duplicitous-fauci-backpedaling/.

813   Eric Boehm. "Anthony Fauci Says Don't Blame Him for COVID Lockdowns and School Closures." *Reason*, April 25, 2023. https://reason .com/2023/04/25/anthony-fauci-says-dont-blame-him-for-covid-lockdowns -and-school-closures/.

814   Ellen Barry. "Many Teens Report Emotional and Physical Abuse by Parents During Lockdown." *New York Times*, March 31, 2022. https://www.nytimes .com/2022/03/31/health/covid-mental-health-teens.html.

815   O.o.t.U.S.S.G. The U.S. Public Health Service, Our Epidemic of Loneliness and Isolation: The U.S. Surgeon General's Advisory on the Healing Effects of Social Connection and Community, by M.D. Dr. Vivek H. Murthy, M.B.A., U.S. Surgeon General. 2023. https://www.hhs.gov/sites/default/files/surgeon -general-social-connection-advisory.pdf

816   Jonas Herby, Lars Jonung, and Steve Hanke, *A Systematic Literature Review and Meta-Analysis of the Effects of Lockdowns on COVID-19 Mortality II.* 2023, medRxiv

817   Sean Cl Deoni, Jennifer Beauchemin, Alexandra Volpe, and Viren D'Sa, *The COVID-19 Pandemic and Early Child Cognitive Development: A Comparison of Development in Children Born During the Pandemic and Historical References.* 2021, Cold Spring Harbor Laboratory.

Chapter 32

818   Peter C. Gøtzsche, Professor, MD, DrMedSci, MSc, "Corporate crime in the pharmaceutical industry is common, serious and repetitive," *BMJ*, 345 no. e8462 (2012).

819   Ibid.

820   BBC News. "Pfizer agrees record fraud fine." September 2, 2009. http://news .bbc.co.uk/2/hi/business/8234533.stm.

821   Janice Hopkins Tanne, "Pfizer pays record fine for off-label promotion of four drugs," *BMJ*, 339 no. (2009): p. b3657, doi: 10.1136/bmj.b3657.

822 Ibid.

823 U.S.D.o. Justice, "GlaxoSmithKline to Plead Guilty and Pay $3 Billion to Resolve Fraud Allegations and Failure to Report Safety Data," by Office of Public Affairs. Consumer Protection, Civil Division, Press Release Number: 12-842: 2012. https://www.justice.gov/opa/pr/glaxosmithkline-plead-guilty -and-pay-3-billion-resolve-fraud-allegations-and-failure-report.

824 Katie Thomas and Michael Schmidt. "Glaxo Agrees to Pay $3 Billion in Fraud Settlement." *New York Times*, July 2, 2012. https://www.nytimes .com/2012/07/03/business/glaxosmithkline-agrees-to-pay-3-billion-in-fraud -settlement.html.

825 U.S.D.o. Justice, "Eli Lilly and Company Agrees to Pay $1.415 Billion to Resolve Allegations of Off-label Promotion of Zyprexa," by Office of Public Affairs. Consumer Protection, Civil Division, Press Release Number: 09-038: 2009. https://www.justice.gov/opa/pr/eli-lilly-and-company-agrees-pay-1415 -billion-resolve-allegations-label-promotion-zyprexa.

826 D.o. Justice, "Aventis Pharmaceutical to Pay U.S. $95.5 Million to Settle False Claims Act Allegations," by Office of Public Affairs. Civil Division, Press Release Number: 09-520: 2009. https://www.justice.gov/opa/pr /aventis-pharmaceutical-pay-us-955-million-settle-false-claims-act-allegations.

827 D.o. Justice, "Pharmaceutical Giant AstraZeneca to Pay $520 Million for Off-label Drug Marketing," by Office of Public Affairs. Civil Division, Press Release Number: 10-487: 2010. https://www.justice.gov/opa/pr /pharmaceutical-giant-astrazeneca-pay-520-million-label-drug-marketing.

828 U.S.D.o. Justice, "Novartis Pharmaceuticals Corp. to Pay More Than $420 Million to Resolve Off-label Promotion and Kickback Allegations," by Office of Public Affairs. Civil Division, Press Release Number: 10-1102: 2010. https://www.justice.gov/opa/pr/novartis-pharmaceuticals-corp-pay-more-420 -million-resolve-label-promotion-and-kickback.

829 U.S.D.o. Justice, "Johnson & Johnson to Pay More Than $2.2 Billion to Resolve Criminal and Civil Investigations," by Office of Public Affairs. Office of the Attorney General, Press Release Number: 13-1170: 2013. https://www .justice.gov/opa/pr/johnson-johnson-pay-more-22-billion-resolve-criminal -and-civil-investigations.

830 Fabien Deruelle, "The pharmaceutical industry is dangerous to health. Further proof with COVID-19," *Surg Neurol Int*, 13 no. (2022): p. 475, doi : 10.25259/sni_377_2022.

831 Ibid.

832 Maryanne Demasi, "FDA oversight of clinical trials is "grossly inadequate," say experts," *BMJ*, 379 no. (2022): p. o2628, doi: 10.1136/bmj.o2628.

833 U.S.F.a.D. Administration, Summary Basis for Regulatory Action, by PhD Sudhakar Agnihothram, Review Committee Chair, DVRPA/OVRR. 2022. https://www.fda.gov/media/155931/download.

834 U.S.F.a.D. Administration, Summary Basis for Regulatory Action, by PhD Sudhakar Agnihothram, Review Committee Chair, DVRPA/OVRR. 2022. https://www.fda.gov/media/155931/download.

835 U.S.P.H.S. Department Of Health And Human Services, Food And Drug Administration, Center For Biologics Evaluation And Research, Office of Inspector General, "The Food and Drug Administration's Oversight of Clinical Trials," by Inspector General Daniel R. Levinson. OEI-01-06-00160, 2007. https://www.oig.hhs.gov/oei/reports/oei-01-06-00160.pdf.

836 Janice Hopkins Tanne, "FDA is failing to oversee human clinical trials, report says," *Bmj*, 335 no. 7622 (2007): p. 691, doi: 10.1136/bmj.39356.358796.DB.

837 Maryanne Demasi, "FDA oversight of clinical trials is "grossly inadequate," say experts," *Bmj*, 379 no. (2022): p. o2628, doi: 10.1136/bmj.o2628.

838 Charles Seife, "Research misconduct identified by the US Food and Drug Administration: out of sight, out of mind, out of the peer-reviewed literature," *JAMA Intern Med*, 175 no. 4 (2015): p. 567–77, doi: 10.1001 /jamainternmed.2014.7774.

839 Charles Piller, "Official Inaction: A Science investigation shows that FDA Oversight of Clinical Trials is lax, slow moving, and secretive - and that enforcement is declining," *Science*, 370 no. 6512 24-29, doi: 10.1126 /science.370.6512.24.

840 Peter Doshi and Fiona Godlee, "The wider role of regulatory scientists," *BMJ*, 357 no. (2017): p. j1991, doi: 10.1136/bmj.j1991.

841 Dr. David Healy, "Disappeared in Argentina," March 1, 2022, https ://davidhealy.org/disappeared-in-argentina/.

842 Ibid.

843 "The Daily Clout," https://dailyclout.io.

844 Pfizer Worldwide Safety, *5.3.6 Cumulative Analysis of Post-Authorization Adverse Event Reports Of Pf-07302048 (Bnt162b2) Received Through 28-Feb-2021.*

845 Naomi Wolf, The WarRoom/DailyClout Pfizer Documents Analysts, Amy Kelly, Stephen K Bannon, *The Pfizer Papers: Pfizer's Crimes Against Humanity* (New York: Skyhorse, 2024), 312.

846 House - Energy and Commerce; Ways and Means | Senate - Labor and Human Resources, H.R.5546 - National Childhood Vaccine Injury Act of 1986, by Rep. Henry A. Waxman. 1986. https://www.congress.gov /bill/99th-congress/house-bill/5546.

847 Norman Fenton and Martin Neil, "The Lancet and the Pfizer Vaccine: A Case Study in Academic Censorship and Deceit in the Covid Era," *ResearchGate. net*, no. (2023): p. 1-18, doi: 10.13140/RG.2.2.29792.56321.

Chapter 33

848 The World Council for Health. "Urgent Expert Hearing on Reports of DNA Contamination in mRNA Vaccines." 2023. https://www. worldcouncilforhealth.org/dna-contamination-covid-19-vaccines/.

849 Michael Palmer, MD and PhD Jonathan Gilthorpe. "COVID-19 mRNA vaccines contain excessive quantities of bacterial DNA: evidence and implications." 2023. https://drjosenasser.substack.com/p/ covid-19-mrna-vaccines-contain-excessive.

850 Peter C. Gøtzsche, Professor, MD, DrMedSci, MSc, "Corporate crime in the pharmaceutical industry is common, serious and repetitive," *BMJ*, 345 no. e8462 (2012): p. doi:

851 Eugene McCarthy, "A Call to Prosecute Drug Company Fraud As Organized Crime," *Syracuse L. Rev.*, no. 69 (2019): p. 51, doi: 10.2139/ssrn.3421124.

852 The World Council for Health. "Urgent Expert Hearing on Reports of DNA Contamination in mRNA Vaccines." 2023. https://www. worldcouncilforhealth.org/dna-contamination-covid-19-vaccines/.

853 Ibid.

854 Ibid.

855 R. Kurth, "Risk potential of the chromosomal insertion of foreign DNA," *Ann N Y Acad Sci*, 772 no. (1995): p. 140–51, doi: 10.1111/j.1749 -6632.1995.tb44739.x.

856 David J. Speicher, Jessica Rose, L. Maria Gutschi, and Kevin McKernan. "DNA fragments detected in monovalent and bivalent Pfizer/BioNTech and

Moderna modRNA COVID-19 vaccines from Ontario, Canada: Exploratory dose response relationship with serious adverse events." (2023). https://www.researchgate.net/publication/374870815_Speicher_DJ_et_al_DNA_fragments_detected_in_COVID-19_vaccines_in_Canada_DNA_fragments_detected_in_monovalent_and_bivalent.

857 Angus G. Dalgleish and Ken O'Byrne, *Inflammation and Cancer*, in *The Link Between Inflammation and Cancer: Wounds that do not heal*, A.G. Dalgleish and B. Haefner, Editors. 2006, Springer US: Boston, MA. p. 1–38.

858 Angus G Dalgleish and Ken J O'Byrne, "Chronic immune activation and inflammation in the pathogenesis of AIDS and cancer," *Adv Cancer Res*, 84 no. (2002): p. 231–76, doi: 10.1016/s0065-230x(02)84008-8

859 The World Council for Health. "Urgent Expert Hearing on Reports of DNA Contamination in mRNA Vaccines." 2023.

860 Robert Kennedy, Jr. , RN Lyn Redwood, MSN, and Member Harold Gielow, *RE: Phase III Moderna mRNA-1273 Vaccine*, D.F. Dr. Steven Hahn, et l.https://childrenshealthdefense.org/wp-content/uploads/RFK_Jr_Letter _toFDA-_CBER-9-25-20.pdf.

861 Children's Health Defense, "FDA Ignores RFK Jr.'s Pleas for Vaccine Safety Oversight Concerning PEG, Suspected To Cause Anaphylaxis," press release, December 14, 2020, https://childrenshealthdefense.org/%20press-release/fda -ignores-rfk-jr-s-pleas-for-vaccine-safety-oversight-concerning-peg-suspected -to-cause-anaphylaxis/

862 C. Li and A. Lieber, "Adenovirus vectors in hematopoietic stem cell genome editing," *FEBS Lett*, 593 no. 24 (2019): p. 3623–3648, doi: 10.1002/1873-3468.13668.

863 Ifigeneia V Mavragani, Zacharenia Nikitaki, Maria P Souli, et al., "Complex DNA Damage: A Route to Radiation-Induced Genomic Instability and Carcinogenesis," *Cancers (Basel)*, 9 no. 7 (2017): p. doi: 10.3390 /cancers9070091.

864 Ibid.

865 The World Council for Health. *Urgent Expert Hearing on Reports of DNA Contamination in mRNA Vaccines*. 2023.

866 Ibid.

Chapter 34

867   N.C.f.I.a.R. Diseases, "Glossary," by Centers for Disease Control and
      Prevention U.S. Department of Health and Human Services. 2022. https
      ://www.cdc.gov/vaccines/terms/glossary.html.

868   Aaron Siri, "Clinical Trial to License RotaTeq, Like Almost All Childhood
      Vaccines, Did Not Use a Placebo Control: Those attacking RFK are wrong.,"
      June 25, 2023, https://aaronsiri.substack.com/p/clinical-trial
      -to-license-rotateq?utm_source=profile&utm_medium=reader2.

869   C.C.A. Branch, RotaTeq® (Rotavirus Vaccine, Live, Oral, Pentavalent), Oral
      Solution, Initial U.S. Approval: 2006, by Food and Drug Administration U.S.
      Department of Health and Human Services, Merck Sharp & Dohme Corp., a
      subsidiary of Merck & Co., Inc. 2006. https://www.fda.gov/media/75718
      /download.

870   Siri, "Clinical Trial to License RotaTeq."

871   U.S.P.H.S. Department Of Health And Human Services, Food And Drug
      Administration, Center For Biologics Evaluation And Research, "Clinical
      Review of New Biologics License Application STN #125122 RotaTeq®,"
      by MD Rosemary Tiernan, MPH, Medical Officer, Vaccines Clinical Trials
      Branch, Division of Vaccines and Related Product Applications (DVRPA).
      2006. http://wayback.archive-it.org/7993/20170723031531/https:/www
      .fda.gov/downloads/BiologicsBloodVaccines/Vaccines/ApprovedProducts
      /UCM142304.pdf.

872   Dr. Paul Offit, "Should Scientists Debate the Undebatable?: Joe Rogan invites
      debate after a misinformation-filled podcast with Robert F. Kennedy Jr.,"
      https://pauloffit.substack.com/p/should-scientists-debate-the-undebatable.

873   Informed Consent Action Network Del Bigtree, *Letter to U.S. Department
      of Health & Human Services, HHS Office of the Secretary, Alex M. Azar II,
      Secretary of Health & Human Services and Tammy R. Beckham, Acting Director,
      National Vaccine Program Office Re: HHS Vaccine Safety Responsibilities and
      Notice Pursuant to 42 U.S.C. § 300aa-31.* 2018.

874   David A Geier, Elizabeth Mumper, Bambi Gladfelter, et al.,
      "Neurodevelopmental disorders, maternal Rh-negativity, and Rho(D)
      immune globulins: a multi-center assessment," *Neuro Endocrinol Lett,* 29 no.
      2 (2008): p. 272-80, doi:

875   M. Tan and J. E. Parkin, "Route of decomposition of thiomersal
      (thimerosal)," *Int J Pharm*, 208 no. 1-2 (2000): p. 23–34, doi: 10.1016
      /s0378-5173(00)00514-7.

876   M. E. Pichichero, E. Cernichiari, J. Lopreiato, and J. Treanor, "Mercury
      concentrations and metabolism in infants receiving vaccines containing
      thiomersal: a descriptive study," *Lancet*, 360 no. 9347 (2002): p. 1737–41,
      doi: 10.1016/s0140-6736(02)11682-5.

877   Michael E Pichichero, Angela Gentile, Norberto Giglio, et al., "Mercury
      levels in newborns and infants after receipt of thimerosal-containing
      vaccines," *Pediatrics*, 121 no. 2 (2008): p. e208-14, doi: 10.1542
      /peds.2006-3363.

878   Thomas M Burbacher, Danny D Shen, Noelle Liberato, et al., "Comparison
      of blood and brain mercury levels in infant monkeys exposed to
      methylmercury or vaccines containing thimerosal," *Environ Health Perspect*,
      113 no. 8 (2005): p. 1015–21, doi: 10.1289/ehp.7712.

879   Ibid.

880   Ibid.

881   J. S. Charleston, R. L. Body, R. P. Bolender, et al., "Changes in the number
      of astrocytes and microglia in the thalamus of the monkey Macaca fascicularis
      following long-term subclinical methylmercury exposure," *Neurotoxicology*, 17
      no. 1 (1996): p. 127–38.

882   J. S. Charleston, R. L. Body, N. K. Mottet, et al., "Autometallographic
      determination of inorganic mercury distribution in the cortex of the calcarine
      sulcus of the monkey Macaca fascicularis following long-term subclinical
      exposure to methylmercury and mercuric chloride," *Toxicol Appl Pharmacol*,
      132 no. 2 (1995): p. 325–33, doi: 10.1006/taap.1995.1114.

883   J. S. Charleston, R. P. Bolender, N. K. Mottet, et al., "Increases in the
      number of reactive glia in the visual cortex of Macaca fascicularis following
      subclinical long-term methyl mercury exposure," *Toxicol Appl Pharmacol*, 129
      no. 2 (1994): p. 196–206, doi: 10.1006/taap.1994.1244.

884   M. E. Vahter, N. K. Mottet, L. T. Friberg, et al., "Demethylation of methyl
      mercury in different brain sites of Macaca fascicularis monkeys during long-
      term subclinical methyl mercury exposure," *Toxicol Appl Pharmacol*, 134 no. 2
      (1995): p. 273–84, doi: 10.1006/taap.1995.1193.

885 M. Vahter, N. K. Mottet, L. Friberg, et al., "Speciation of mercury in the primate blood and brain following long-term exposure to methyl mercury," *Toxicol Appl Pharmacol,* 124 no. 2 (1994): p. 221–9, doi: 10.1006 /taap.1994.1026.

886 Vahter, Mottet, Friberg, et al., "Demethylation of methyl mercury."

887 Vahter, Mottet, Friberg, et al., "Speciation of mercury in the primate."

888 Diana L Vargas, Caterina Nascimbene, Chitra Krishnan, et al., "Neuroglial activation and neuroinflammation in the brain of patients with autism," *Ann Neurol,* 57 no. 1 (2005): p. 67–81, doi: 10.1002/ana.20315.

889 Thomas M Burbacher, Danny D Shen, Noelle Liberato, et al., "Comparison of blood and brain mercury levels in infant monkeys exposed to methylmercury or vaccines containing thimerosal," *Environ Health Perspect,* 113 no. 8 (2005): p. 1015–21, doi: 10.1289/ehp.7712.

890 M. Aschner and J. L. Aschner, "Mercury neurotoxicity: mechanisms of blood-brain barrier transport," *Neurosci Biobehav Rev,* 14 no. 2 (1990): p. 169–76, doi: 10.1016/s0149-7634(05)80217-9.

891 Alessia Carocci, Nicola Rovito, Maria Stefania Sinicropi, and Giuseppe Genchi, "Mercury toxicity and neurodegenerative effects," *Rev Environ Contam Toxicol,* 229 no. (2014): p. 1–18, doi: 10.1007/978-3-319-03777-6_1.

892 Peter N Alexandrov, Aileen I Pogue, and Walter J Lukiw, "Synergism in aluminum and mercury neurotoxicity," *Integr Food Nutr Metab,* 5 no. 3 (2018): p. doi: 10.15761/ifnm.1000214.

893 Diana L Vargas, Caterina Nascimbene, Chitra Krishnan, et al., "Neuroglial activation and neuroinflammation in the brain of patients with autism," *Ann Neurol,* 57 no. 1 (2005): p. 67–81, doi: 10.1002/ana.20315.

894 Vasco Branco, Michael Aschner, and Cristina Carvalho, "Neurotoxicity of mercury: an old issue with contemporary significance," *Adv Neurotoxicol,* 5 no. (2021): p. 239–262, doi: 10.1016/bs.ant.2021.01.001.

895 Sabrina Llop, Ferran Ballester, and Karin Broberg, "Effect of Gene-Mercury Interactions on Mercury Toxicokinetics and Neurotoxicity," *Curr Environ Health Rep,* 2 no. 2 (2015): p. 179–94, doi: 10.1007/s40572-015-0047-y.

896 Centers for Disease Control and Prevention U.S. Department of Health and Human Services, "Notice to Readers: Thimerosal in Vaccines: A Joint Statement of the American Academy of Pediatrics and the Public

Health Service," *Morbidity and Mortality Weekly Report*, 48 no. 26 (1999): p. 563–565.

897 I. Knezevic, E. Griffiths, F. Reigel, and R. Dobbelaer, "Thiomersal in vaccines: a regulatory perspective WHO Consultation, Geneva, 15-16 April 2002," *Vaccine*, 22 no. 15–16 (2004): p. 1836-41, doi: 10.1016/j.vaccine.2003.11.051.

898 Institute of Medicine (US) Immunization Safety Review Committee, *Immunization Safety Review: Thimerosal-Containing Vaccines and Neurodevelopmental Disorders.*, ed. G.A. Stratton K, McCormick MC. 2001, Washington (DC): National Academies Press (US).

899 Robert F. Kennedy Jr., *The Wuhan Cover-Up: And the Terrifying Bioweapons Arms Race*. 986. (New York: Skyhorse, 2023).

900 Anna Malkova, Dmitriy Kudlay, Igor Kudryavtsev, et al., "Immunogenetic Predictors of Severe COVID-19," *Vaccines (Basel)*, 9 no. 3 (2021): p. doi : 10.3390/vaccines9030211.

901 Berend Jan Bosch, Byron E Martina, Ruurd Van Der Zee, et al., "Severe acute respiratory syndrome coronavirus (SARS-CoV) infection inhibition using spike protein heptad repeat-derived peptides," *Proc Natl Acad Sci U S A*, 101 no. 22 (2004): p. 8455–60, doi: 10.1073/pnas.0400576101.

902 Daniel Wrapp, Nianshuang Wang, Kizzmekia S Corbett, et al., "Cryo-EM structure of the 2019-nCoV spike in the prefusion conformation," *Science*, 367 no. 6483 (2020): p. 1260–1263, doi: 10.1126/science.abb2507.

903 Yuan Hou, Junfei Zhao, William Martin, et al., "New insights into genetic susceptibility of COVID-19: an ACE2 and TMPRSS2 polymorphism analysis," *BMC Med*, 18 no. 1 (2020): p. 216, doi: 10.1186/s12916-020-01673-z.

904 André Salim Khayat, Paulo Pimentel de Assumpção, Bruna Claudia Meireles Khayat, et al., "ACE2 polymorphisms as potential players in COVID-19 outcome," *PLoS One*, 15 no. 12 (2020): p. e0243887, doi: 10.1371/journal.pone.0243887.

905 Naoki Yamamoto, Nao Nishida, Rain Yamamoto, et al., "Angiotensin-Converting Enzyme (ACE) 1 Gene Polymorphism and Phenotypic Expression of COVID-19 Symptoms," *Genes (Basel)*, 12 no. 10 (2021): p. doi: 10.3390/genes12101572.

906 Tasnim H Beacon, Genevieve P Delcuve, and James R Davie, "Epigenetic regulation of ACE2, the receptor of the SARS-CoV-2 virus(1)," *Genome*, 64 no. 4 (2021): p. 386–399, doi: 10.1139/gen-2020-0124.

907 Zafer Yildirim, Oyku Semahat Sahin, Seyhan Yazar, and Vildan Bozok Cetintas, "Genetic and epigenetic factors associated with increased severity of Covid-19," *Cell Biol Int*, 45 no. 6 (2021): p. 1158–1174, doi: 10.1002 /cbin.11572.

908 Osama A Badary, "Pharmacogenomics and COVID-19: clinical implications of human genome interactions with repurposed drugs," *Pharmacogenomics J*, 21 no. 3 (2021): p. 275-284, doi: 10.1038/s41397-021-00209-9.

909 Marion Nyamari, Kauthar Omar, Ayorinde Fayehun, et al., *Expression Level Analysis of ACE2 Receptor Gene in African-American and Non-African-American COVID-19 Patients*. 2023.

910 Virna Margarita Martín Giménez, León Ferder, Felipe Inserra, et al., "Differences in RAAS/vitamin D linked to genetics and socioeconomic factors could explain the higher mortality rate in African Americans with COVID-19," *Therapeutic Advances in Cardiovascular Disease*, 14 no. (2020): p. 1753944720977715, doi: 10.1177/1753944720977715.

911 Mart De La Cruz, David P Nunes, Vaishali Bhardwaj, et al., "Colonic Epithelial Angiotensin-Converting Enzyme 2 (ACE2) Expression in Blacks and Whites: Potential Implications for Pathogenesis Covid-19 Racial Disparities," *J Racial Ethn Health Disparities*, 9 no. 2 (2022): p. 691–697, doi: 10.1007/s40615-021-01004-9.

912 Jiawei Chen, Quanlong Jiang, Xian Xia, et al., "Individual variation of the SARS-CoV-2 receptor ACE2 gene expression and regulation," *Aging Cell*, 19 no. 7 (2020): p. doi: 10.1111/acel.13168.

913 Aileen Graef, Sarah Fortinsky. "Robert F. Kennedy Jr. invokes Nazi Germany in offensive anti-vaccine speech." *CNN*, January 24, 2022. https://www.cnn .com/2022/01/23/politics/robert-f-kennedy-nazi-germany-offensive-anti -vaccine-speech/index.html.

914 https://archive.vanityfair.com/issue/20210501.

915 https://time.com/5585702/robert-kennedy-vaccines/.

Chapter 35

916 BMC Public Health, "Research and Publication Ethics to Consider,"
     BMC Public Health. https://bmcpublichealth.biomedcentral.com/
     equity-in-academic-publishing/research-and-publication-ethics-to-consider.

917 Baruch Green, "WATCH: Pediatrician Says 'Massive Increase' in Myocarditis,
     Asks What We Are Injecting in Children," December 9, 2022, https
     ://vinnews.com/2022/12/09/watch-pediatrician-says-massive-increase-in
     -myocarditis-asks-what-we-are-injecting-in-children/.

918 Parth Shah, Imani Thornton, Danielle Turrin, and John E Hipskind, *Informed
     Consent*, in *StatPearls*. 2024, StatPearls Publishing Copyright © 2024,
     StatPearls Publishing LLC.: Treasure Island (FL).

919 Neal Halfon, Gregory D Stevens, Kandyce Larson, and Lynn M Olson,
     "Duration of a well-child visit: association with content, family-centeredness,
     and satisfaction," *Pediatrics*, 128 no. 4 (2011): p. 657–64, doi: 10.1542
     /peds.2011-0586.

920 Ibid.

921 Centers for Disease Control and Prevention, U.S. Department of Health
     and Human Services, National Center for Immunization and Respiratory
     Diseases, "Current VISs," December 7, 2023, https://www.cdc.gov/vaccines
     /hcp/vis/current-vis.html.

922 E. Shuster, "Fifty years later: the significance of the Nuremberg Code,"
     N Engl J Med, 337 no. 20 (1997): p. 1436–40, doi: 10.1056
     /nejm199711133372006.

923 Muriel Bowser Government of the District of Columbia, Executive Office of
     the Mayor, "Mayor Bowser and DC Health Announce Vaccination Plans for
     Children 5-11 Years Old," press release, October 29, 2021, https://mayor
     .dc.gov/release/mayor-bowser-and-dc-health-announce-vaccination-plans
     -children-5-11-years-old.

924 Children's Health Defense Team. "CHD Lawsuit Seeks to Overturn D.C. Law
     Allowing Kids to Be Vaccinated Without Parents' Knowledge or Consent."
     Children's Health Defense, July, 16 2021. https://childrenshealthdefense.org
     /defender/chd-lawsuit-kids-vaccinated-without-parents-knowledge-consent/.

925 "Children's Health Defense Wins Federal District Court Injunction On
     District of Columbia's Minor Consent for Vaccinations Act," press release,
     March 21, 2022 https://childrenshealthdefense.org/press-release/childrens

-health-defense-wins-federal-district-court-injunction-on-district-of-columbias-minor-consent-for-vaccinations-act/.

926 Jessica Rose, "What exactly is suddenly killing the young who've been injected with no history, no meds, and no previous illness?," in *Unacceptable Jessica*.

927 House - Energy and Commerce; Ways and Means | Senate - Labor and Human Resources, H.R.5546 - National Childhood Vaccine Injury Act of 1986, by Rep. Henry A. Waxman. 1986. https://www.congress.gov /bill/99th-congress/house-bill/5546.

928 Peter C. Gøtzsche, Professor, MD, DrMedSci, MSc, "Corporate crime in the pharmaceutical industry is common, serious and repetitive," *BMJ*, 345 no. e8462 (2012): p. doi:

929 Eugene McCarthy, "A Call to Prosecute Drug Company Fraud As Organized Crime," *Syracuse L. Rev.* , no. 69 (2019): p. 51, doi: 10.2139/ssrn.3421124.

930 Joel Lexchin, *Medicines and Money: The Corruption of Clinical Information*, in *Crime and Corruption in Organizations*, E.C.T. Ronald J. Burke, Editor. 1984, Routledge.

931 M.D. Sammy Almashat, M.P.H. and M.D Sidney Wolfe, *Pharmaceutical Industry Criminal and Civil Penalties: An Update*. Public Citizen.

Chapter 36

932 Barbara Loe Fisher, "Sen. Ron Johnson Roundtable on Feb 26, 2024," February 26, 2024, https://www.nvic.org/newsletter/feb-2024 /sen-ron-johnson-roundtable-on-feb-26,-2024.

933 https://www.ifm.org.

934 https://www.a4m.com/pediatrics.

935 https://www.medmaps.org/.

Index